Lecture Notes in Computer Science   14395

Founding Editors

Gerhard Goos
Juris Hartmanis

The series Lecture Notes in Computer Science (LNCS), including its subseries Lecture Notes in Artificial Intelligence (LNAI) and Lecture Notes in Bioinformatics (LNBI), has established itself as a medium for the publication of new developments in computer science and information technology research, teaching, and education.

LNCS enjoys close cooperation with the computer science R & D community, the series counts many renowned academics among its volume editors and paper authors, and collaborates with prestigious societies. Its mission is to serve this international community by providing an invaluable service, mainly focused on the publication of conference and workshop proceedings and postproceedings. LNCS commenced publication in 1973.

Nicholas Olenev · Yuri Evtushenko ·
Milojica Jaćimović · Michael Khachay ·
Vlasta Malkova

Editors

Optimization and Applications

14th International Conference, OPTIMA 2023
Petrovac, Montenegro, September 18–22, 2023
Revised Selected Papers

Springer

Editors
Nicholas Olenev (iD)
FRC CSC RAS
Moscow, Russia

Yuri Evtushenko (iD)
FRC CSC RAS
Moscow, Russia

Milojica Jaćimović (iD)
University of Montenegro
Podgorica, Montenegro

Michael Khachay (iD)
Krasovsky Institute of Mathematics
and Mechanics
Ekaterinburg, Russia

Vlasta Malkova (iD)
FRC CSC RAS
Moscow, Russia

ISSN 0302-9743   ISSN 1611-3349 (electronic)
Lecture Notes in Computer Science
ISBN 978-3-031-47858-1   ISBN 978-3-031-47859-8 (eBook)
https://doi.org/10.1007/978-3-031-47859-8

This Springer imprint is published by the registered company Springer Nature Switzerland AG
The registered company address is: Gewerbestrasse 11, 6330 Cham, Switzerland

Paper in this product is recyclable.

Preface

This volume contains the first part of the refereed proceedings of the XIV International Conference on Optimization and Applications (OPTIMA 2023)[1].

Organized annually since 2009, the conference has attracted a significant number of researchers, academics, and specialists in many fields of optimization, operations research, optimal control, game theory, and their numerous applications in practical problems of data analysis and software development.

The broad scope of OPTIMA has made it an event where researchers involved in different domains of optimization theory and numerical methods, investigating continuous and discrete extremal problems, designing heuristics and algorithms with theoretical bounds, developing optimization software, and applying optimization techniques to highly relevant practical problems can meet together and discuss their approaches and results. We strongly believe that this facilitates collaboration between researchers working in optimization theory, methods, and applications, to advance optimization theory and methods and employ them on valuable practical problems.

The conference was held during September 18–22, 2023, in Petrovac, Montenegro, on the picturesque Budvanian riviera on the azure Adriatic coast. For those who were not able to come to Montenegro this year, an online session was organized. The main organizers of the conference were the Montenegrin Academy of Sciences and Arts, Montenegro, FRC CSC RAS, Russia, and the University of Évora, Portugal. This year, the key topics of OPTIMA were grouped into eight tracks:

    (i) Mathematical Programming
   (ii) Global Optimization
  (iii) Continuous Optimization
  (iv) Discrete and Combinatorial Optimization
   (v) Optimal Control
  (vi) Game Theory and Mathematical Economics
 (vii) Optimization in Economics and Finance
(viii) Applications

The Program Committee and the reviewers of the conference included more than one hundred well-known experts in continuous and discrete optimization, optimal control and game theory, data analysis, mathematical economics, and related areas from leading institutions of 26 countries including Argentina, Australia, Austria, Belgium, China, Finland, France, Germany, Greece, India, Israel, Italy, Lithuania, Kazakhstan, Mexico, Montenegro, The Netherlands, Poland, Portugal, Russia, Serbia, Sweden, Taiwan, the UAE, the UK, and the USA. This year we received 107 submissions mostly from Russia but also from Algeria, Armenia, Azerbaijan, Belarus, China, Egypt, France, Germany, Iran, Kazakhstan, Montenegro, Poland, Turkey, United Arab Emirates, the UK, and the USA. Each submission was single-blind reviewed by at least three PC members or invited

---

[1] http://agora.guru.ru/display.php?conf=OPTIMA-2023

reviewers, experts in their fields, to supply detailed and helpful comments. Out of 68 qualified submissions, the Program Committee decided to accept 27 papers to the first volume of the proceedings. Thus the acceptance rate for this volume was about 40%.

In addition, the Program Committee proposed about 20 papers to be included in the second volume of the proceedings after a short presentation of the candidate papers, discussion at the conference, and subsequent revision.

The conference featured three invited lecturers, and several plenary and keynote talks. The invited lectures included:

– Alexey Tret'yakov, Systems Research Institute, Polish Academy of Sciences, Warsaw, Poland, "The pth-Order Karush-Kuhn-Tucker Type Optimality Conditions for Nonregular Inequality Constrained Optimization Problems"
– Panos M. Pardalos, University of Florida, USA, "Diffusion in Networks"
– Yurii Nesterov, UCLouvain, Belgium, "Universality, the New Trend in Development of Optimization Schemes"

We would like to thank all the authors for submitting their papers and the members of the PC for their efforts in providing exhaustive reviews. We would also like to express special gratitude to all the invited lecturers and plenary speakers.

September 2023

Nicholas Olenev
Yuri Evtushenko
Milojica Jaćimović
Michael Khachay
Vlasta Malkova

Organization

Program Committee Chairs

Milojica Jaćimović	Montenegrin Academy of Sciences and Arts, Montenegro
Yuri G. Evtushenko	FRC CSC RAS, Russia
Michael Yu. Khachay	Krasovsky Institute of Mathematics and Mechanics, Russia
Vlasta U. Malkova	FRC CSC RAS, Russia
Nicholas N. Olenev	CEDIMES-Russie, FRC CSC RAS, Russia

Program Committee

Majid Abbasov	St. Petersburg State University, Russia
Samir Adly	University of Limoges, France
Kamil Aida-Zade	Institute of Control Systems of ANAS, Azerbaijan
Alexander P. Afanasiev	Institute for Information Transmission Problems, RAS, Russia
Yedilkhan Amirgaliyev	Suleyman Demirel University, Kazakhstan
Anatoly S. Antipin	FRC CSC RAS, Russia
Adil Bagirov	Federation University, Australia
Artem Baklanov	International Institute for Applied Systems Analysis, Austria
Evripidis Bampis	LIP6 UPMC, France
Olga Battaïa	ISAE-SUPAERO, France
Armen Beklaryan	National Research University Higher School of Economics, Russia
Nikolay Belotelov	FRC CSC RAS, Russia
Vladimir Beresnev	Sobolev Institute of Mathematics, Russia
Anton Bondarev	Xi'an Jiaotong-Liverpool University, China
Sergiy Butenko	Texas A & M University, USA
Vladimir Bushenkov	University of Évora, Portugal
Igor A. Bykadorov	Sobolev Institute of Mathematics, Russia
AlexanderChichurin	The John Paul II Catholic University of Lublin, Poland
Duc-Cuong Dang	INESC TEC, Portugal
Tatjana Davidovic	Mathematical Institute of Serbian Academy of Sciences and Arts, Serbia
Stephan Dempe	TU Bergakademie Freiberg, Germany
Alexandre Dolgui	IMT Atlantique, LS2N, CNRS, France

Anton Eremeev	Omsk Division of Sobolev Institute of Mathematics, SB RAS, Russia
Adil Erzin	Novosibirsk State University, Russia
Francisco Facchinei	Sapienza University of Rome, Italy
Denis Fedyanin	V. A. Trapeznikov Institute of Control Sciences, Russia
Tatiana Filippova	Krasovsky Institute of Mathematics and Mechanics, Russia
Anna Flerova	FRC CSC RAS, Russia
Manlio Gaudioso	Università della Calabria, Italy
Victor Gorelik	FRC CSC RAS, Russia
Alexander Yu. Gornov	Inst. System Dynamics and Control Theory, SB RAS, Russia
Edward Kh. Gimadi	Sobolev Institute of Mathematics, SB RAS, Russia
Alexander Grigoriev	Maastricht University, The Netherlands
Mikhail Gusev	N. N. Krasovskii Institute of Mathematics and Mechanics, Russia
Vladimir Jaćimović	University of Montenegro, Montenegro
Vyacheslav Kalashnikov	ITESM, Campus Monterrey, Mexico
Maksat Kalimoldayev	Institute of Information and Computational Technologies, Kazakhstan
Valeriy Kalyagin	Higher School of Economics, Russia
Igor E. Kaporin	FRC CSC RAS, Russia
Alexander Kazakov	Matrosov Institute for System Dynamics and Control Theory, SB RAS, Russia
Oleg V. Khamisov	L. A. Melentiev Energy Systems Institute, Russia
Andrey Kibzun	Moscow Aviation Institute, Russia
Donghyun Kim	Kennesaw State University, USA
Roman Kolpakov	Moscow State University, Russia
Igor Konnov	Kazan Federal University, Russia
Yury A. Kochetov	Sobolev Institute of Mathematics, Russia
Dmitri E. Kvasov	University of Calabria, Italy
Alexander A. Lazarev	V. A. Trapeznikov Institute of Control Sciences, Russia
Vadim Levit	Ariel University, Israel
Bertrand M. T. Lin	National Chiao Tung University, Taiwan
Alexander V. Lotov	FRC CSC RAS, Russia
Vladimir Mazalov	Institute of Applied Mathematical Research, Karelian Research Center, Russia
Nevena Mijajlović	University of Montenegro, Montenegro
Mikhail Myagkov	University of Oregon, USA
Angelia Nedich	Arizona State University, USA

Yuri Nesterov	Université Catholique de Louvain, Belgium
Yuri Nikulin	University of Turku, Finland
Evgeni Nurminski	Far Eastern Federal University, Russia
Natalia K. Obrosova	FRC CSC RAS, Russia
Victor Orlov	Moscow State University of Civil Engineering, Russia
Panos Pardalos	University of Florida, USA
Dmitry Pasechnyuk	Mohammed bin Zayed University of Artificial Intelligence, United Arab Emirates
Alexander V. Pesterev	V. A. Trapeznikov Institute of Control Sciences, Russia
Alexander Petunin	Ural Federal University, Russia
Stefan Pickl	University of the Bundeswehr Munich, Germany
Leonid Popov	IMM UB RAS, Russia
Mikhail A. Posypkin	FRC CSC RAS, Russia
Alexander N. Prokopenya	Warsaw University of Life Sciences, Poland
Artem Pyatkin	Novosibirsk State University; Sobolev Institute of Mathematics, Russia
Ioan Bot Radu	University of Vienna, Austria
Soumyendu Raha	Indian Institute of Science, India
Leonidas Sakalauskas	Institute of Mathematics and Informatics, Lithuania
Sergei Semakov	FRC CSC RAS, Russia
Yaroslav D. Sergeyev	University of Calabria, Italy
Natalia Shakhlevich	University of Leeds, UK
Alexander A. Shananin	Moscow Institute of Physics and Technology, Russia
Bismark Singh	University of Southampton, UK
Angelo Sifaleras	University of Macedonia, Greece
Mathias Staudigl	Maastricht University, The Netherlands
Fedor Stonyakin	V. I. Vernadsky Crimean Federal University, Russia
Alexander Strekalovskiy	Matrosov Institute for System Dynamics & Control Theory, SB RAS, Russia
Vitaly Strusevich	University of Greenwich, UK
Michel Thera	University of Limoges, France
Tatiana Tchemisova	University of Aveiro, Portugal
Anna Tatarczak	Maria Curie-Skłodowska University, Poland
Alexey A. Tretyakov	Siedlce University of Natural Sciences and Humanities, Poland
Stan Uryasev	University of Florida, USA
Frank Werner	Otto von Guericke University Magdeburg, Germany

Adrian Will	National Technological University, Argentina
Anatoly A. Zhigljavsky	Cardiff University, UK
Aleksandra Zhukova	FRC CSC RAS, Russia
Julius Žilinskas	Vilnius University, Lithuania
Yakov Zinder	University of Technology, Australia
Tatiana V. Zolotova	Financial University under the Government of the Russian Federation, Russia
Vladimir I. Zubov	FRC CSC RAS, Russia
Anna V. Zykina	Omsk State Technical University, Russia

Organizing Committee Chairs

Milojica Jaćimović	Montenegrin Academy of Sciences and Arts, Montenegro
Yuri G. Evtushenko	FRC CSC RAS, Russia
Nicholas N. Olenev	FRC CSC RAS, Russia

Organizing Committee

Anna Flerova	FRC CSC RAS, Russia
Alexander Gasnikov	National Research University Higher School of Economics, Russia
Alexander Gornov	Institute of System Dynamics and Control Theory, SB RAS, Russia
Vesna Dragović	Montenegrin Academy of Sciences and Arts, Montenegro
Vladimir Jaćimović	University of Montenegro, Montenegro
Michael Khachay	Krasovsky Institute of Mathematics and Mechanics, Russia
Yury Kochetov	Sobolev Institute of Mathematics, Russia
Vlasta Malkova	FRC CSC RAS, Russia
Anton Medennikov	FRC CSC RAS, Russia
Oleg Obradovic	University of Montenegro, Montenegro
Natalia Obrosova	FRC CSC RAS, Russia
Mikhail Posypkin	FRC CSC RAS, Russia
Kirill Teymurazov	FRC CSC RAS, Russia
Yulia Trusova	FRC CSC RAS, Russia
Svetlana Vladimirova	FRC CSC RAS, Russia
Ivetta Zonn	FRC CSC RAS, Russia
Vladimir Zubov	FRC CSC RAS, Russia

Invited Talks

The pth-Order Karush-Kuhn-Tucker Type Optimality Conditions for Nonregular Inequality Constrained Optimization Problems

Alexey Tret'yakov

Systems Research Institute, Polish Academy of Sciences, Warsaw, Poland
https://www.researchgate.net/profile/Alexey_Tretyakov

Abstract. In this paper, we present necessary and sufficient optimality conditions for optimization problems with inequality constraints in the finite dimensional spaces. We focus on the degenerate (nonregular) case when the classical constraint qualifications are not satisfied at a solution of the optimization problem. We present optimality conditions of the Karush-Kuhn-Tucker type under new regularity assumptions. To formulate the optimality conditions, we use the p-factor operator, which is the main construction of the p-regularity theory.

The approach of p-regularity used in the paper can be applied to various degenerate nonlinear optimization problems due to its flexibility and generality.

This is joint work with Yuri Evtushenko, Olga Brezhneva, and Vlasta Malkova.

Diffusion in Networks

Panos M. Pardalos

University of Florida, USA
http://www.ise.ufl.edu/pardalos/
https://nnov.hse.ru/en/latna/

Abstract. This lecture addresses the significant challenge of comprehending diffusive processes in networks in the context of complexity. Networks possess a diffusive potential that depends on their topological configuration, but diffusion also relies on the process and initial conditions. The lecture introduces the concept of Diffusion Capacity, a measure of a node's potential to diffuse information that incorporates a distance distribution considering both geodesic and weighted shortest paths and the dynamic features of the diffusion process. This concept provides a comprehensive depiction of individual nodes' roles during the diffusion process and can identify structural modifications that may improve diffusion mechanisms.

The lecture also defines Diffusion Capacity for interconnected networks and introduces Relative Gain, a tool that compares a node's performance in a single structure versus an interconnected one. To demonstrate the concept's utility, we apply the methodology to a global climate network formed from surface air temperature data, revealing a significant shift in diffusion capacity around the year 2000. This suggests a decline in the planet's diffusion capacity, which may contribute to the emergence of more frequent climatic events. Our goal is to gain a deeper understanding of the complexities of diffusive processes in networks and the potential applications of the Diffusion Capacity concept.

Universality, the New Trend in Development of Optimization Schemes

Yurii Nesterov

UCLouvain, Belgium
https://uclouvain.be/fr/repertoires/yurii.nesterov

Abstract. In the early years of Optimization, the first classical schemes were derived from an abstract concept of approximation (e.g. Gradient method, Newton's methods, etc.). However, since the development of Complexity Theory for Convex Optimization (Nemirovsky, Yudin), the most powerful approaches for constructing efficient (optimal) methods are based on the model of the objective function. This model incorporates the characteristic properties of the corresponding problem class and provides us with comprehensive information on the behavior of the objective. At the same time, it helps in deriving theoretically unimprovable complexity bounds for the target class.

However, this framework completely neglects the fact that every objective function belongs, at the same time, to many different problem classes. Hence, it should be treated by a method developed for the most appropriate class of problems. However, for real-life problems, such a choice is seldom feasible, at least in advance.

In this talk, we discuss several ideas for constructing universal methods which automatically ensure the best possible convergence rate among appropriate problem classes. The simplest methods of this type adjust to the best power in the Holder condition for the target derivative. Our most promising super-universal Regularized Newton's Method works properly for a wide range of problems, starting from functions with bounded variation of Hessian up to functions with Lipschitz continuous third derivative. Thus, being a second-order scheme, it covers a diversity of problems, from problems traditionally treated by the first-order methods, up to problems which are usually attributed to third-order schemes. For its proper work, no preliminary information on the objective function is needed.

(Some of the results are obtained jointly with N. Doikov, G. Grapiglia, and K. Mishchenko.)

In Memory of Igor Germogenovich Pospelov (12.06.1950-30.12.2022) (Epitaph)

Colleagues

FRC CSC RAS, Russia

After a serious long illness, Igor Germogenovich Pospelov, a brilliant scientist, the greatest expert in the field of mathematical modeling, corresponding member of the Russian Academy of Sciences, Doctor of Physical and Mathematical Sciences, Chief Scientific Associate of the FRC CSC RAS, passed away in the 73rd year of his life.

Igor Ghermogenovich worked at the Computing Center of the USSR Academy of Sciences since 1976, which he entered after graduation from the MIPT graduate school. He organically joined the staff of the "Mathematical Economics" sector headed by A.A. Petrov. With his arrival, research began in the framework of a new direction of modeling economic systems, called "System Analysis of Developing Economies". Igor Germogenovich and his students obtained important results in the field of modeling economic systems, including the theory of intertemporal equilibrium. Under his leadership, the ECOMOD system was created to provide intellectual support for mathematical modeling of the economy. This system was further utilized in important applied research. In recent years, Igor Germogenovich together with his students proposed a new approach to the construction of a general equilibrium model.

Igor Germogenovich is the author of numerous works, including monographs, which are well known among Russian economists. His scientific work was combined with great organizational activity. After the establishment of the Federal Research Center "Computer Sciences and Control" of the Russian Academy of Sciences, he headed the department of "Mathematical Modeling". As a teacher and then head of the department of MIPT, Igor Germogenovich paid great attention to the education of young specialists. He was also a professor at Lomonosov Moscow State University and the National Research University Higher School of Economics. Igor Germogenovich raised many talented scientists, including doctors and candidates of sciences. In 2008 Igor Germogenovich was elected a corresponding member of the Russian Academy of Sciences.

Igor Germogenovich was an amazingly versatile person. He possessed truly encyclopedic knowledge, could read to his friends whole lectures on philosophy, history, poetry. Igor Germogenovich Pospelov's benevolence and responsiveness, his diligence and organizational skills earned him the respect of his colleagues and friends. His passing is a great loss not only for the staff of FRC CSC RAS, but also for the whole country. The memory of this remarkable man will long remain in the hearts of his colleagues and students.

Contents

Applications

Mathematical Programming

Representation of the Eternal Vertex Cover Problem as a Dynamic Stackelberg Game

Vladimir Beresnev⊙, Andrey Melnikov⊙, and Stepan Utyupin⁽⊠⁾

Sobolev Institute of Mathematics, 4 Acad. Koptyug avenue, 630090 Novosibirsk, Russia
stepan.utyupin@gmail.com

Abstract. Eternal vertex cover problem is a variant of the graph vertex cover problem that can be considered as a dynamic game between two players (Attacker and Defender) with an infinite number of turns. At each turn, there is an arrangement of guards over the vertices of the graph, forming a vertex cover. Attacker attacks one of the graph's edges, and Defender must move the guard along the attacked edge from one vertex to another. Additionally, Defender can move other guards from their current vertices to some adjacent ones to obtain a new vertex cover. Attacker wins if Defender cannot build a new vertex cover after an attack. The eternal vertex cover problem is to determine the smallest number of guards and their placement such that Attacker has no sequence of attacks leading to a win.

We consider the eternal vertex cover problem as a dynamic Stackelberg game. At each turn of the game, a vertex cover is considered and the optimal solution of the bi-level programming problem is calculated, which allows to determine whether the given vertex cover is stable against a single attack (1-stable). The results of the computational experiment on checking the 1-stability of various vertex covers of random graphs are presented.

Keywords: 1-stable vertex cover · Protection · Bi-level programming

1 Introduction

A family of eternal graph protection models originates in 1990 s as an abstraction inspired by Emperor Constantine's strategy to protect the Roman Empire [1]. In these models, vertices or edges of a graph are secured against an infinite sequence of attacks. The protection is performed by means of guards placed in vertices of a graph and able to move along its edges after the attack happens.

The present paper considers the eternal cover problem formulated in [2]. This problem can be interpreted as an infinite game of two parties, Attacker and

The study was carried out within the framework of the state contract of the Sobolev Institute of Mathematics (project FWNF-2022-0019).

Defender. Initially, Defender places guards at the vertices of the given graph so that the vertices, where the guards are placed, form a vertex cover. Then, Attacker chooses an edge to attack. In response, Defender must move guards from the vertices of the attacked edge along it. When both vertices of the attacked edge are occupied by guards, the guards are simply interchange their positions. Additionally, Defender can move other guards to their neighboring vertices so that there are no vertices occupied by more than one guard, and no guards making more than one move. If the new placement of guards forms a vertex cover, the next turn begins. Oppositely, if it's impossible to reconstruct a vertex cover by using a feasible guards move, that is, at least one edge stays unprotected, then Attacker wins and the game ends. Defender wins if they do not lose to any sequence of attacks.

The eternal cover problem aims to place the minimum number of guards forming a vertex cover being a winning strategy of Defender. In [3] Fedor V. Fomin et al. state that this problem is NP-hard, and in [4] Jasine Babu et al. show that the problem remains NP-hard for bipartite graphs. In the paper [5], Jasine Babu et al. propose a new lower bound for the eternal vertex cover number of an arbitrary graph in terms of the cardinality of a vertex cover of minimum size, containing all its cut vertices. Also they present a quadratic time algorithm for computing the eternal vertex cover number of chordal graphs. In [6] for some classes of graphs, Jasine Babu et al. obtain polynomial algorithms for calculating the least number of guards for constructing an eternal vertex cover. In the paper [7], Hisashi Araki et al. propose polynomial algorithm for specialized tree graphs. In [8], Kaustav Paul et al. obtain a polynomial algorithm to find the eternal vertex cover number in chain graphs and cographs. Also they propose a linear-time algorithm to find the eternal vertex cover number for split graphs, a subclass of chordal graphs. In [3], the authors show that Eternal Vertex Cover Problem has a polynomial time 2-approximation algorithm.

One could notice that the listed works and many related ones either propose constructions for very specific graphs families or provide qualitative results or estimations for general or special cases. In this paper, we address the situation, when a graph is given, but no restrictions on its structure are imposed in advance. Additionally, some vertex cover with no assumptions about it is given as well. Then, the Attacker-Defender game could start from this setup, and one may wonder, which player wins in a given situation.

Answering this question demands for some kind of analysis of all infinite sequences of attacks. Actually, it is enough to consider sequences of some suffi-ciently large length due to finiteness of the set of possible game configurations, but one may expect the sufficient length to be unmanageable. In this work, we propose a first step in this analysis, where an existence of Defender's winning strategy on a given finite planning horizon is verified by an enumeration scheme.

To construct a strategy, the game is represented as a repeated or dynamic Stackelberg game. At each turn, the interaction of the opposing sides is for-malized as a two-level mathematical programming problem. The solution of the upper-level problem (Attacker's problem) determines the attack that maximizes the number of unprotected edges after Defender reconstructs the cover. The solution of the lower-level problem (the Defender's problem) with a given Attacker's solution changes the placement of the guards to minimize the number of unprotected edges.

The paper proposes a procedure for constructing the best solutions of the parties in the considered two-level problem. The results of computational experiments using the proposed procedure are presented. For randomly generated graphs and various vertex covers, the stability of vertex covers against a single-turn attack (*1-stability*) is studied.

2   Formulation of the Single-Turn Problem

Let $G = (I, J)$ be a directed graph without loops with a set of vertices I and set of arcs J. For any $j \in J$, denote by $i(j)$ and $k(j)$, respectively, the starting and ending vertices of the arc j and call these vertices *neighbors*. Note that consideration of a directed graph does not reduce the generality of the formulated two-level problem considered at each turn of the game. Indeed, in an undirected graph, each edge can be replaced by two oppositely oriented parallel arcs. We assume that a guard always moves along the arc from its starting to the ending vertex.

A vertex cover in a graph G is any subset of the set I that contains a starting or ending vertex of any arc from the set J.

Let the planning horizon has length of T, what means that the game consists of T turns, where each turn is represented by an attack followed by a Defender's response. To formally write the two-level problem considered at a turn t of the game under study, we introduce the following notation.

x_{i0}, $i \in I$, are given values that determine the initial vertex cover; the value x_{i0} is equal to one if a guard is placed at the vertex $i \in I$, and equals zero otherwise.

x_{it}, $i \in I$, are variables determining the placement of guards upon completion of the turn t, $1 \leq t \leq T$; the value of x_{it} is equal to 1 if a guard is placed at the vertex $i \in I$, and equals 0 otherwise.

y_{jt}, $j \in J$, are variables determining the arc attacked at turn t, $1 \leq t \leq T$; the value y_{jt} is equal to one if the arc $j \in J$ is attacked, and equals to zero otherwise.

u_{jt}, $j \in J$, are variables determining the arcs that, after the completion of turn t, stay unprotected; the value u_{jt} is equal to one if no guard is placed at vertices of the arc $j \in J$, and equals zero otherwise.

v_{jt}, $j \in J$, are variables determining the movement of the guards along the arcs; the value v_{jt} is equal to one if the guard moves from the vertex $i(j)$ to the vertex $k(j)$, and equals to zero otherwise.

Using the introduced notation, a single-turn problem could be written in terms of bi-level mathematical programming as follows:

$$\max_{(y_{jt})} \sum_{j \in J} u_{jt}^*; \tag{1}$$

$$\sum_{j \in J} y_{jt} \leq 1; \tag{2}$$

$$x_{i(j)t-1} \geq y_{jt}, \qquad j \in J; \tag{3}$$

$$x_{i(j)t-1} + x_{k(j)t-1} + y_{jt} \leq 2, \qquad j \in J; \tag{4}$$

$(v_{jt}^*), (u_{jt}^*)$ - optimal solution of the problem:

$$\min_{(x_{it}),(v_{jt}),(u_{jt})} \sum_{j \in J} u_{jt}; \tag{5}$$

$$v_{jt} \geq y_{jt}, \qquad j \in J; \tag{6}$$

$$x_{it-1} \geq \sum_{j|i(j)=i} v_{jt}, \qquad i \in I; \tag{7}$$

$$1 - x_{it-1} \geq \sum_{j|k(j)=i} v_{jt} - \sum_{j|i(j)=i} v_{it}, \qquad i \in I; \tag{8}$$

$$x_{it} = x_{it-1} + \sum_{j|k(j)=i} v_{jt} - \sum_{j|i(j)=i} v_{it}, \qquad i \in I; \tag{9}$$

$$x_{i(j)t} + x_{k(j)t} + u_{jt} \geq 1, \qquad j \in J; \tag{10}$$

$$x_{it}, v_{jt}, u_{jt} \in \{0,1\}, \qquad j \in J, i \in I; \tag{11}$$

This model, like any two-level optimization problem, has an upper-level problem (1)–(4) (the Attacker's problem) and a lower-level problem (5)-(11) (the Defender's problem).

The objective functions (1) and (5) of both problems show the number of arcs left unprotected after the Defender's response on turn t. The inequality (2) means that at turn t at most one arc can be attacked. And from the inequalities (3) and (4) it follows that only an arc with a guard placed at the starting vertex and with no guard at the ending vertex can be attacked.

The constraints (6)–(8) of the lower-level problem formalize the rules for possible movement of guards along the corresponding arcs. The constraint (6) means that if the arc is attacked, the guard moves to the ending vertex of the arc. From the inequality (7), it follows that if a guard is placed at vertex i, then their movement is possible along at most one arc starting from this vertex. If there is no guard at vertex i, then the guards cannot move along the arcs starting from this vertex. Similarly, from the inequality (8), subject to the condition (7), we obtain that if there is no guard at vertex i in the vertex cover considered at turn t, then $\sum_{j|k(j)=i} v_{jt} \leq 1$. It means that the guard can be moved along at most one arc entering the vertex i. If a guard is placed at the vertex i, then from the inequalities (7) and (8) we obtain

$$1 \geq \sum_{j|i(j)=i} v_{jt} \geq \sum_{j|k(j)=i} v_{jt}.$$

It means that a guard can be moved along one of the arcs entering the i vertex, but only if the guard located at the i vertex also moves. The equality (9) determines the new placement of guards as a result of their movement at turn t. The inequality (10) forces the variable u_{jt} to take a non-zero value when there are no guards at the starting and ending vertices of the arc j.

A feasible solution to a two-level problem at turn t of the game is a pair $((y_{jt}), (v_{jt}, u_{jt}))$, where (y_{jt}) is a feasible solution to the upper-level problem, and (v_{jt}, u_{jt}) is the optimal solution of the lower-level problem. A feasible solution will be optimal if there are no feasible solutions with a higher value of the objective function. Calculation of the optimal solution enables one to check if the current vertex cover is *stable* against an attack. If the optimal value of the objective function is zero, then the cover is stable. If this value is greater than zero, then there exists a feasible solution of the upper-level problem (an attacked arc) for which there is no vertex cover reconstruction leading to a new vertex cover.

3   Stability of Vertex Covers

The problem considered at each turn of the game allows us to determine whether the vertex cover obtained by turn t is stable against an attack. If, for a given feasible solution of the Attacker's problem (y_{jt}), for which $y_{jt} = 1$, an optimal value of the objective function of the Defender's problem is equal to zero, then the considered vertex cover will be called *stable against an attack on arc j*. If the vertex cover is stable against attack on any arc, i.e. if the optimal value of the objective function of the Attacker's problem is equal to zero, then the considered vertex cover will be called *1-stable*, or simply *stable*.

Consider a sequence of attacks of length L. Let at turns $1, 2, ..., L-1$, the optimal values of objective functions are equal to zero and let this value becomes strictly positive at turn L, i.e. at turn L, when attacking the L-th arc from the sequence, a successful reconstruction of the vertex cover fails. Note that this does not mean that Defender always loses after L turns. Indeed, the vertex cover obtained up to the turn L is the result of successive reconstructions of vertex covers considered at previous turns. These reconstructions are determined by optimal solutions of the corresponding Defender's problems, which may not be unique. Therefore, with an appropriate choice of optimal solutions of the Defender's problems at previous turns, it may be possible to construct a vertex cover that is stable against the attack at the L-th turn.

A vertex cover is called *L-stable* if it is resistant against any sequence of L attacks by Attacker, i.e. if, for any feasible solution of the Attacker's problem, the considered vertex cover is stable and there exists a reconstruction (an optimal solution of the Defender's problem) that defines an $(L-1)$-stable vertex cover.

The *1-stability* of a cover can be established by enumerating arcs, that can be attacked by Attacker, and verifying if the cover can be reconstructed. Such an enumeration scheme is represented by an algorithm 1.

Algorithm 1. Procedure for determining the 1-stability

Input: Di-graph $G = (I, J)$, vertex cover C encoded by vector (x_{i0}), $i \in I$.
Output: $result \in \{\text{TRUE}, \text{FALSE}\}$
1: $result \leftarrow \text{TRUE}$
2: **for each** $j \in J$ **do**
3:     **if** $x_{i(j)0} + x_{k(j)0} = 1$ **then**
4:         Check the stability of C against an attack on j
5:         **if** Cover C is not stable to attack on the arc j **then**
6:             $result \leftarrow \text{FALSE}$
7:             **break**

Capabilities of the Algorithm 1 can be extended by relying on useful properties of the problem such that one formulated in the following lemma. For the sake of brevity, we omit the index t of the variables.

Lemma 1. *Consider a feasible solution (y_j) of the Attacker's problem, and let there exists a solution (v_j), (u_j) of the corresponding Defender's problem such that the objective function value equals to zero. Let $J_0 = \{j \in J | v_j = 1\}$, and let feasible solution (y'_j) of the Attacker's problem be such that $y'_{j_0} = 1$ for some $j_0 \in J_0$. Then the optimal objective function value of the Defender's problem, corresponding to the Attacker's solution (y'_j), equals to zero.*

Proof. Consider a feasible solution (y'_j) of the Attacker's problem such that $y'_{j_0} = 1$, $j_0 \in J_0$. Notice that the solution (v_j), (u_j) is feasible in the Defender's problem, corresponding the Attacker's solution (y'_j). By definition of (v_j), (u_j), the objective function value is zero, what finishes the proof.

Lemma 1 allows to significantly speed up the enumeration Algorithm 1 when checking 1-stability. Indeed, when a stability check with an attack on some arc is performed within the Algorithm 1, a solution (v_j), (u_j) of the corresponding Defender's problem must be computed. In the case, when the objective function value on this solution is zero, by Lemma 1, all the checks considering attacks on edges from the set J_0 from the lemma's formulation can be skipped, since they are proven to be passed successfully.

4   Computational Experiments

In the experiments, we study the performance of the Algorithm 1 and demonstrate its capabilities as an analytical tool. Firstly, we have measured a computational time the Algorithm 1 needs to finish the computations for problem instances of different sizes. All the calculations were performed on a computer with a 3.3 GHz quad-core processor and 16 gigabytes of RAM. The Algorithm 1 was written in Julia programming language [9], and the COIN-OR Branch-and-Cut solver [10] was used to find the optimal value of arising MIP problems.

4.1 Graph Generation

In the experiments we use randomly generated connected graphs of different density. The density is defined as the ratio between the number of edges in the graph and the number of edges in a clique having the same number of vertices. The graphs are constructed as follows: given the number n of vertices in a graph, we start from a spanning tree having n vertices and called *initial tree*. The initial tree could have either special structure (path, star, etc.) or be a randomly generated tree constructed by looking through the list of vertices and connecting a vertex under consideration with a randomly chosen one considered earlier. Then, we choose a number r, $r \in [1, \frac{(n-1)(n-2)}{2}]$ and supplement the tree generated with new r edges by connecting r randomly chosen pairs of unconnected vertices.

4.2 Cover Generation

Since finding a minimal cover is an NP-hard problem, the covers of sub-optimal sizes could be of practical interest as well. Such covers, that none of their proper subsets remain to be a cover, would be called *irreducible* ones. A graph could have multiple irreducible covers of different sizes, and generation of different irreducible covers in our experiments is organized as a randomized greedy heuristic represented by Algorithm 2.

Algorithm 2. Randomized greedy heuristic to find irreducible vertex covers

Input: $G = (V, E)$
Output: $C \subset V$
 1: **while** $V \neq \emptyset$ **do**
 2:    $V' \leftarrow$ random subset of V of size 3
 3:    $v \leftarrow$ vertex from V' with the highest degree
 4:    $V \leftarrow V \setminus \{v\}$
 5:    $C \leftarrow C \cup \{v\}$
 6:    Remove all the edges adjacent to v from E
 7: **for each** v from randomly ordered C **do**
 8:    **if** $C \setminus \{v\}$ is a vertex cover **then**
 9:       $C \leftarrow C \setminus \{v\}$

The algorithm constructs an irreducible cover in two phases. On the first phase, a vertex cover is built within a loop, where a randomly chosen triplet of graph vertices, not yet taken into a cover, is considered on each pass. Among the vertices of the triplet, the one, covering the highest number of yet uncovered edges, is taken into a cover, and the loop body repeats. When the first phase is finished, a vertex cover, which is not guaranteed to be irreducible, is built.

On the second phase, a random order of vertices of the cover is chosen. The vertices are considered in this order, and if a vertex under consideration can be removed from the cover so that all the graph edges remain covered, the vertex is removed.

4.3   Running Time

Figure 1 demonstrates a dependency of computational time on the number of additional edges r for a graph with 20 vertices. Instances with $r = 108$ additional edges (density equals to 0.67) appeared to be the most time-consuming. To calculate the time cost for a given r, we have generated 5 random graphs using the scheme described above. For each graph, we generated 5 random smallest vertex covers using the Algorithm 2. For each cover, we checked whether it is 1-stable and measured the elapsed time. The result for each r is the average time over 25 runs.

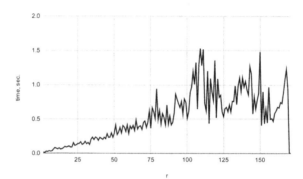

Fig. 1. Dependence of computational time of Algorithm 1 on the number r of edges added to a tree with 20 vertices.

Further, we have considered a sequence of graphs with edges density 0.67 and measured the computational time of the Algorithm 1 on graphs with different number of vertices. The results are summarized on Fig. (2). Here, to calculate the time costs for each N, 5 graphs were also generated, according to the scheme described above (for each dimension, a random initial tree was taken); for each graph, 5 random smallest vertex covers were generated using the algorithm 2, and the average time costs for 25 runs were taken. As it shows, while the graph size does not exceed 100, the computational time does not show a dramatic growth, and it takes around 6 min to check 1-stability for a cover within a graph with 128 vertices.

4.4   Proportion of 1-Stable Covers

To demonstrate analytical capabilities of the Algorithm 1, we have performed a statistical study of proportion of 1-stable vertex covers among the irreducible covers of a graph. In the experiment, given the number of vertices n, for each

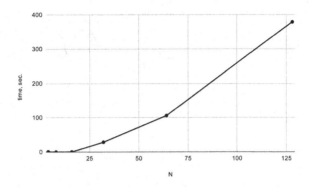

Fig. 2. Dependence of Algorithm 1's computational time on the number of vertices in the graph with density 0.67.

number of additional edges $r \in [1, \frac{(n-1)(n-2)}{2}]$, we have generated N_g random connected graphs with n vertices and $n - 1 + r$ edges, using the graph generation routine described earlier. For each graph generated, we have computed N_c random irreducible vertex covers using the Algorithm 2. Each vertex cover was checked by the Algorithm 1 on the subject if it is 1-stable or not. For each r, we got a value $S(r)$ showing a proportion of irreducible covers, which are 1-stable (out of $N_g N_c$ generated ones), and $C(r)$, showing the average size of the cover.

Figure 3 demonstrates the results of these computations with parameters $n = 10$, $N_g = 100$, $N_c = 10$. The lines show the dependency $S(r)$ for random graphs, obtained from different initial trees. The dashed line corresponds to the initial tree being a path, the dotted line corresponds to a star-graph, and the solid line corresponds to a randomly generated initial tree. The histogram bars indicate the values of $C(r)$ computed for randomly generated spanning tree.

As one could notice, the value $C(r)$ demonstrates a monotone growth resulting from an increase of vertex cover size caused by increase of the number of edges in a graph. The value $S(r)$ shows an undulating growth, which is hard to be explained. The line's shape does not depend on the topology of the initial graph.

Figure 4 shows analogous results for $n = 20$; $N_g = 100$, $N_c = 10$. A random spanning tree was chosen to be an initial tree here. It is noticeable, that the undulation is not affected by the size of a graph and is present here as well. The range of undulation increases while r grows.

We assume that the undulation is observed due to the discrete nature of irreducible cover size. While the graph density grows but irreducible covers remain of the same size, a proportion of stable covers decreases, since a larger set of edges must be covered with the same number of guards. This situation corresponds to intervals of decline of $S(r)$. When the graph density forces the irreducible cover to increase, the proportion of stable covers increases. At this moment, we observe an increase of $S(r)$. Unfortunately, the hypothesis does not answers on

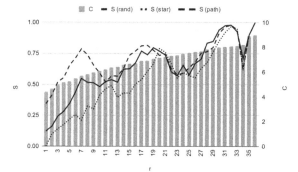

Fig. 3. Dependence of the share of stable covers $S(r)$ and the number of guards $C(r)$ on the number r of edges added to a tree with $n = 10$ vertices.

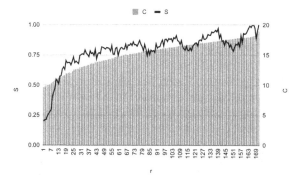

Fig. 4. Dependence of the share of stable covers and the number of guards on the number r of edges, added to a tree with $n = 20$ vertices

why the undulation amplitude visibly grows, and why the waves on the graph of $S(r)$ have a moderate increasing part and a steeper decrease.

5   Conclusion and Future Work

The present work considers an eternal cover problem and its representation as a dynamic Stackelberg game between an Attacker and Defender with an infinite number of turns. On each of the game turns, a vertex cover is considered, and an optimal solution of a bi-level programming problem is computed to verify if the cover is stable against an attack or not.

To answer the question on cover stability, a testing procedure was developed and tested. The results show that the procedure takes a reasonable time to finish the computations for graphs of sizes up to 100 vertices, what allows to consider a perspective of using the procedure to test L-stability of covers, i.e. a stability against attacks composed of L turns.

Another intriguing result obtained within the computational experiments and demanding for further study is a behavior of the function $S(r)$ showing the proportion of stable covers depending on the graph's density. The nature of undulation is to be examined further.

References

1. Klostermeyer, W.F., Mynhardt, C.M.: Protecting a graph with mobile guards. Appl. Anal. Discrete Math. **10**(1), 1–29 (2016)
2. Klostermeyer, W.F., Mynhardt, C.M.: Edge protection in graphs. Australas. J Comb. **45**, 235–250 (2009)
3. Fomin, F.V., Gaspers, S., Golovach, P.A., Kratsch, D., Saurabh, S.: Parameterized algorithm for eternal vertex cover. Inf. Process. Lett. **110**(16), 702–706 (2010)
4. Babu, J., Misra, N., Nanoti, S.G.: Eternal vertex cover on bipartite graphs, Computer Science - Theory and Applications. In: 17th International Computer Science Symposium in Russia, CSR 2022, Virtual Event, June 29 - July 1, 2022, Proceedings, pp. 64–76 (2022)
5. Babu, J., Prabhakaran, V.: A new lower bound for the eternal vertex cover number of graphs. J. Comb. Optim. **44**, 2482–2498 (2022)
6. Babu, J., Veena, P., Arko, S.: A substructure based lower bound for eternal vertex cover number. Theor. Comput. Sci. **890**, 87–104 (2021)
7. Araki, H., Fujito, T., Inoue, S.: On the eternal vertex cover numbers of generalized trees IEICE transactions on fundamentals of electronics. Commun. Comput. Sci. **E98.A**, 1153–1160 (2015)
8. Paul, K., Pandey, A.: Some algorithmic results for eternal vertex cover problem in graphs. In: Lin, CC., Lin, B.M.T., Liotta, G. (eds.) WALCOM: Algorithms and Computation. WALCOM 2023. Lecture Notes in Computer Science, vol. 13973, pp. 242–253. Springer, Cham (2023). https://doi.org/10.1007/978-3-031-27051-2_21
9. Bezanson, J., Edelman, A., Karpinski, S., Shah, V.B.: Julia: a fresh approach to numerical computing. SIAM Rev. **59**(1), 65–98 (2017)
10. Cbc code repository. htttps://github.com/coin-or/Cbc. Accessed Jun 2023

The pth-Order Karush-Kuhn-Tucker Type Optimality Conditions for Nonregular Inequality Constrained Optimization Problems

Olga Brezhneva[1], Yuri Evtushenko[2,3] , Vlasta Malkova[2(✉)] ,
and Alexey Tret'yakov[2,4,5]

[1] Department of Mathematics, Miami University, Oxford, OH 45056, USA
`brezhnoa@miamioh.edu`
[2] FRC CSC RAS, Vavilov Street 40, 119333 Moscow, Russia
`vmalkova@yandex.ru`
[3] Moscow Institute of Physics and Technology, Moscow, Russia
[4] System Res. Inst., Polish Acad. Sciences, Newelska 6, 01-447 Warsaw, Poland
[5] Faculty of Exact and Natural Sciences, Siedlce University, 08-110 Siedlce, Poland
`tret@ap.siedlce.pl`

Abstract. In this paper, we present necessary and sufficient optimality conditions for optimization problems with inequality constraints in the finite dimensional spaces. We focus on the degenerate (nonregular) case when the classical constraint qualifications are not satisfied at a solution of the optimization problem. We present optimality conditions of the Karush-Kuhn-Tucker type under new regularity assumptions. To formulate the optimality conditions, we use the p-factor operator, which is the main construction of the p-regularity theory. The approach of p-regularity used in the paper can be applied to various degenerate nonlinear optimization problems due to its flexibility and generality.

Keywords: Nonregular problems · Constrained optimization · Optimality conditions · p-regularity

1 Introduction

Consider the following optimization problem with inequality constraints

$$\begin{array}{ll} \underset{x \in \mathbb{R}^n}{\text{minimize}} & f(x) \\ \text{subject to } g(x) = (g_1(x), \ldots, g_m(x)) \leq 0 \end{array} \qquad (1)$$

where f and g_i, $i = 1, \ldots, m$, are sufficiently smooth functions from $\mathbb{R}^n$ to $\mathbb{R}$. We are interested in the case when the classical regularity assumptions (constraint qualifications (CQs)) known in the literature (see, for example, [19]) are not

---

This work was supported by the Russian Foundation for Basic Research, project No. 21-71-30005.

satisfied at the solution $\bar{x}$ of the problem (1) and, therefore, classical Karush-Kuhn-Tucker (KKT) optimality conditions do not hold at $\bar{x}$. In this case, we call problem (1) degenerate (nonregular, singular) at $\bar{x}$.

The Karush-Kuhn-Tucker (KKT) Theorem states that if $\bar{x}$ is a local solution of problem (1) and a regularity assumption holds, then there exist Lagrange multipliers $\bar{\lambda}_1, \ldots, \bar{\lambda}_m$ such that

$$f'(\bar{x}) + \sum_{j=1}^{m} \bar{\lambda}_j g_j'(\bar{x}) = 0, \ g(\bar{x}) \leq 0, \ \bar{\lambda}_j \geq 0, \ \bar{\lambda}_j g_j(\bar{x}) = 0, \ j = 1, \ldots, m. \quad (2)$$

In a more general form, the KKT conditions (2) can be written as

$$\bar{\lambda}_0 f'(\bar{x}) + \sum_{j \in I(\bar{x})} \bar{\lambda}_j g_j'(\bar{x}) = 0, \quad \bar{\lambda}_0 = 1, \quad \bar{\lambda}_j \geq 0, \quad j \in I(\bar{x}), \quad (3)$$

where $I(\bar{x}) = \{i \in \{1, \ldots, m\} \mid g_i(\bar{x}) = 0\}$ is the set of indices of active at $\bar{x}$ constraints. However, in the degenerate case, one cannot guarantee that $\bar{\lambda}_0 \neq 0$ in (3). For example, in the following optimization problem

$$\begin{aligned} \text{minimize } & -(x_1 + x_2) \\ \text{subject to } & g_1(x) = (x_1^2 + x_2^2)x_1 \leq 0, \\ & g_2(x) = (x_1^2 + x_2^2)x_2 \leq 0, \end{aligned} \quad (4)$$

a solution is $\bar{x} = (0,0)^T$, but conditions (3) are not satisfied at $\bar{x}$ with $\bar{\lambda}_0 = 1$. In this paper, we derive generalizations of the KKT conditions for optimization problem (4).

Our main purpose is to obtain necessary and sufficient conditions of optimality in the degenerate case under new regularity assumptions. While careful comparison and classification of the existing constraint qualifications are given in [19], all CQs are stated using at most the first derivatives of the constraints. In contrast, the regularity assumptions proposed in this paper expand constraint qualifications known in the literature to new classes of optimization problems by using higher-order derivatives. As a result, our approach allows us to analyze problems, where, for example, the first-order derivatives of all constraints are equal to zero, which are not covered by any regularity assumptions given in [2,3,18], and [19]. Also, in addition to the classical KKT-type optimality conditions analyzed in the literature, our approach is applicable to cases when the KKT Theorem fails but generalized forms of the KKT conditions proposed in the paper hold.

There are several methods to overcome the difficulty of degeneracy, for instance, those in [4,11,12,15–17,20]. Here we pursue an approach based on the construction of p-regularity introduced earlier in [24–26,29]. The main idea of the p-regularity approach is in using higher order derivatives of the constraints $g_i(x)$ to replace the gradients of the active constraints which are linearly dependent. To compare our approach with others, we would like to note that Ledzewicz and Schättler in [16] introduced a concept of the p-regular mapping, but in a different sense. A mapping is called p–regular at a point $\bar{x}$ with respect to an element h_1 in our sense if it is p–regular in the direction of the sequence $H_{p-1} = (h_1, 0, \ldots, 0)$ in the sense of Ledzewicz and Schättler [16]. However, both our definition and the definition from [16] reduce to the same definition of 2–regularity for $p = 2$. Ledzewicz and Schättler [16] also analyze p–regular problems, but they require the functions to be $(2p - 1)$-times continuously differentiable while we assume that the functions are $(p + 1)$-times continuously differentiable. Other results of this type are obtained in the work of Izmailov [13, 14] and of Izmailov and Solodov [15]. We compare our results with ones obtained in [13–15] and other relevant work in Sect. 6.

This paper continues the series of publications devoted to optimality conditions for degenerate optimization problems. In [5], we considered problems with equality constraints given in the operator form as $F(x) = 0$. The focus of the paper was on the case when the constraints are not regular at the solution $\bar{x}$ in the sense that the operator $F'(\bar{x})$ is not surjective. In [5], we derived new sufficient conditions for p-regular problems and necessary optimality conditions for problems satisfying the generalized condition of p–regularity. In [6] and [7], we turned our attention to problems with inequality constraints in the finite dimensional spaces in the completely degenerate case, when the following holds for any index i from the set $I(\bar{x})$ of indices of active at $\bar{x}$ constraints:

$$g_i^{(k)}(\bar{x}) = 0, \quad k = 1, \ldots, p - 1, \quad p \geq 2, \tag{5}$$

where p denotes a number for which (5) holds. Then work on optimality conditions for problems with inequality constraints was continued in [8–10, 21–23, 27], and [28]. Usually the completely degenerate case plays a key role in studying a degenerate situation.

The following assumption was made in [5–7, 21–23], and [25]: In the completely degenerate case (5), there exists a vector h, $\|h\| = 1$, such that

$$g_i^{(p)}(\bar{x})[h]^p = 0, \quad \forall\, i \in I(\bar{x}). \tag{6}$$

In addition, it was assumed that vectors $g_i^{(p)}(\bar{x})[h]^{p-1}$, $i \in I(\bar{x})$, are linearly independent. Under these assumptions (including (5) and (6)), necessary optimality conditions were obtained in the form

$$f'(\bar{x}) + \sum_{i \in I(\bar{x})} \lambda_i(h) \left(g_i^{(p)}(\bar{x})[h]^{p-1} \right)^T = 0, \quad \lambda_i(h) \geq 0, \quad i \in I(\bar{x}). \tag{7}$$

However, there exists a class of problems for which there is no vector $h \neq 0$ satisfying condition (6), and therefore, optimality conditions (7) do not hold. For example, if the constraints are defined by

$$g_1(x) = \left(x_1^2 + x_2^2 \right) x_1 \leq 0,$$
$$g_2(x) = \left(x_1^2 + x_2^2 \right) x_2 \leq 0,$$

then with $\bar{x} = 0$ and $p = 3$, there is no vector h, $\|h\| = 1$ such that $g_i'''(0)[h]^3 = 0$, $i = 1, 2$. This case will be considered in details in Example 1.

A similar situation can occur for a not completely degenerate case. For example, if the constraints in an optimization problem are defined by

$$g_1(x) = \left(x_1^2 + x_2^2\right) x_1 \leq 0,$$
$$g_2(x) = \left(x_1^2 + x_2^2\right) x_2 \leq 0, \tag{8}$$
$$g_3(x) = (x_2 - x_1)(x_1 - 2x_2) \leq 0,$$

or if the last constraint is $g_3(x) = x_1 x_2 \leq 0$. This case will be considered in details in Example 2. For the case of constraints defined in (8), optimality conditions similar to (7) were obtained in [27] and [28] under some additional assumptions.

In this paper, we extend our consideration to the general case and propose optimality conditions for degenerate problems that are not necessarily satisfy assumptions (5) and (6) (or similar assumptions in the not completely degenerate case). The optimality conditions presented in the paper resemble the structure of the classical optimality conditions. Necessary conditions reduce to the KKT conditions in the regular case. The optimality conditions can be applied to discretization of calculus of variations and optimal control problems. While our consideration covers the case when $p = 2$, there are some important applications that require $p > 2$ in such areas as geometric programming, quantum physics, singular optimal controls and others. Moreover, the examples given in the paper illustrate that there exist problems for which the proposed p-regularity concept works. The consideration in the paper does not intend to cover all degenerate (nonregular, abnormal) problems, but it covers some classes of the nonregular problems and allows us to get some theoretical results that were not obtained earlier for these classes of problems. The results presented in the paper can be viewed as extension of the earlier work published in [10] and [27].

The paper is organized as follows. In Sect. 2, we consider some definitions and results of the p-regularity-theory that we are going to use in this paper. In Sect. 3, we consider the general case of mapping $g(x)$ when condition (16) may fail. In Sect. 4, we present sufficient conditions for optimality in the general case. We illustrate the obtained results by several examples in Sect. 5.

Notation: The set $S = \{x \in \mathbb{R}^n \mid g_i(x) \leq 0, i = 1, \ldots, m\}$ is the feasible set for problem (1). For some set C, we denote the span of C by $\text{span}(C)$, and the set of all nonnegative combinations of vectors in C by $\text{cone}\, C$. We let $g_i^{(p)}(x)$ be the pth derivative of $g_i : \mathbb{R}^n \to \mathbb{R}^n$ at the point x; the associated p-form is $g_i^{(p)}(x)[h]^p = g_i^{(p)}(x)(h, h, \ldots, h)$. The notation $g_i^{(p)}(x)[h]^{p-1}$ means $\left(g_i^{(p-1)}(x)[h]^{p-1}\right)'_x$ (see [1] for additional details). We denote the scalar (dot) product between two elements x and y in $\mathbb{R}^n$ by $\langle x, y \rangle$. The other notation will be introduced below as needed.

2   Definitions and Results of the p-Regularity Theory

Consider some definitions and results of the p-regularity theory, which we are going to use in this paper. Without loss of generality, we can assume that all constraints of problem (1) are active at a solution $\bar{x}$ or $I(\bar{x}) = \{1, \ldots, m\}$. Assume that the gradients of the active constraints $g'_1(\bar{x}), \ldots, g'_m(\bar{x})$ are linearly dependent, so the solution $\bar{x}$ is nonregular. If, for example, r gradients $g'_1(\bar{x}), \ldots, g'_r(\bar{x})$, $r < m$, are linearly independent, then the other vectors $g'_{r+1}(\bar{x}), \ldots, g'_m(\bar{x})$ can be expressed as a linear combination of the linearly independent gradients:

$$g'_j(\bar{x}) = -\alpha_{j_1} g'_1(\bar{x}) - \ldots - \alpha_{j_r} g'_r(\bar{x}), \ \alpha_{j_k} \geq 0, \ k = 1, \ldots, r, \ j = r+1, \ldots, m. \quad (9)$$

Otherwise, there exists a vector $\xi \neq 0$ such that $\langle \xi, g'_i(\bar{x}) \rangle < 0$, $i \in I(\bar{x})$, and classical KKT conditions (3) hold with $\bar{\lambda}_0 = 1$. Therefore, we assume that (9) holds. Define $\tilde{g}_j(x)$ by

$$\tilde{g}_j(x) = g_j(x) + \alpha_{j_1} g_1(x) + \ldots + \alpha_{j_r} g_r(x), \quad j = r+1, \ldots, m. \quad (10)$$

Then the set of functions $\{g_1(x), \ldots, g_r(x), \tilde{g}_{r+1}(x), \ldots, \tilde{g}_m(x)\}$ satisfies

$$\tilde{g}'_j(\bar{x}) = 0, \quad j = r+1, \ldots, m. \quad (11)$$

Choosing different initial sets $\{g'_{k_1}(\bar{x}), \ldots, g'_{k_r}\}$ of r linearly independent gradients, $k = 1, \ldots, l$ ($l \leq m$), and applying transformations similar to (10), we get l different sets of functions G_k that satisfy (11) and have the form:

$$G_k = \{g_{k_1}(x), \ldots, g_{k_r}(x), \tilde{g}_j(x) \mid j \in I(\bar{x}) \backslash \{k_1, \ldots k_r\}\}, \quad k = 1, \ldots, l. \quad (12)$$

Note that since initial sets of linearly independent gradients are all different, the resulting sets of functions G_k are also different.

For every $k = 1, \ldots, l$, define a set S_k as follows:

$$S_k = \{x \in \mathbb{R}^n \mid g_{k_1}(x) \leq 0, \ldots, g_{k_r}(x) \leq 0, \ \tilde{g}_j(x) \leq 0, \ j \in I(\bar{x}) \backslash \{k_1, \ldots k_r\}\}. \quad (13)$$

We will need the following lemma.

Lemma 1. *For the functions $g_1(x), \ldots, g_m(x)$, there exist $l \leq m$ sets of functions G_k defined in (12) such that $S = \bigcap\limits_{k=1}^{l} S_k$, where sets S_k are defined in (13).*

Proof. The proof of the lemma follows from the fact that, for any $k = 1, \ldots, l-1$, the definition of the set of functions G_k with the number $k+1$ implies that at least one function $g_j(x)$, $j \in I(\bar{x})$, is used in the definition of the set S_{k+1} and is not in S_k. The process of defining the sets G_k also implies that each index j from the set $I(\bar{x})$ is used at least once. Then, by the definition of S_k, $\bigcap\limits_{k=1}^{l} S_k \subseteq \bigcap\limits_{i=1}^{m} A_i = S$, where $A_i = \{x \in \mathbb{R}^n \mid g_i(x) \leq 0\}$, $i = 1, \ldots, m$. At the same time, for every $k = 1, \ldots, l$, $S \subseteq S_k$, and hence, $S \subseteq \bigcap\limits_{k=1}^{l} S_k$. Thus $S = \bigcap\limits_{k=1}^{l} S_k$ holds.

Lemma 1 implies that problem (1) can be written as

$$\text{minimize} \ f(x)$$
$$\scriptstyle x \in \mathbb{R}^n$$

$$\text{subject to } x \in \bigcap_{k=1}^{l} S_k.$$

2.1   General Case with $p \geq 2$

Similarly to consideration above with sets of functions G_K satisfying (9)–(10) and with some number $p \geq 2$, problem (1) can be transformed into the following one

$$\min f(x), \quad x \in S = \bigcap_{k=1}^{l'} S'_k, \tag{14}$$

where

$$S'_k = \{x \in \mathbb{R}^n \mid g_{k_1}(x) \leq 0, \ldots, g_{k_{r_k}}(x) \leq 0,$$
$$\tilde{g}_{k_{r_k+1}}(x) \leq 0, \ldots, \tilde{g}_{k_{r_{k+1}}}(x) \leq 0,$$
$$\ldots$$
$$\tilde{g}_{k_{r_{k+p-1}+1}}(x) \leq 0, \ldots, \tilde{g}_{k_{r_{k+p}}}(x) \leq 0\}.$$

and

$$
\begin{aligned}
g'_j(\bar{x}) &= 0, \quad j \in I(\bar{x}) \backslash \{k_1, \ldots, k_{r_k}\} \\
\tilde{g}''_j(\bar{x}) &= 0, \quad j \in I(\bar{x}) \backslash \{k_1, \ldots, k_{r_k}, k_{r_k+1}, \ldots, k_{r_{k+1}}\} \\
&\ldots\ldots\ldots\ldots\ldots\ldots\ldots\ldots\ldots\ldots\ldots \\
\tilde{g}_j^{(p-1)}(\bar{x}) &= 0, \quad j \in I(\bar{x}) \backslash \{k_1, \ldots, k_{r_k}, \ldots, k_{r_{k+p-1}+1}, \ldots, k_{r_{k+p}}\}
\end{aligned}
\tag{15}
$$

The transformation of the constraints is possible under an assumption that the operators of higher derivatives $\left\{g_i^{(k)}(\bar{x})\right\}_{i=r_{k-1}+1,\ldots,r_k}$, $k = 1, \ldots, p$, $r_0 = 0$, are linearly independent. As a result of the problem transformation, necessary and sufficient conditions of optimality for degenerate problem (1) can be obtained by considering problem (14) instead.

First of all, under some additional p–regularity conditions, we will obtain descriptions of critical cones to every set S'_k, $k = 1, \ldots, l'$, with specific properties (15) and of the critical cone to the set S at the point $\bar{x}$.

Without loss of generality, we use notation S_k instead of S'_k and l instead of l'. Also, to simplify the consideration, let us assume that the set S_k is defined using a set of functions

$$G_k = \{g_1(x), \ldots, g_{r_1}(x), g_{r_1+1}(x), \ldots, g_{r_2}(x), \ldots, g_{r_{p-1}+1}(x), \ldots, g_{r_p}(x)\},$$

where $r_p = m$ and

$$
\begin{aligned}
g'_i(\bar{x}) &= 0, \quad i = r_1 + 1, \ldots, m, \\
g''_i(\bar{x}) &= 0, \quad i = r_2 + 1, \ldots, m, \\
&\ldots\ldots\ldots\ldots\ldots\ldots\ldots\ldots\ldots\ldots\ldots \\
g_i^{(p-1)}(\bar{x}) &= 0, \quad i = r_{p-1} + 1, \ldots, r_p = m.
\end{aligned}
\tag{16}
$$

Conditions (16) can be guaranteed as a result of special transformations of the constraints of the problem (1) using linear transformations of the mapping $g(x)$ similar to the one defined in (9).

Introduce a new set of vectors as follows with $r_0 = 0$:

$$H_p(\bar{x}) = \{h \in \mathbb{R}^n \mid g_i^{(k)}(\bar{x})[h]^k \leq 0, \ i = r_{k-1} + 1, \ldots, r_k, \ k = 1, \ldots, p\}. \tag{17}$$

The set $H_p(\bar{x})$ is a generalization of the set $H_2(\bar{x})$ introduced in [10].

Define additional sets of indices for some vector $h \in H_p(\bar{x})$ as follows:

$$I_0^k(h) = \left\{ i \in \{r_{k-1}+1,\ldots,r_k\} \mid g_i^{(k)}(\bar{x})[h]^k = 0 \right\}, \quad k = 1,\ldots,p, \quad r_0 = 0.$$

and $I_0(h) = \bigcup\limits_{k=1}^{p} I_0^k(h)$.

To give the following definition, we will introduce some additional assumptions and notation. Let $\mathcal{K} \in \mathbb{R}^n$ be a closed cone with a nonempty interior and let $\partial\mathcal{K}$ be the boundary of the cone $\mathcal{K}$. Assume that a set Σ consists of indices σ_i such that for each $\sigma_i \in \Sigma$, a part of the boundary $\partial\mathcal{K}$ divides $\mathbb{R}^n$ into two cones $\mathcal{K}_{\sigma_i}^+ \subset \mathbb{R}^n$ and $\mathcal{K}_{\sigma_i}^- \subset \mathbb{R}^n$ with boundary $\partial\mathcal{K}_{\sigma_i}^+ \subset \partial\mathcal{K}$ and $\partial\mathcal{K}_{\sigma_i}^- \subset \partial\mathcal{K}$ in such a way that $\mathbb{R}^n = \mathcal{K}_{\sigma_i}^+ \cup \mathcal{K}_{\sigma_i}^-$, and the following holds:

1. $\mathcal{K} \cap \mathrm{int}(\mathcal{K}_{\sigma_i}^+) = \emptyset$;
2. $\mathcal{K} \subset \mathcal{K}_{\sigma_i}^-$;
3. $\mathrm{int}(\mathcal{K}_{\sigma_i}^+) \cap \mathrm{int}(\mathcal{K}_{\sigma_j}^+) = \emptyset$ for any two indices $\sigma_i \in \Sigma$ and $\sigma_j \in \Sigma$, $\sigma_i \neq \sigma_j$.

Definition 1. *Let* $\overline{\mathcal{K}} = \mathbb{R}^n \backslash \mathrm{conv}(\mathcal{K})$, *where* $\mathcal{K} \in \mathbb{R}^n$ *is a closed cone with a nonempty interior. If for some* $\sigma_i \in \Sigma$, $\overline{\mathcal{K}} \subset \mathcal{K}_{\sigma_i}^+$, *then we define the exterior boundary* $\partial_E\mathcal{K}$ *of the cone* $\mathcal{K}$ *as the boundary of* $\partial\mathcal{K}_{\sigma_i}^+$.

Remark 1. The boundary $\partial\mathcal{K}$ divides the space $\mathbb{R}^n$ into several parts, in general. However, it will be important to consider only the parts (cones) approximating the feasible set S. The remaining parts (cones) can be consider as not essential. (For the better understanding of the concepts, see Examples 1 and 2.)

Let us introduce the concept of p-regularity of the constraints $g_i(x)$, $i = 1,2,\ldots,m$. For some $h \in \mathbb{R}^n$, define the set of vectors:

$$G(h) = \{g'_{i_1}(\bar{x}), g''_{i_2}(\bar{x})[h],\ldots,g_{i_p}^{(p)}(\bar{x})[h]^{p-1} \mid i_k \in I_0^k(h), \ k = 1,\ldots,p\}. \quad (18)$$

The vectors $g_{i_k}^{(k)}(\bar{x})[h]^{k-1}$, $k = 1,\ldots,p$, are called k-factor operators.

Definition 2. *The mapping* $g(x) = (g_1(x),\ldots,g_m(x))$ *is called p-regular at* $\bar{x} \in S$ *with respect to some vector* $h \in \mathbb{R}^n$ *if the set of vectors* $G(h)$ *defined in (18) is linearly independent.*

Definition 3. *The mapping* $g(x) = (g_1(x),\ldots,g_m(x))$ *is called p-regular at* $\bar{x} \in S$ *if the set of vectors* $G(h)$ *defined in (18) is linearly independent for any* $h \in H_p(\bar{x})$, $\|h\| = 1$, *or* $H_p(\bar{x}) = \{0\}$.

Lemma 2. *Let mapping* $g(x) = (g_1(x),\ldots,g_m(x))$ *be p-regular at the point* $\bar{x} \in S$. *Then any vector* $h \in H_p(\bar{x})$ *such that* $\|h\| \neq 0$ *and* $I_0(h) \neq \emptyset$ *belongs to the boundary* $\partial H_p(\bar{x})$, *i.e.* $h \in \partial H_p(\bar{x})$.

Proof. Since for any vector $h \in H_p(\bar{x})$, the set of vectors $G(h)$ defined in (18) is linearly independent, then there exists a vector $\bar{h}$ such that $g_i^{(k)}(\bar{x})[h]^{k-1}\bar{h} < 0$, $i = 1,\ldots,m$, $k = 1,\ldots,p$, and the proof follows from the property of the p-form.

Corollary 1. *Let mapping $g(x) = (g_1(x), \ldots, g_m(x))$ be p-regular at the point $\bar{x} \in S$.*

Let $h \in H_p(\bar{x})$, $h \neq 0$, and $I_0(h) \neq \emptyset$. Then these conditions on the vector h guarantee existence of a nonempty interior $\mathrm{Int}_h H_p(\bar{x}) \neq \emptyset$ and a boundary element $h \in \partial H_p(\bar{x})$.

Corollary 2. *Assume that mapping $g(x) = (g_1(x), \ldots, g_m(x))$ is p-regular at the point $\bar{x} \in S$. Consider a set $\bar{H}$ of all vectors $h \in H_p(\bar{x})$ such that $\|h\| = 1$ and $I_0(h) \neq \emptyset$. Then $\mathrm{Int}_h H_p(\bar{x})$ contains a ball D_h of the same radius $\delta > 0$ uniformly for all h from the set $\bar{H}$.*

Definition 4. *Let J be a set of indices j. We say that the cone $H_p(\bar{x})$ is decomposable into elementary cones $H_p^j(\bar{x})$, $j \in J$, if the following holds:*

1. $H_p(\bar{x}) = \bigcup\limits_{j \in J} H_p^j(\bar{x})$.
2. *Each cone $H_p^j(\bar{x})$ has a nonempty interior $\mathrm{int}(H_p^j(\bar{x}))$.*
3. $\mathrm{int}(H_p^{j_1}(\bar{x})) \cap \mathrm{int}(H_p^{j_2}(\bar{x})) = \emptyset$ *for any $j_1, j_2 \in J$, $j_1 \neq j_2$.*
4. $H_p^j(\bar{x})$ *is not decomposable into elementary cones for any $j \in J$.*

Lemma 3. *Let mapping $g(x) = (g_1(x), \ldots, g_m(x))$ be p-regular at the point $\bar{x} \in S$. Assume that there exist vectors $h \in H_p(\bar{x})$ such that $\|h\| \neq 0$ and $I_0(h) \neq \emptyset$.*

Then the cone $H_p(\bar{x})$ is decomposable into a finite sum of elementary cones with a nonempty interior:

$$H_p(\bar{x}) = \bigcup_{j=1}^{s} H_p^j(\bar{x}), \tag{19}$$

i.e. $J = \{1, \ldots, s\}$ is a finite set.

2.2   The pth Order Optimality Conditions for Inequality Constrained Optimization Problems in a Special Case

In this section, we prove the pth order optimality conditions for problem (1) under assumptions (16).

Assume that conditions of Lemma 3 hold and, for some $j \in J = \{1, \ldots, s\}$, consider an elementary cone $H_p^j(\bar{x})$ defined in (19). Introduce a set of indices:

$$\Omega_j = \{\omega \mid h_\omega^j \in \partial_E H_p^j(\bar{x})\}. \tag{20}$$

Note that $\bigcup\limits_{\omega \in \Omega_j} h_\omega^j = \partial_E H_p^j(\bar{x})$ and the set Ω_j may be continuum.

Theorem 1. *(Necessary Conditions for Optimality) Let $f \in C^2(\mathbb{R}^n)$, $g_i \in C^{p+1}(\mathbb{R}^n)$, $i = 1, \ldots, m$, and assume that the mapping $g(x) = (g_1(x), \ldots, g_m(x))$ is p-regular at the solution $\bar{x}$ of the problem (1). Then, for every $j \in \{1, \ldots, s\}$,* there exists a finite number of indices $\{\omega_{\nu_1}, \ldots, \omega_{\nu_j}\} \subset \Omega_j$ and numbers $\lambda_i^k(h_{\omega_\nu}^j) \leq 0$, $i \in I_0^k(h_{\omega_\nu}^j)$, $k = 1, \ldots, p$, $\nu = 1, \ldots, \nu_j$, such that

$$
f'(\bar{x}) = \sum_{\nu=1}^{\nu_j} \left\{ \sum_{i \in I_0^1(h_{\omega_\nu}^j)} \lambda_i^1(h_{\omega_\nu}^j) g_i'(\bar{x}) + \sum_{i \in I_0^2(h_{\omega_\nu}^j)} \lambda_i^2(h_{\omega_\nu}^j) \left(g''_i(\bar{x})[h_{\omega_\nu}^j] \right)^T + \ldots \right.
$$
$$
\left. + \sum_{i \in I_0^p(h_{\omega_\nu}^j)} \lambda_i^p(h_{\omega_\nu}^j) \left(g_i^{(p)}(\bar{x})[h_{\omega_\nu}^j]^{p-1} \right)^T \right\}.
$$

(21)

Remark 2. If $\partial H_p(\bar{x}) = \emptyset$ then $\partial_E H_p(\bar{x}) = \emptyset$ and $f'(\bar{x}) = 0$, i.e. $H_p(\bar{x}) = \mathbb{R}^n$.

Remark 3. If both $H_p(\bar{x}) = \{0\}$ and $\partial_E H_p(\bar{x}) = \{0\}$, then the right-hand side of (21) is assumed to be equal to $\mathbb{R}^n$, i.e. $f'(\bar{x}) \in \mathbb{R}^n$.

Proof. (See [28].) Since the mapping $g(x) = (g_1(x), \ldots, g_m(x))$ is p-regular at the solution $\bar{x}$ of the problem (1) and (16) holds, the constraints $g_i(x)$, $i = 1, \ldots, m$ can be rewritten in the following form:

$$
g_i(x) = g_i'(\bar{x})[x - \bar{x}], \quad i = 1, \ldots, r_1;
$$
$$
\ldots
$$
$$
g_i(x) = g_i^{(p)}(\bar{x})[x - \bar{x}]^p, \quad i = r_{p-1} + 1, \ldots, r_p.
$$

This follows from the fact that under the assumptions of the theorem, the feasible set S can be approximated by the set

$$
S_{app} = \{x \in \mathbb{R}^n \mid g_i^{(k)}(\bar{x})[x - \bar{x}] \leq 0, \ i = r_{k-1}, \ldots, r_k, \ k = 1, \ldots, p, \ r_0{=}1\}.
$$

The approximation follows from the Taylor expansion of functions $g_i(x)$, $i = 1, \ldots, m$, at the point $\bar{x}$ up to the first nonzero term.

We choose and fix an index $j \in \{1, \ldots, s\}$. To prove the theorem, it is sufficient to prove the following for the chosen index j:

$$
f'(\bar{x}) \in \bigcup_{h^j \in \partial_E H_p^j(\bar{x})} \text{Cone} \left(g_{i_1}'(\bar{x}), g_{i_2}''(\bar{x})[h^j], \ldots, g_{i_p}^{(p)}(\bar{x})[h^j]^{p-1} \right),
$$

(22)

where $i_k \in I_0^k(h^j)$, $k = 1, \ldots, p$.

Recall that we introduced the exterior boundary $\partial_E H_p^j(\bar{x})$ in Definition 1. We consider two cases:

Case 1. In the first case, assume that there is no supporting hyperplane for the cone $H_p(\bar{x})$ that passes through the origin. Thus, there is no hyperplane such that the cone $H_p(\bar{x})$ is entirely contained in one of the two closed half-spaces bounded by the hyperplane.

In this case, consider any hyperplane π that is formed by a normal vector ξ and passes through the origin so that $\pi = \{x \in \mathbb{R}^n \mid \langle \xi, x \rangle = 0\}$. Then there exist vectors $h_1, h_2 \in H_p(\bar{x})$ such that $h_1 \in \text{int}(\pi_+)$ and $h_2 \in \text{int}(\pi_-)$, so that

$\langle h_1, \xi \rangle > 0$ and $\langle h_2, \xi \rangle < 0$. This means that $f'(\bar{x}) = 0$ and the conditions (21) are satisfied with

$$\lambda_i^k(h_\omega^j) = 0, \quad i \in I_0^k(h_\omega^j), \quad k = 1, \ldots, p, \quad h_\omega^j \in \partial_E H_p^j(\bar{x}), \quad j \in 1, \ldots, s.$$

Case 2. In this case, assume that there exists a supporting hyperplane π for the cone $H_p(\bar{x})$ that passes through the origin and one of the two closed half-spaces bounded by the hyperplane contains $H_p^j(\bar{x})$, $j = 1, \ldots, s$. Without loss of generality, assume that the half-space is π_+, that is for all $h \in H_p^j(\bar{x})$ the following holds: $\langle h, \xi \rangle \geq 0$. If $\theta_{h_1 h_2}$ is an angle between vectors h_1 and h_2, then obviously, in this case, $\theta_{h_1 h_2} \leq \pi$ for any $h_1, h_2 \in H_p^j(\bar{x})$.

Without loss of generality, in the next part of the proof, we consider the completely degenerate case, similarly to how it was done in previous publications [5,7,8], and [6]. Indeed, consideration in the general case is similar to the degenerate case but requires more technical details and additional notation.

Assume, that

$$g_i^{(k)}(\bar{x}) = 0, \quad i \in I(\bar{x}), \quad k = 1, \ldots, p - 1.$$

It follows from the p-regularity of the constraints and Lemma 3 that there exists nonempty interior, $\text{int}(H_p^j(\bar{x})) \neq \emptyset$, and the closed exterior boundary $\partial_E H_p^j(\bar{x})$, which consists of smooth $(p-1)$-dimensional manifolds

$$\{h \in \mathbb{R}^n | g_i^{(p)}(\bar{x})[h]^p = 0, \quad h \in H_p^j(\bar{x}), \quad i \in I_0^p(\bar{x})\}.$$

Consider the conjugate to $H_p^j(\bar{x})$ defined as

$$(H_p^j(\bar{x}))^* = \{h^* \in \mathbb{R}^n \,|< h^*, h >\leq 0, \quad \forall h \in H_p^j(\bar{x})\}.$$

Obviously, necessary optimality conditions can be written as

$$-f'(\bar{x}) \in (H_p^j(\bar{x}))^*. \tag{23}$$

Now our goal is to show that

$$(H_p^j(\bar{x}))^* \subset \{\lambda_i g_i^{(p)}(\bar{x})[h^j]^{p-1} \,|\, \lambda_i \geq 0, \quad h^j \in \partial_E H_p^j(\bar{x}), \quad i \in I_0^p(h^j)\}. \tag{24}$$

Let $\widehat{h} \in \text{int}(H_p^j(\bar{x}))$ and let $\widehat{l}$ be a vector orthogonal to $\widehat{h}$. Define a line l in the direction of vector $\widehat{l}$. Let $\widehat{h}_1(l)$ and $\widehat{h}_2(l)$ be two points of intersection of the line l with the boundary $\partial_E H_p^j(\bar{x})$.

Note that in the case of $H_p^j(\bar{x})$ being a half-space, we can replace the line l with an arc

$$\xi(\beta_1, \beta_2) = \beta_1 \widehat{h} + \beta_2 \widehat{l}, \quad \beta_1^2 + \beta_2^2 = 1.$$

Without loss of generality, assume that the boundary of $H_p^j(\bar{x})$ is limited by one of the constraints $g_{i_0}(x)$. As the result, we have

$$g_{i_0}^{(p)}(\bar{x})[\widehat{h}_1(l)]^p = 0 = g_{i_0}^{(p)}(\bar{x})[\widehat{h}_2(l)]^p,$$

$$I_0^p(\widehat{h}_1(l)) = I_0^p(\widehat{h}_2(l)) = i_0.$$

Consider vectors $g_{i_0}^{(p)}(\bar{x})[\widehat{h}_1(l)]^{p-1}$ and $g_{i_0}^{(p)}(\bar{x})[\widehat{h}_2(l)]^{p-1}$, which, in this case, are also the p-factor operators.

Obviously, the following inclusion holds:

$$(H_p^j(\bar{x}))^* \cap \mathrm{span}(g_{i_0}^{(p)}(\bar{x})[\widehat{h}_1(l)]^{p-1}, g_{i_0}^{(p)}(\bar{x})[\widehat{h}_2(l)]^{p-1})$$

$$\subset \{\alpha_1 g_{i_0}^{(p)}(\bar{x})[\widehat{h}_1(l)]^{p-1} + \alpha_2 g_{i_0}^{(p)}(\bar{x})[\widehat{h}_2(l)]^{p-1}, \alpha_1 \geq 0, \alpha_2 \geq 0\}.$$

By the definition of the conjugate cone in the two-dimensional space (or Farkas Theorem) we can make the following observation. When we move the line l along the hyperplane γ defined by $\gamma = \{z \in \mathbb{R}^n | \langle \widehat{h}, z \rangle = 0\}$, we obtain all points of intersection of the hyperplane γ with the boundary $\partial_E H_p^j(\bar{x})$ and come back to the original points of intersection $\widehat{h}_1(l)$ and $\widehat{h}_2(l)$.

On the other hand, considering the span defined as

$$\mathrm{span}(g_{i_0}^{(p)}(\bar{x})[\widehat{h}_1(l)]^{p-1}, g_{i_0}^{(p)}(\bar{x})[\widehat{h}_2(l)]^{p-1})$$

and taking the full displacement of l along γ gives the points of intersections $(H_p^j(\bar{x}))^*$ with space $\mathbb{R}^n$, so that

$$(H_p^j(\bar{x}))^* \subset \bigcup_{l \in \gamma} \{\alpha_1 g_{i_0}^{(p)}(\bar{x})[\widehat{h}_1(l)]^{p-1} + \alpha_2 g_{i_0}^{(p)}(\bar{x})[\widehat{h}_2(l)]^{p-1}, \quad \alpha_1 \geq 0, \quad \alpha_2 \geq 0\}.$$

Thus, we get

$$\bigcup_{l \in \gamma} \{\alpha_1 g_{i_0}^{(p)}(\bar{x})[\widehat{h}_1(l)]^{p-1} + \alpha_2 g_{i_0}^{(p)}(\bar{x})[\widehat{h}_2(l)]^{p-1}, \quad \alpha_1 \geq 0, \alpha_2 \geq 0\}$$

$$\subset \{\alpha_1 g_{i_0}^{(p)}(\bar{x})[\widehat{h}_1(l)]^{p-1}, \alpha_1 \geq 0, l \in \gamma) = \{\alpha_1 g_{i_0}^{(p)}(\bar{x})[h]^{p-1} \mid h \in \partial_E H_p^j(\bar{x})\}.$$

Taking into account (23) we obtain (24).

Corollary 3. *Let the assumptions of Theorem 1 hold. Then for $h_{\omega_l}^j \in \partial_E H_p^j(\bar{x})$, $l = 1, \ldots, l_j$, the following holds with $\lambda_i^k \leq 0$, $k = 1, \ldots, p$, and any i:*

$$f'(\bar{x}) \in \bigcap_{j=1}^{s} \left\{ \sum_{l=1}^{l_j} \left(\sum_{i \in I_0^1(h_{\omega_l}^j)} \lambda_i^1 g_i^1(\bar{x}) + \ldots + \sum_{i \in I_0^p(h_{\omega_l}^j)} \lambda_i^p g_i^{(p)}(\bar{x})[h_{\omega_l}^j]^{p-1} \right) \right\}.$$

3   The pth Order Optimality Conditions in the General Case of Degeneracy

In this section, we consider the general case of degeneracy, when condition (16) may fail.

For every $k = 1, \ldots, s$, we choose a different set of indices $\{k_1, \ldots, k_r\} \subset \{1, \ldots, m\}$, such that $g'_{k_1}(\bar{x}), \ldots, g'_{k_r}(\bar{x})$ are linearly independent and define $\tilde{g}_l(x)$ using Equation (10). Note that there might be various sets of indices $\{k_1, \ldots, k_r\}$ corresponding to the same value of k.

Similarly to notation in (17), we denote by $H_k(x)$, $k = 1, \ldots, s$, the cone corresponding to the set of constraints,

$$\{g_{k_1}(x), \ldots, g_{k_r}(x), \tilde{g}_l(x) \mid l \in I(\bar{x}) \backslash \{k_1, \ldots, k_r\}\}, \tag{25}$$

so that with $l \in I(\bar{x}) \backslash \{k_1, \ldots, k_r\}$,

$$H_k(\bar{x}) = \{h \in \mathbb{R}^n \mid g'_{k_1}(\bar{x})h \le 0, \ldots, g'_{k_r}(\bar{x})h \le 0, \tilde{g}''_l(\bar{x})[h]^2 \le 0\}.$$

For each $H_k(x)$, $k = 1, \ldots, s$, we define l_k elementary cones $H_k^{j_r}(x)$, $j_r \in \{1, \ldots, l_k\}$ corresponding to system (25).

Let $\Omega_k^{j_r}$ be a set of indexes ω such that

$$h_\omega^{j_r} \in \partial_E H_k^{j_r}(x) \qquad \text{and} \qquad \bigcup_{\omega \in \Omega_k^{j_r}} h_\omega^{j_r} = \partial_E H_k^{j_r}(x).$$

To simplify the consideration, we continue with the case $p = 2$ in the rest of the section.

Definition 5. *We say that the mapping $g(x)$ is tangent $2-$regular at the point $\bar{x}$ if for any $h \in H_k(\bar{x})$, there exists a set of feasible points $x(\alpha) \in S$ in the form*

$$x(\alpha) = \bar{x} + \alpha h + r(\alpha), \quad \alpha \in (0, \varepsilon),$$

where $\varepsilon > 0$ is sufficiently small and $\|r(\alpha)\| = o(\alpha^{3/2})$. Obviously, $x(\alpha)$ and $r(\alpha)$ depend on h.

Remark. If there exists an element ξ such that

$$< g'_{k_i}(\bar{x}), \xi > < 0, \quad i = 1, \ldots, r$$
$$< \tilde{g}''_i(\bar{x}) h_\omega^{j_r}, \xi > < 0, \quad i \in I(\bar{x}) \backslash \{k_1, \ldots, k_r\}.$$

then $g(x)$ is tangent $2-$regular at the point $\bar{x}$. Moreover, any point in the arc $\bar{x}(\alpha)$ defined by

$$\bar{x}(\alpha) = \bar{x} + \alpha h + \alpha^{3/2}\xi + r(\alpha), \quad \alpha \in (0, \varepsilon),$$

is a feasible point, where $\varepsilon > 0$ is sufficiently small and $\|r(\alpha)\| = o(\alpha^{3/2})$.

Now we are ready to formulate the main result for the case $p = 2$.

Theorem 2. *Let $f \in C^2(R^n)$ and the mapping $g(x) \in C^3(R^n)$ be tangent $2-$regular at $\bar{x}$. Then for any $j_r \in \{1, \ldots, l_k\}$, $k = 1, \ldots, s$, there exist $h_{\omega_l}^{j_r}$, $l = 1, \ldots, l_r$, and multipliers $\lambda_k^1(h_{\omega_l}^{j_r}) \le 0$, $\lambda_k^2(h_{\omega_l}^{j_r}) \le 0$, such that*

$$f'(\bar{x}) = \sum_{l=1}^{l_r} \left(\sum_{i=k_1}^{k_r} \lambda_i^1(h_{\omega_l}^{j_r}) g'_i(\bar{x}) + \sum_{i \in I(\bar{x}) \backslash \{k_1, \ldots, k_r\}} \lambda_i^2(h_{\omega_l}^{j_r}) \tilde{g}''_i(\bar{x})[h_{\omega_l}^{j_r}] \right).$$

The proof of the theorem is similar to the proof in the absolutely degenerate case.

Corollary 4. *Let* $f \in C^2(R^n)$ *and the mapping* $g(x) \in C^3(R^n)$ *be tangent* 2−*regular at* $\bar{x}$. *Then for any* $j_r \in \{1, \ldots, l_k\}$, $k = 1, \ldots, s$, *there exist* $h_{\omega_l}^{j_r}$, $l = 1, \ldots, l_r$, *and multipliers* $\lambda_k^1(h_{\omega_l}^{j_r}) \leq 0$, $\lambda_k^2(h_{\omega_l}^{j_r}) \leq 0$, *such that* $\lambda_i^1 \leq 0, \lambda_i^2 \leq 0, \forall i$ *and*

$$f'(\bar{x}) \in \bigcap_{k=1}^{s} \bigcap_{r=1}^{k} \left\{ \sum_{l=1}^{l_r} \left(\sum_{i=k_1}^{k_r} \lambda_i^1 g_i'(\bar{x}) + \sum_{i \in I(\bar{x}) \backslash \{k_1, \ldots, k_r\}} \lambda_i^2 \tilde{g}_i''(\bar{x})[h_{\omega_l}^{j_r}] \right) \right\}.$$

4   Sufficient Optimality Conditions

In this section we present sufficient conditions for optimality in the general case ($p \geq 0$). For the sake of simplicity we consider the optimization problem

$$\min_{x \in \mathbb{R}^n} f(x) \tag{26}$$

subject to

$$g(x) = (g_1(x), \ldots, g_{r_1}(x), g_{r_1+1}(x), \ldots, g_m(x)) \leq 0, \tag{27}$$

where $g(x)$ satisfies (16).

Let us introduce a generalization of the Lagrange function:

$$L_p(x, h, \lambda(h)) = f(x) + \sum_{i \in I_0^1(h)} \lambda_i^1 g_i(x) + \ldots + \sum_{i \in I_0^p(h)} \lambda_i^p g_i^{(p-1)}(x)[h]^{p-1},$$

where $\lambda(h) = (\lambda^1(h), \ldots, \lambda^p(h))$ and $\lambda^j(h) = (\lambda_i^j(h))_{i \in I_0^j(h)}, j = 1, \ldots, p$.

Theorem 3. *(Sufficient Conditions for Optimality)*
   Let $\bar{x}$ *be a feasible point for the problem* (26)–(27), $f(x) \in C^2(\mathbb{R}^n, R)$ *and* $g(x) \in C^{p+1}(\mathbb{R}^n, \mathbb{R}^m)$. *Assume that* $g(x)$ *is p-regular at the point* $\bar{x}$. *Assume also that for any* $h \in H_p(\bar{x})$, *either* $\langle f'(\bar{x}), h \rangle > 0$ *or there exists* $\bar{\lambda}(h)$, $\bar{\lambda}(h) \leq 0$, *and* $\beta > 0$ *such that* $L_p'(\bar{x}, h, \bar{\lambda}(h)) = 0$ *and*

$$\langle L_p''(\bar{x}, h, \tilde{\lambda}(h))h, h \rangle \geq \beta \|h\|^2, \tag{28}$$

where

$$\tilde{\lambda}(h) = (\bar{\lambda}^1(h), \frac{1}{3}\bar{\lambda}^2(h), \ldots, \frac{2}{i(i+1)}\bar{\lambda}^i(h), \ldots, \frac{2}{p(p+1)}\bar{\lambda}^p(h)). \tag{29}$$

Then $\bar{x}$ *is an isolated local minimizer of* (26)–(27).

Proof. We give a proof of this theorem for the general case of degeneracy. The proof of the completely degenerate case is considered in [9].

Assume on the contrary that $\bar{x}$ is not a local minimizer. Then there exists a sequence $\{x_k\} \to \bar{x}$ such that $f(x_k) < f(\bar{x})$ and $g(x_k) \leq 0$. Consider a sequence (we use the same notation for the sequence and its convergent subsequence) $\left\{\frac{x_k - \bar{x}}{\|x_k - \bar{x}\|}\right\}$ that converges to some element $\tilde{h}$. Then

$$x_k = \bar{x} + \|x_k - \bar{x}\|\tilde{h} + \omega(x^k) = \bar{x} + t\tilde{h} + \xi_k, \tag{30}$$

where $\|\omega(x^k)\| = o(\|x_k - \bar{x}\|)$. Observe that $\tilde{h} \in H_p(\bar{x})$ and consider two cases.

Case 1. If $\langle f'(\bar{x}), \tilde{h} \rangle > 0$, then

$$f(x_k) = f(\bar{x}) + \langle f'(\bar{x}), \|x_k - \bar{x}\|\tilde{h} \rangle + \xi_k > f(\bar{x}),$$

which leads to a contradiction, so this case does not hold.

Case 2. If $\langle f'(\bar{x}), \tilde{h} \rangle = 0$, then there exists $\lambda(\tilde{h}) \leq 0$ such that

$$f(x_k) - f(\bar{x}) \geq f(x_k) - f(\bar{x}) + \sum_{i \in I_0^1(\tilde{h})} \bar{\lambda}^1(\tilde{h})g_i(x_k) + \ldots + \frac{(p-1)!}{t^{p-1}} \sum_{i \in I_0^p(\tilde{h})} \bar{\lambda}^p(\tilde{h})g_i(x_k)$$

$$= f'(\bar{x})(x_k - \bar{x}) + \frac{1}{2}f''(\bar{x})(x_k - \bar{x})^2 + \sum_{i \in I_0^1(\tilde{h})} \bar{\lambda}^1(\tilde{h})g_i'(\bar{x})[x_k - \bar{x}]$$

$$+ \frac{1}{2} \sum_{i \in I_0^1(\tilde{h})} \bar{\lambda}^1(\tilde{h})g_i''(\bar{x})[x_k - \bar{x}]^2 + \frac{1}{2t} \sum_{i \in I_0^2(\tilde{h})} \bar{\lambda}^2(\tilde{h})g_i''(\bar{x})[x_k - \bar{x}]^2 + \ldots$$

$$+ \frac{1}{p(p+1)t^{p-1}} \sum_{i \in I_0^p(\tilde{h})} \bar{\lambda}^p(\tilde{h})g_i^{(p+1)}(\bar{x})[x_k - \bar{x}]^{p+1} + o(\|x_k - \bar{x}\|^2) =$$

$$= \left\langle f'(\bar{x}) + \sum_{i \in I_0^1(\tilde{h})} \bar{\lambda}^1(\tilde{h})g_i'(\bar{x}) + \ldots + \sum_{i \in I_0^p(\tilde{h})} \bar{\lambda}^p(\tilde{h})g_i^{(p)}(\bar{x})[\tilde{h}]^{p-1}, \xi_k \right\rangle +$$

$$+ \frac{1}{2}\left\langle \left(f''(\bar{x}) + \sum_{i \in I_0^1(\tilde{h})} \bar{\lambda}^1(\tilde{h})g_i''(\bar{x}) + \ldots \right. \right.$$

$$\left. \left. + \frac{1}{p(p+1)} \sum_{i \in I_0^p(\tilde{h})} \bar{\lambda}^p(\tilde{h})g_i^{(p+1)}(\bar{x})[\tilde{h}]^{p-1} \right) t\tilde{h}, t\tilde{h} \right\rangle + o(t^2)$$

$$= \langle L_p'(\bar{x}, \tilde{h}, \bar{\lambda}(\tilde{h})), \xi_k \rangle + \frac{1}{2}\langle L_p''(\bar{x}, \tilde{h}, \bar{\lambda}(\tilde{h}))t\tilde{h}, t\tilde{h} \rangle + o(t^2) \geq$$

$$\geq \frac{\beta}{2} \|t\tilde{h}\|^2 + o(t^2) > 0,$$

which contradicts the assumption $f(x_k) < f(\bar{x})$. Hence, $\bar{x}$ is a strict local minimizer.

5   Examples

In this section we illustrate the obtained results by several examples.

Example 1. Let us consider the example of the completely degenerate case up to order $p = 3$ with $\bar{x} = 0$ and

$$g_1(x) = \left(x_1^2 + x_2^2\right) x_1 \le 0, \quad g_2(x) = \left(x_1^2 + x_2^2\right) x_2 \le 0.$$

Here $\mathcal{K} = \{x \in \mathbb{R}^2 \mid x_1 \le 0, x_2 \le 0\}$ and

$$\partial_E \mathcal{K} = \{x \in \mathbb{R}^2 \mid x_1 = 0, x_2 \le 0\} \cup \{x \in \mathbb{R}^2 \mid x_2 = 0, x_1 \le 0\}.$$

Obviously, $f'(0)$ belongs to the third quadrant, i.e.,

$$f'(0) \in \left\{ \lambda_1 \begin{bmatrix} 1 \\ 0 \end{bmatrix} + \lambda_2 \begin{bmatrix} 0 \\ 1 \end{bmatrix} \mid \lambda_1 \le 0, \lambda_2 \le 0 \right\}. \tag{31}$$

The constraints $g_1(x)$ and $g_2(x)$ are 3–regular at the point $\bar{x} = 0$. Indeed,

$$\begin{aligned}
\mathcal{K} = H_3(0) &= \{h \in \mathbb{R}^2 \mid g_i^{(3)}(0)[h]^3 \le 0, \ i = 1,2\} = \\
&= \{h \in \mathbb{R}^2 \mid h_1(h_1^2 + h_2^2) \le 0, \ h_2(h_1^2 + h_2^2) \le 0\} = \\
&= \{h \in \mathbb{R}^2 \mid h_1 \le 0, \ h_2 \le 0\}.
\end{aligned}$$

In this example, there is one elementary cone $H_3^1(0)$ and, by Lemma 3, $s = 1$ and $H_3(0) = H_3^1(0)$, so

$$\partial_E H_3^1(0) = \{h \in \mathbb{R}^2 \mid h_1 = 0, \ h_2 \le 0\} \cup \{h \in \mathbb{R}^2 \mid h_1 \le 0, \ h_2 = 0\},$$

We can choose the following vectors from $\partial_E H_3^1(0)$:

$$h_1^1 = \begin{bmatrix} 0 \\ -1 \end{bmatrix}, \quad h_2^1 = \begin{bmatrix} -1 \\ 0 \end{bmatrix}.$$

Recall that we define

$$I_0^k(h) = \left\{ i \in \{r_{k-1} + 1, \dots, r_k\} \mid g_i^{(k)}(\bar{x})[h]^k = 0 \right\}, \quad k = 1, \dots, p, \quad r_0 = 0.$$

With $p = 3$, we get the following:

$$I_0^3(h_1^1) = \{1\}, \quad I_0^3(h_2^1) = \{2\}, \quad g_1^{(3)}(0)[h_1^1]^2 = \begin{bmatrix} 2 \\ 0 \end{bmatrix}, \quad g_2^{(3)}(0)[h_2^1]^2 = \begin{bmatrix} 0 \\ 2 \end{bmatrix}.$$

Therefore, by Corollary 3,

$$f'(0) \in \left\{ \lambda_1 \begin{bmatrix} 2 \\ 0 \end{bmatrix} + \lambda_2 \begin{bmatrix} 0 \\ 2 \end{bmatrix}, \quad \lambda_1 \le 0, \lambda_2 \le 0 \right\},$$

which coincides with (31).

Example 2. Consider the not completely degenerate case with $p = 3$, $\bar{x} = 0$ and the following constrains:

$$g_1(x) = (x_1^2 + x_2^2)x_1 \leq 0, \quad g_2(x) = (x_1^2 + x_2^2)x_2 \leq 0, \quad g_3(x) = (x_2 - x_1)(x_1 - 2x_2) \leq 0.$$

The constraints $g_i(x)$, $i = 1, 2, 3$, are 3-regular at the point $\bar{x} = 0$.
Here $\mathcal{K} = \{x \in \mathbb{R}^2 \mid x_1 \leq 0, x_2 \leq 0\}$ and

$$\partial_E \mathcal{K} = \{x \in \mathbb{R}^2 \mid x_1 = 0, x_2 \leq 0\} \cup \{x \in \mathbb{R}^2 \mid x_2 = 0, x_1 \leq 0\}.$$

We obtain

$$\mathcal{K} = H_3(0) = \left\{ h \in \mathbb{R}^2 \mid g_1^{(3)}(0)[h]^3 \leq 0, \ g_2^{(3)}(0)[h]^3 \leq 0, \ g_3''(0)[h]^2 \leq 0 \right\}$$

$$= \left\{ h \in \mathbb{R}^2 \mid h_2 \leq 0, \ h_2 \geq \frac{1}{2} h_1 \right\} \cup \{h \in \mathbb{R}^2 \mid h_2 \leq h_1, \ h_1 \leq 0\}.$$

Obviously,

$$f'(0) \in \left\{ \lambda_1 \begin{bmatrix} 1 \\ 0 \end{bmatrix} + \lambda_2 \begin{bmatrix} 0 \\ 1 \end{bmatrix} \mid \lambda_1 \leq 0, \lambda_2 \leq 0 \right\}. \tag{32}$$

By Lemma 3, there are two elementary cones and relevant vectors defined by

$$H_p^1(0) = \left\{ h \in \mathbb{R}^2 \mid h_1 \leq 0, \ h_2 \geq \frac{1}{2} h_1 \right\}, H_p^2(0) = \{h \in \mathbb{R}^2 \mid h_1 \leq 0, \ h_2 \leq h_1\},$$

$$h_1^1 = \begin{bmatrix} -1 \\ 0 \end{bmatrix}, \quad h_2^1 = \begin{bmatrix} -1 \\ -1/2 \end{bmatrix}, \quad h_1^2 = \begin{bmatrix} 0 \\ -1 \end{bmatrix}, \quad h_2^2 = \begin{bmatrix} -1 \\ -1 \end{bmatrix}.$$

Then we can define the sets

$$\partial_E H_p^1(0) = \left\{ h \in \mathbb{R}^2 \mid t \begin{bmatrix} -1 \\ 0 \end{bmatrix} \cup t \begin{bmatrix} -1 \\ -1/2 \end{bmatrix}, \ t \geq 0 \right\},$$

$$\partial_E H_p^2(0) = \left\{ h \in \mathbb{R}^2 \mid t \begin{bmatrix} 0 \\ -1 \end{bmatrix} \cup t \begin{bmatrix} -1 \\ -1 \end{bmatrix}, \ t \geq 0 \right\},$$

so that $I_0^2(h_2^1) = \{3\}$, $I_0^3(h_1^1) = \{2\}$, $I_0^2(h_2^2) = \{3\}$, $I_0^3(h_1^2) = \{1\}$,

$$g_3''(0)[h_2^1] = \begin{bmatrix} 1/2 \\ -1 \end{bmatrix}, \quad g_3''(0)[h_2^2] = \begin{bmatrix} -1 \\ 1 \end{bmatrix}.$$

Therefore, by Corollary 3,

$$f'(0) \in \left\{ \bar{\lambda}_1 \begin{bmatrix} 0 \\ 2 \end{bmatrix} + \bar{\lambda}_2 \begin{bmatrix} 1/2 \\ -1 \end{bmatrix} \mid \lambda_i \leq 0 \right\} \cap \left\{ \bar{\bar{\lambda}}_1 \begin{bmatrix} 2 \\ 0 \end{bmatrix} + \bar{\bar{\lambda}}_2 \begin{bmatrix} -1 \\ 1 \end{bmatrix} \mid \bar{\bar{\lambda}}_i \leq 0 \right\}.$$

The last inclusion is equivalent to

$$f'(0) = \left\{ \lambda_1 \begin{bmatrix} 1 \\ 0 \end{bmatrix} + \lambda_2 \begin{bmatrix} 0 \\ 1 \end{bmatrix} \mid \lambda_1 \leq 0, \lambda_2 \leq 0 \right\}$$

and coincides with (32).

6   Comparison with Other Work and Concluding Remarks

In this paper we extended the results from [6] and [7] to new classes of nonregular optimization problems. The closest results to ours are those obtained in the work of Izmailov [13,14] and of Izmailov and Solodov [15]. Papers [14] and [15] consider the case of $p = 2$ only, while we are considering the case of $p \geq 2$. Even for the case of $p = 2$, our results are derived under assumptions that are weaker than those in [13–15]. For example, our Theorems 1 can be viewed as a generalization of the results obtained in [13]. Namely, there is an additional restrictive assumption in [13] that the objective function and its derivatives up to some order are equal to zero, i.e., $f^{(k)}(\bar{x}) = 0$, $k = 1, 2, \ldots, q - 1$. As follows from consideration in [13], k has to be greater than or equal to 1. We do not make this assumption, so our theorems cover classes of problems that are not subsumed by the theorems proposed in [13]. Another additional assumption in [13] is one on the constraint functions, which is $g_i^{(k)}(\bar{x}) = 0$, $k = 1, \ldots, p_i - 1$. Having such an assumption would restrict classes of nonregular problems under consideration and would not cover the general case that is considered in our paper.

Moreover, papers [13–15] present only necessary conditions for optimality and do not consider sufficient ones. In addition, Theorem 3 covers the case of any $p \geq 2$ and also subsumes the case when the set $I_0(h)$ is empty.

The optimality conditions given in papers [4,11,12] can be used to analyze some degenerate optimization problems in case $p = 2$. However, those optimality conditions cannot be applied in the case of $p > 2$, which is the main focus of this paper. In addition, necessary conditions given in [12] allow the coefficient λ_0 of the objective function to be zero. In the contrast, the conditions given in our paper provide $\lambda_0 \neq 0$. Paper [16] considers the case of $p > 2$. However, it requires the functions to be $(2p - 1)$-times continuously differentiable in the case of degeneracy of order p. At the same time, in this paper, we only require the functions to be $(p + 1)$-times continuously differentiable.

Our results will also be true in the Banach spaces, but under some additional assumptions (see, for example, [8]).

References

1. Alekseev, V.M., Tikhomirov, V.M., Fomin, S.V.: Optimal Control. Consultants Bureau, New York (1987)
2. Andreani, R., Haeser, G., Schuverdt, M.L., Silva, P.J.S.: A relaxed constant positive linear dependence constraint qualification and applications. Math. Program. Ser. A. **135**, 255–273 (2012)
3. Andreani, R., Haeser, G., Schuverdt, M.L., Silva, P.J.S.: Two new weak constraint qualifications and applications. SIAM J. Control. Optim. **22**, 1109–1135 (2012)
4. Ben-Tal, A.: Second order and related extremality conditions in nonlinear programming. J. Optim. Theory Appl. **31**, 143–165 (1980)
5. Brezhneva, O.A., Tret'yakov, A.A.: Optimality conditions for degenerate extremum problems with equality constraints. SIAM J. Control. Optim. **42**, 729–745 (2003)

6. Brezhneva, O., Tret'yakov, A.: The p-th order necessary optimality conditions for inequality—constrained optimization problems. In: Kumar, V., Gavrilova, M.L., Tan, C.J.K., L'Ecuyer, P. (eds.) ICCSA 2003. LNCS, vol. 2667, pp. 903–911. Springer, Heidelberg (2003). https://doi.org/10.1007/3-540-44839-X_95
7. Brezhneva, O.A., Tret'yakov, A.A.: The pth order optimality conditions for inequality constrained optimization problems. Nonlinear Anal. **63**, e1357–e1366 (2005)
8. Brezhneva, O.A., Tret'yakov, A.A.: The pth order optimality conditions for non-regular optimization problems. Dokl. Math. **77**, 1–3 (2008)
9. Brezhneva, O., Tret'yakov, A.A.: The p -th order optimality conditions for degenerate inequality constrained optimization problems. TWMS J. Pure Appl. Math. **1**, 198–223 (2010)
10. Brezhneva, O., Tret'yakov, A.A.: When the karush-kuhn-tucker theorem fails constraint qualifications and higher-order optimality conditions for degenerate optimization problems. J. Optim. Theory Appl. **174**(2), 367–387 (2017)
11. Gfrerer, H.: Second-order optimality conditions for scalar and vector optimization problems in Banach spaces. SIAM J. Control. Optim. **45**, 972–997 (2006)
12. Ioffe, A.D.: Necessary and sufficient conditions for a local minimum 3: second order conditions and augmented duality. SIAM J. Control. Optim. **17**, 266–288 (1979)
13. Izmailov, A.F.: Degenerate extremum problems with inequality-type constraints. Comput. Math Math Phys. **32**, 1413–1421 (1992)
14. Izmailov, A.F.: Optimality conditions for degenerate extremum problems with inequality-type constraints. Comput. Math Math Phys. **34**, 723–736 (1994)
15. Izmailov, A.F., Solodov, M.V.: Optimality conditions for irregular inequality-constrained problems. SIAM J. Control. Optim. **40**, 1280–1295 (2001)
16. Ledzewicz, U., Schättler, H.: High-order approximations and generalized necessary conditions for optimality. SIAM J. Control. Optim. **37**, 33–53 (1999)
17. Levitin, E.S., Milyutin, A.A., Osmolovskii, N.P.: Conditions of higher order for a local minimum in problems with constraints. Russian Math Surveys. **33**, 97–168 (1978)
18. Moldovan, A., Pellegrini, L.: On regularity for constrained extremum problems. Part 1: sufficient optimality conditions. J Optim Theory Appl. **142**, 147–163 (2009)
19. Moldovan, A., Pellegrini, L.: On regularity for constrained extremum problems. Part 2: necessary optimality conditions. J. Optim. Theory Appl. **142**, 165–183 (2009)
20. Penot, J.-P.: Second-order conditions for optimization problems with constraints. SIAM J. Control. Optim. **37**, 303–318 (1999)
21. Prusińska, A., Tretýakov, A.A.: On the existence of solutions to nonlinear equations involving singular mappings with non-zero p-kernel. Set-Valued Variational Anal. **19**, 399–416 (2011)
22. Szczepanik, E., Tret'yakov, A.A.: P-factor methods for nonregular inequality-constrained optimization problems. Nonlinear Anal. Theory Methods Appl. **69**(12), 4241–4251 (2008)
23. Szczepanik, E., Tret'yakov, A.A., Prusińska, A.: The p-factor method for nonlinear optimization. Schedae Informaticae **21**, 141–157 (2012)
24. Tret'yakov, A.A.: Necessary conditions for optimality of pth order. Control and Optimization, Moscow, MSU, pp. 28–35 (1983). (in Russian)
25. Tret'yakov, A.A.: Necessary and sufficient conditions for optimality of pth order. USSR Comput. Math. Math. Phys. **24**, 123–127 (1984)
26. Tret'yakov, A.A.: The implicit function theorem in degenerate problems. Russ. Math. Surv. **42**, 179–180 (1987)

27. Tret'yakov, A.A.: The p-th order optimality conditions of Kuhn-Tucker type for degenerate inequality constraints optimization problems. Doklady Academii Nauk. **434**(5), 591–594 (2010)
28. Tret'yakov, A.A., Szczepanik, E.: Irregular optimization models and p-order Kuhn-Tucker optimality conditions. J. Comput. Syst. Sci. Int. **53**(3), 384–391 (2014)
29. Tret'yakov, A.A., Marsden, J.E.: Factor-analysis of nonlinear mappings: p-regularity theory. Commun. Pure Appl. Anal. **2**, 425–445 (2003)

Convergence Rate
of Gradient-Concordant Methods
for Smooth Unconstrained Optimization

Alexey Chernov[ID] and Anna Lisachenko[✉][ID]

Moscow Institute of Physics and Technology,
9 Institutskiy per., Dolgoprudny, Moscow Region 141701, Russian Federation
lisachenko.am@phystech.edu

Abstract. The article discusses the class of gradient-concordant numerical methods for smooth unconstrained minimization where the descent direction is restricted to a subset of the descent cone. This class covers a wide range of well-known optimization methods, such as gradient descent, conjugate gradient method, and Newton's method. While previous research has demonstrated the linear convergence rate of gradient-concordant methods for strongly convex functions, many practical functions do not meet this criterion. Our research explores the convergence of gradient-concordant methods for a broader class of functions. We prove that the Polyak-Łojasiewicz condition is sufficient for the linear convergence of the gradient-concordant method. Additionally, we show sublinear convergence for convex functions that are not necessarily strongly convex.

Keywords: Gradient-concordant methods · Polyak-Łojasiewicz condition · Non-convex optimization · Convex optimization · Convergence rate · Gradient methods

1 Introduction

The concept of gradient-concordant methods was first introduced by J. Ortega and W. Reinbolt in 1970 [1]. Later [2] gave a slightly different definition, which we will operate in this paper. This class covers a wide range of well-known optimization methods, such as gradient descent, conjugate gradient method, and Newton's method. Previous research has demonstrated the linear convergence rate of gradient-concordant methods for strongly convex functions [2]. However, practical machine learning problems, such as least-squares and logistic regression, are not necessarily strongly convex but meet the weaker Polyak-Łojasiewicz condition [3,4]. In 1963, B.T. Polyak showed that this condition is sufficient for linear convergence of the gradient method [3]. In recent works, the Polyak-Łojasiewicz has been used to analyze the convergence of other methods as well,

Supported by Russian Science Foundation (project No. 21-71-30005) https://rscf.ru/en/project/21-71-30005/.

such as the stochastic gradient method [5,6], the coordinate descent method [7,8], and the heavy ball method [9,10]. In addition, the Frank-Wolfe method [11] was investigated using the Polyak-Łojasiewicz condition.

Our research explores the convergence of gradient-concordant methods for a broader class of functions. We prove that the Polyak-Łojasiewicz condition is sufficient for the linear convergence of the gradient-concordant method. Additionally, we show sublinear convergence for convex functions that are not necessarily strongly convex.

2   Gradient-Concordant Methods

Consider the smooth unconstrained optimization problem

$$f(x) \to \min, \quad x \in \mathbb{R}^n \tag{1}$$

which is solved numerically

$$x^k = x^{k-1} + \lambda_k p^k, \quad k = 1, 2, \ldots. \tag{2}$$

It is assumed that p^k is the descent direction of the objective function $f(x)$ at $x^k - 1$, i.e., there exists $\overline{\lambda} \in \mathbb{R}_+$ such that for any $\lambda \in (0, \overline{\lambda})$, $f(x^{k-1} + \lambda p^k) < f(x^{k-1})$, and the step $\lambda_k \in (0, \overline{\lambda})$. It follows that, instead of the initial unconstrained problem (1), the following constrained problem defined by the initial point x_0 is solved:

$$f(x) \to \min_{x \in G}, \quad G = \left\{ x \in \mathbb{R}^n : f(x) \le f(x^0), x^0 \in \mathbb{R}^n \right\}. \tag{3}$$

The solution sets of problems (1) and (3) obviously coincide.

Next, we will discuss the class of numerical methods where certain constraints are imposed on the sequence $\{p^k\}$ [2].

Definition 1. *A sequence of vectors $\{p^k\}$ is called gradient-concordant with a sequence of points $\{x^k\}$ if there are numbers $\alpha > 0$ and $\beta > 0$ such that for any $k > 0$*

$$\left\| p^k \right\| \le \alpha \left\| \nabla f(x^{k-1}) \right\|, \tag{4}$$

$$\beta \left\| \nabla f(x^{k-1}) \right\|^2 \le -\left\langle p^k, \nabla f(x^{k-1}) \right\rangle \tag{5}$$

Geometrically, the inequality (4) constrains the vector p^k in length, and together with the inequality (5) constrains p^k in direction, so that the cosine of the angle between these vectors is strictly negative and does not exceed $-\frac{\beta}{\alpha}$ in magnitude. This property of gradient-concordant methods is represented in Fig. 1.

Methods for solving the unconstrained minimization problem for which the conditions (4) and (5) are satisfied will be called gradient-concordant methods. Since the above conditions are natural, many known unconstrained minimization methods satisfy them. For example, it is easy to see that the gradient method is gradient-concordant with $\alpha = \beta = 1$. The conjugate gradient and Newton's method are also gradient-concordant.

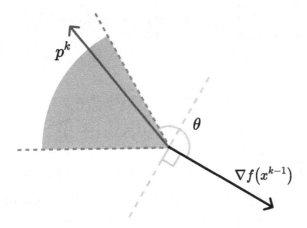

Fig. 1. Property of gradient-concordant methods: the vector p^k is bounded by a subset of the descent cone, $cos\theta = -\frac{\beta}{\alpha}$

Remark 1. Note the direct corollary of the definition 1, which defines the relationship between the coefficients α and β [2]:

$$\beta \leq \alpha \tag{6}$$

Before proceeding to the convergence of gradient-concordant methods, let us introduce the two ways of choosing the step size, which we will consider hereafter—the backtracking line search and the exact line search.

1. We will say that the step λ_k is chosen according to the backtracking line search if it satisfies the Armijo-Goldstein condition

$$f(x^k) - f(x^{k-1}) \leq \varepsilon\lambda_k\left\langle p^k, \nabla f(x^{k-1})\right\rangle, \tag{7}$$

where $\varepsilon \in (0,1)$ and $\left\langle p^k, \nabla f(x^{k-1})\right\rangle < 0$. If the condition (7) is not met, the step length is reduced until this condition is satisfied.
2. In the exact line search, the step length λ_k is chosen to minimize the value of the target function on the ray:

$$\lambda_k = \arg\min_{\lambda > 0} f(x^{k-1} + \lambda p^k) \tag{8}$$

These ways of choosing step size are often used in practice. For them, the following statement was proved in [2]:

Theorem 1. *Let the function $f(x)$ on the set G be bounded from below, differentiable, and the first derivative of f is L-Lipschitz continuous with $L > 0$:*

$$\left\|\nabla f(x) - \nabla f(y)\right\| \leq L\|x - y\|.$$

Let the sequence of vectors $\{p^k\} \in \mathbb{R}^n \backslash \{0_n\}$ be gradient-concordant and the step size be chosen according to (7) or (8). Then

$$\lim_{k \to \infty} \left\| \left\| \nabla f(x^k) \right\| \right\| = 0.$$

Additionally, if a step is chosen according to (7), then

$$f(x^k) - f(x^{k-1}) \leq -\varepsilon \tilde{\lambda} \beta \left\| \nabla f(x^{k-1}) \right\|^2, \tag{9}$$

where $\varepsilon \in (0,1)$, $\lambda_k = \tilde{\lambda} \in (0, \overline{\lambda}]$, $\overline{\lambda} = \dfrac{2(1-\varepsilon)\beta}{L\alpha^2}$.

Also, [2] shows the linear convergence rate of gradient-concordant methods in the case of a strongly convex function.

3 Polyak-Łojasiewicz Condition

Next, we will assume that the optimization problem (1) has a non-empty solution set χ^*, on which the optimal value of the function f^* is achieved. Let us provide the definition [3].

Definition 2. *A function $f(x)$ satisfies the Polyak-Łojasiewicz condition on a set G if there exists $\mu > 0$ such that for any $x \in G$*

$$\left\| \nabla f(x) \right\|^2 \geq 2\mu \left(f(x) - f^* \right). \tag{10}$$

In this case, μ is called the Polyak-Łojasiewicz constant.

This condition means that the gradient grows faster than the quadratic function as it moves away from the optimal value. It is easy to show that this condition is a corollary of strong convexity [12], but not all functions satisfying the Polyak-Łojasiewicz condition are strongly convex. Moreover, there are examples of nonconvex functions that nevertheless satisfy the Polyak-Łojasiewicz condition [8].

Remark 2. The condition (10) implies that any stationary point is a global minimum. However, unlike the strongly convex case, the uniqueness of the solution cannot be guaranteed.

Let us now prove the main claim of this paper, the linear convergence of gradient-concordant methods for functions satisfying the Polyak-Łojasiewicz condition.

Theorem 2. *Let the function $f(x)$ satisfy the Polyak-Łojasiewicz condition on the set G and the conditions of Theorem 1 are satisfied. Then the gradient-concordant method converges to the solution set χ^* with linear rate, that is, there exists a number $q \in (0,1)$ such that*

$$f(x^k) - f^* \leq q^2 \left(f(x^{k-1}) - f^* \right), \qquad \left\| x^k - x^{k*} \right\| \leq Cq^k, \tag{11}$$

where x^{k*} is the projection of the point x^k on the solution set χ^*,

$$C = \sqrt{\frac{2\left(f(x^0) - f^*\right)}{\mu}}, \quad q = \sqrt{1 - 2\mu\varepsilon\tilde{\lambda}\beta}, \quad \tilde{\lambda} \in (0, \overline{\lambda}], \quad \overline{\lambda} = \frac{2(1 - \varepsilon)\beta}{L\alpha^2}.$$

Proof. Let us first perform the proof for the step choice according to the backtracking line search (7). The conditions of Theorem 1 are satisfied, so

$$f(x^k) - f(x^{k-1}) \leq -\varepsilon\tilde{\lambda}\beta \left\|\nabla f(x^{k-1})\right\|^2$$

From here, using the Polyak-Lojasiewicz condition:

$$f(x^k) - f^* \leq f(x^{k-1}) - f^* - 2\mu\varepsilon\tilde{\lambda}\beta\left(f(x^{k-1}) - f^*\right) =$$
$$= \left(f(x^{k-1}) - f^*\right)(1 - 2\mu\varepsilon\tilde{\lambda}\beta) = q^2\left(f(x^{k-1}) - f^*\right), \quad (12)$$

where $q = \sqrt{1 - 2\mu\varepsilon\tilde{\lambda}\beta}$. If we choose $\tilde{\lambda} = \overline{\lambda} = \dfrac{2(1 - \varepsilon)\beta}{L\alpha^2}$, then

$$q^2 = 1 - 4\varepsilon(1 - \varepsilon)\frac{\mu}{L}\left(\frac{\beta}{\alpha}\right)^2. \quad (13)$$

Let us evaluate the obtained expression. First, if $\varepsilon \in (0; 1)$ holds

$$0 < 4\varepsilon(1 - \varepsilon) \leq 1. \quad (14)$$

Second, it follows from the property (6) that

$$\frac{\beta}{\alpha} \leq 1. \quad (15)$$

And finally,

$$\frac{\mu}{L} \leq 1. \quad (16)$$

This can be shown by combining the Polyak-Lojasiewicz condition and the property of functions having a Lipschitz gradient [13]:

$$2\mu\left(f(x) - f^*\right) \leq \left\|\nabla f(x)\right\|^2 \leq 2L\left(f(x) - f^*\right).$$

Thus, substituting (14), (15), and (16) into (13), we obtain an estimate $q \in [0; 1)$. Since $q < 1$, $f(x)$ converges with linear rate to f^* at $k \to \infty$.

Now let us estimate the distance from the point obtained in the k-step of the gradient method to the solution set, that is, the value $\left\|x^k - x^{k*}\right\|$, where x^{k*} is the projection of the point x^k on the solution set χ^*. To do this, we use the resulting inequality (12), and the quadratic growth condition

$$f(x) - f^* \geq \frac{\mu}{2}\left\|x - x^*\right\|^2,$$

where x^* is the projection of the point x on the solution set. This condition is a consequence of the Polyak-Lojasiewicz condition [8]. We obtain:

$$\left\|x^k - x^{k*}\right\|^2 \leq \frac{2\left(f\left(x^k\right) - f^*\right)}{\mu} \leq \frac{2q^2\left(f\left(x^{k-1}\right) - f^*\right)}{\mu} \leq \frac{2q^{2k}}{\mu}\left(f\left(x^0\right) - f^*\right).$$

Hence,

$$\left\|x^k - x^{k*}\right\| \leq Cq^k, \qquad C = \sqrt{\frac{2\left(f\left(x^0\right) - f^*\right)}{\mu}}$$

Thus, for backtracking line search, the assertion is proved. Now consider the case of the exact line search (8). Denote by $\left\{x^{k,1}\right\}$ and $\left\{x^{k,2}\right\}$ the sequences of points obtained by the step choice methods (7) and (8) respectively. Obviously, $f\left(x^{k,2}\right) \leq f\left(x^{k,1}\right)$. Therefore, from (12):

$$f\left(x^{k,2}\right) - f\left(x^*\right) \leq f\left(x^{k,1}\right) - f\left(x^*\right) \leq q^2\left(f\left(x^{k-1}\right) - f^*\right)$$

Consequently, the convergence rate estimates obtained above also carry over to the case of exact line search.

4   Convex Functions

In this section we will assume objective function f to be convex on the set G, which implies

$$f(y) - f(x) \geq \langle \nabla f(x), \, y - x \rangle. \tag{17}$$

and

$$f\left(x^k\right) - f^* \leq \left\langle \nabla f\left(x^k\right), \, x^k - x^* \right\rangle. \tag{18}$$

Convexity is weaker than strong convexity, but it does not follow from the Polyak-Lojasiewicz condition. We will show that in this case sublinear convergence takes place.

Theorem 3. *Let the function $f(x)$ be convex on the set G. Then under the assumptions of Theorem 1, gradient-concordant methods have sublinear convergence $O\left(\frac{1}{\sqrt{N}}\right)$. Specifically,*

$$f\left(x^N\right) - f^* \leq \sqrt{\frac{f\left(x^0\right) - f\left(x^N\right)}{\varepsilon\beta\lambda_{min}N}} \cdot R, \tag{19}$$

where

$$\lambda_{min} = \min_{k=1,\ldots,N} \lambda_k, \quad R = \operatorname{diam} G.$$

Proof. From the Armijo-Goldstein condition (7) and (5) we obtain

$$f(x^k) - f(x^{k-1}) \leq \varepsilon \lambda_k \langle p^k, \nabla f(x^{k-1}) \rangle \leq -\varepsilon \lambda_k \beta \left\| \nabla f(x^{k-1}) \right\|^2, \qquad (20)$$

which gives

$$\varepsilon \lambda_k \beta \left\| \nabla f(x^{k-1}) \right\|^2 \leq f(x^{k-1}) - f(x^k). \qquad (21)$$

Summarizing Eq. (21) from $k = 1$ to N we get

$$\sum_{k=0}^{N-1} \varepsilon \lambda_{k+1} \beta \left\| \nabla f(x^k) \right\|^2 \leq f(x^0) - f(x^N) \leq f(x^0) - f(x^*). \qquad (22)$$

Letting $N \to \infty$ we can assert that

$$\lim_{N \to \infty} \left\| \nabla f(x^k) \right\| = 0. \qquad (23)$$

Now let us denote by $\tilde{r}$ the gradient vector with the minimum norm of the first N points and by $\tilde{k}$ the corresponding index

$$\tilde{k} = \arg \min_{k=0,\ldots,N-1} \left\| \nabla f(x^k) \right\|. \qquad (24)$$

$$\tilde{r} = \nabla f\left(x^{\tilde{k}}\right). \qquad (25)$$

It follows from (22) that

$$\|\tilde{r}\|^2 \leq \frac{f(x^0) - f(x^N)}{\varepsilon \beta \sum_{k=0}^{N-1} \lambda_{k+1}}. \qquad (26)$$

By the convexity of f,

$$f\left(x^{\tilde{k}}\right) - f^* \leq \langle \tilde{r}, x^{\tilde{k}} - x^* \rangle \leq \|\tilde{r}\| \cdot \left\| x^{\tilde{k}} - x^* \right\| \leq$$

$$\leq \sqrt{\frac{f(x^0) - f(x^N)}{\varepsilon \beta \sum_{k=0}^{N-1} \lambda_{k+1}}} \cdot \left\| x^{\tilde{k}} - x^* \right\| \leq \sqrt{\frac{f(x^0) - f(x^N)}{\varepsilon \beta \sum_{k=0}^{N-1} \lambda_{k+1}}} \cdot R, \qquad (27)$$

where R is the diameter of set G defined in (3). It is finite because of the convexity of f.

In view of Theorem 1, we have $\lambda_k \in \left(0, \frac{2(1-\varepsilon)\beta}{L\alpha^2}\right]$. Therefore, there exists $\lambda_{min} > 0$ which is the minimum of the first N step sizes

$$\lambda_{min} = \min_{k=1,\ldots,N} \lambda_k, \qquad (28)$$

and the expression (27) can be further estimated

$$f(x^N) - f^* \leq f\left(x^{\tilde{k}}\right) - f^* \leq \sqrt{\frac{f(x^0) - f(x^N)}{\varepsilon \beta \lambda_{min} N}} \cdot R, \qquad (29)$$

which completes the proof.

5 Experiments

To compare the obtained theoretical estimates with the experiment, let us examine the convergence of the gradient descent for the function

$$f(x) = \|Ax - b\|^2, \tag{30}$$

where A is matrix $m \times n$, $x \in R^n, b \in R^m$. This objective function is widely used in solving a linear system by the least squares. Its Hessian $H = 2A^\top A$ is degenerate in the case $m < n$, so the function (30) is not strongly convex. However, for any m and n it satisfies the Polyak-Lojasiewicz condition as a composition of strongly convex $g(z) = \|z - b\|^2$ and linear function. The Polyak-Lojasiewicz constant μ can be found as the product of the strong convexity constant of the function g ($\sigma = 2$) and the square of the least nonzero singular value of the matrix A—θ [8]:

$$\mu = \sigma\theta^2 = 2\theta^2,$$

and the Lipschitz constant L—as the largest eigenvalue of the Hessian. As noted earlier, the gradient method belongs to the gradient-concordant method with constants $\alpha = \beta = 1$. Assuming $\varepsilon = 0.5$, we find

$$\overline{\lambda} = \frac{2(1 - \varepsilon)\beta}{L\alpha^2} = \frac{1}{L} \tag{31}$$

And by choosing the step size $\tilde{\lambda} = \overline{\lambda}$, we get an estimate of the convergence parameter

$$q = \sqrt{1 - 2\mu\varepsilon\tilde{\lambda}\beta} = \sqrt{1 - 2\frac{\theta^2}{L}} \tag{32}$$

For the experiments, the algorithm was implemented in C++, and the analysis was performed using Python. The results are shown in Figs. 2, 3 and Table 1.

In the case considered $m < n$ the solution of the equation $Ax = b$ exists, so the optimal value of the target function $f^* = 0$. Figure 2b shows that the theoretical estimate of the square of the convergence parameter q^2 does indeed bound the ratio

$$\frac{f(x^k) - f^*}{f(x^{k-1}) - f^*} = \frac{f(x^k)}{f(x^{k-1})},$$

with the estimate becoming increasingly accurate as the iteration number increases. At a constant step $\frac{1}{L}$, the experimental convergence parameter tends monotonically to the theoretical one, whereas when the step size is chosen by the backtracking line search, there are regions where the convergence rate significantly exceeds the estimate. In Fig. 3 similar results are shown for more complex problem. In this experiment the rate $\frac{\theta^2}{L}$ is ten times smaller, and, therefore, q is closer to 1 and the convergence is slower. Table 2 summarizes the experiments.

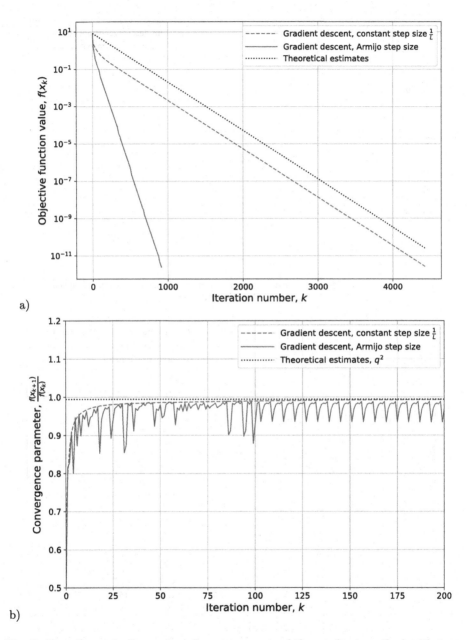

a)

b)

Fig. 2. Experiment 1. Comparing the convergence of the gradient method with two ways of choosing the step length for the function (30) with the theoretical estimate at $n = 30, m = 25, L = 62.67, \theta^2 = 0.18$.

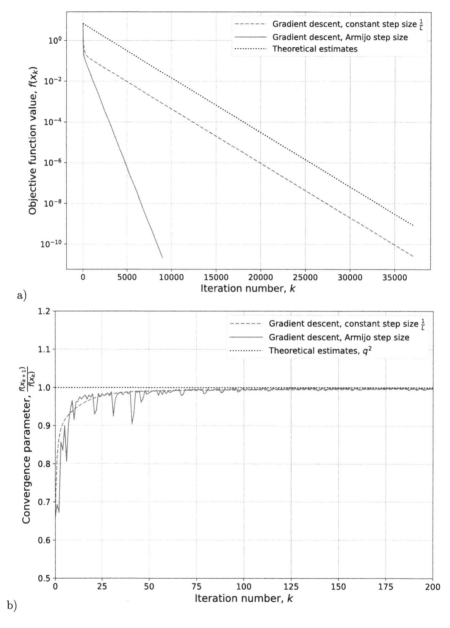

a)

b)

Fig. 3. Experiment 2. Comparing the convergence of the gradient method with two ways of choosing the step length for the function (30) with the theoretical estimate at $n = 30, m = 29, L = 66.86, \theta^2 = 0.02$.

Table 1. Comparison of the number of iterations for which the method converges to the optimal value with δ precision, with the theoretically obtained upper bound. k^1—number of steps of gradient descent with constant step size; k^2—gradient descent with step size (7); k^{ub}—upper bond for given accuracy.

	δ	10^{-1}	10^{-2}	10^{-3}	10^{-4}	10^{-5}	10^{-6}	10^{-7}	10^{-8}	10^{-9}	10^{-10}
Exp. 1	k^1	358	743	1129	1514	1900	2285	2671	3056	3442	3828
	k^2	57	148	273	398	518	643	768	888	1016	1105
	k^{ub}	739	1124	1509	1894	2279	2664	3049	3434	3819	4204
Exp. 2	k^1	1197	4922	8665	12407	16149	19891	23634	27376	31118	34860
	k^2	279	1149	2081	3013	3890	4792	5694	6613	7523	8423
	k^{ub}	6862	10603	14345	18087	21828	25570	29311	33053	36795	40536

6   Conclusion

Gradient-concordant methods are a crucial class of numerical methods for smooth unconstrained minimization. Due to their general properties, it is possible to obtain convergence estimates uniformly for many methods simultaneously. Earlier, these methods were shown to have linear convergence for strongly convex functions. In this paper, we extended the range of functions under consideration and demonstrated that the weaker Polyak-Łojasiewicz condition is adequate for the linear convergence of these methods. Furthermore, we proved that for convex functions sublinear convergence takes place. Finally, we provided theoretical estimates of the convergence parameters and compared them with experimental results. Consequently, we derived universal convergence estimates for established gradient-concordant methods and other possible members of this class.

Acknowledgements. We would like to thank the late Dr. Alexander Birjukov for his considerable help that made this work possible.

References

1. Ortega, J.M., Rheinbolt, W.G.: Iterative Solution of Nonlinear Equations in Several Variables. Academic Press, New York (1970)
2. Birjukov, A.G., Chernov, A.V.: Gradient-concordant methods for solving unconstrained minimization problems. In: Proceedings of MIPT, pp. 113–125 (2017). (in Russian)
3. Polyak, B.T.: Gradient methods for minimizing functionals. Zh. Vychisl. Mat. Mat. Fiz. 643–653 (1963). (in Russian)
4. Łojasiewicz., S.: A topological property of real analytic subsets. Coll. du CNRS, Les 'equations aux d'eriv'ees partielles, 87–89 (1963). (in French)
5. Ajalloeian, A., Stich, S.: Analysis of SGD with biased gradient estimators. In: Workshop on "Beyond First Order Methods in ML Systems" at the 37th ICML, Vienna, Austria (2020)
6. Kim, S., Madden, L., Dall'Anese, E.: Online stochastic gradient methods under sub-weibull noise and the Polyak-Łojasiewicz condition. In: 2022 IEEE 61st Conference on Decision and Control (CDC), Cancun, Mexico, pp. 3499–3506 (2022)

7. Nesterov, Y.: Efficiency of coordinate descent methods on huge-scale optimization problems. SIAM J. Optim. 341–362 (2012)
8. Karimi, H., Nutini, J., Schmidt, M.: Linear convergence of gradient and proximal-gradient methods under the Polyak-Łojasiewicz condition. In: Frasconi, P., Landwehr, N., Manco, G., Vreeken, J. (eds.) ECML PKDD 2016. LNCS (LNAI), vol. 9851, pp. 795–811. Springer, Cham (2016). https://doi.org/10.1007/978-3-319-46128-1_50
9. Apidopoulos, V., Ginatta, N., Villa, S.: Convergence rates for the heavy-ball continuous dynamics for non-convex optimization, under Polyak-Lojasiewicz condition. J. Glob. Optim. **84**, 563–589 (2022)
10. Wang, J.K., et al.: Provable acceleration of heavy ball beyond quadratics for a class of Polyak-Lojasiewicz functions when the non-convexity is averaged-out. In: ICML, pp. 22839–22864. PMLR (2022)
11. Garber, D., Hazan, E.: Faster rates for the Frank-Wolfe method over strongly-convex sets. In: ICML, pp. 541–549 (2015)
12. Sukharev, A.G., Timokhov, A.V., Fedorov, V.V.: Course in Optimization Methods. Physmatlit, Moscow (2005)
13. Polyak, B.T.: Introduction to Optimization, 2nd edn. Lenand, Moscow (2014). (in Russian)

A Derivative-Free Nonlinear Least Squares Solver for Nonsmooth Functions

Igor Kaporin[(✉)](ID)

Federal Research Center "Computer Science and Control"
of the Russian Academy of Sciences, Vavilova 40, Moscow, Russia
igorkaporin@mail.ru

Abstract. Applicability of appropriately modified derivative-free nonlinear least squares iterative solver to nonsmooth problems is considered. The main assumption is that the residual function is continuous and piecewise smooth with integrable Jacobian matrix. The preconditioner is taken as a rectangular matrix depending on pseudorandom parameters. The search subspaces are generated using Arnoldi type recurrences with additional basis orthonormalization. Numerical results are given for several representative test problems with piecewise smooth residual functions to demonstrate the effectiveness of the method.

Keywords: Nonlinear least squares · Derivative-free optimization · Zero-order optimization · Nonsmooth problems · Pseudorandom preconditioning · Structured finite-difference directions · Preconditioned subspace descent

1 Introduction

Many important applications involve difficult optimization problems such as unconstrained minimization

$$x_* = \arg \min_{x \in R^n} \varphi(x). \tag{1}$$

Frequently, the black-box problem setting arises, when only evaluations of objective function $\varphi(x)$ are permitted. In such cases, the use of derivative-free methods, also known as zeroth-order (ZO) ones, is necessary. An excellent introduction into the topic is the monograph [6].

In other cases, the development of software for the gradient evaluation may involve more expenses than the use of ZO optimization methods despite their potentially slower convergence. ZO methods can be viewed as derivative-free counterparts of first-order optimization methods, since they approximate the full gradients or stochastic gradients through function value based gradient estimates. Hence, the interest in ZO optimization has grown rapidly in the past few years.

N. Olenev et al. (Eds.): OPTIMA 2023, LNCS 14395, pp. 45–59, 2023.
https://doi.org/10.1007/978-3-031-47859-8_4

Quite often, the objective function can be presented as a sum of squares of m nonlinear functions $f_i(x)$ in n variables x,

$$\varphi(x) = \frac{1}{2}\|f(x)\|^2 \equiv \frac{1}{2}f^{\mathsf{T}}(x)f(x), \tag{2}$$

where $f(x)$ is a nonlinear mapping $f : R^n \to R^m$, $m \geq n$, thus giving rise to a nonlinear least squares problem. The latter also may arise when one wants to solve a system of m nonlinear equations in n variables

$$f(x) = 0; \tag{3}$$

in such cases, the objective value is typically set to zero.

The present paper continues the research started in [11] where a derivative-free version of the method [10] was presented. Similar to [3–5], we use the inexact Newton/Krylov subspace framework, however with search subspaces augmented by several previous directions, and using a different stepsize choice rule. A critically important feature of the proposed solver is the use of quasirandomly parametrized rectangular preconditioner involving quite moderate computational expenses.

Moreover, comparing the definition of search subspaces used in [11] with the construction of structured gradients used, e.g., in [15], has suggested to apply orthogonalization for the search subspace basis in the proposed method (see Sect. 5.1) which improved both its practical robustness and theoretical justification.

2 Motivation of the Research

The development of the nonlinear least squares solver [11] was mainly motivated by the need to solve the set of trilinear equations (3) proposed by R.Brent in [2]. Related to the problem of evaluating the product of two $q \times q$ matrices in $r < q^3$ essential multiplications (note that $r = 7$ for $q = 2$, as was discovered by V.Strassen in [16]), Brent equations contain $3rq^2$ unknowns and q^6 equations. Already for $q > 3$, an exact solution of these equations cannot be easily found using contemporary solvers and computers. The main problem here is that Brent equations have no isolated solutions. Consequently, the Jacobian at the solution is always rank deficient. Moreover, a typical situation is that the residual norm $\|f\|$ decreases as iterations progress, but the solution norm tends to infinity.

In double precision calculations, we refer x as an "exact solution" of Brent equations if

$$\|f(x)\| < 10^{-15} \qquad \text{and} \qquad \|x\|_\infty < 1.$$

In order to limit the possible growth of the solution norm in the course of iteration, the original target function was replaced by the Euclidean norm of the modified residual function, for instance,

$$\widetilde{f}(x) = \frac{f(x)}{\sqrt{1 - 0.1\|x\|_\infty}}.$$

Note that the latter is no longer a smooth function. Moreover, when we are getting closer to the exact solution, the components of $\tilde{f}(x)$, additionally to their nonsmoothness, start to demonstrate discontinuities due to the floating-point round-off. Nevertheless, the solvers presented in [10,11] have demonstrated quite satisfactory performance in solving small-sized Brent equations [10] with high precision.

If an exact solution to the Brent equations is found, one can use recursion to derive the corresponding algorithm for the evaluation of $n \times n$ square matrix product in $O(n^\omega)$ multiplications, where $\omega = \log r / \log q$. For Strassen algorithm one has $\omega = \log_2 7 < 2.8074$.

In order to make the problem easier to solve, in [11] a parametrization (in general, complex-valued) of the matrix coefficients in Brent equations was proposed which provides for a 12 times reduction in the number of unknowns. A lot of various parametrizations have been considered for Brent equations, so the use of Jacobian-free solvers made it possible to minimize the related coding efforts. Several new exact matrix multiplication algorithms were discovered using nonlinear solver [11]. In particular, it was found that two 4×4 matrices can be multiplied using only 48 multiplications, which results in an algorithm for matrix product with $O(n^{2.792})$ complexity bound.

3   Preliminary Remarks on Gradient Methods

Let $x_0, x_1, \ldots, x_t, \ldots$ be the sequence of approximations to the stationary point x_* of $\varphi(x)$ constructed in the course of iterations, where t is the iteration index. Assume for a moment that φ is sufficiently smooth. For its minimization, one can apply the gradient descent method

$$x_{t+1} = x_t - \alpha_t \operatorname{grad} \varphi(x_t), \qquad t = 0, 1, \ldots, \qquad \alpha_t > 0,$$

with an appropriately chosen sequence of stepsizes α_t. A finite-difference estimate for the gradient vector

$$g = \operatorname{grad} \varphi(x)$$

can be obtained using an arbitrary orthogonal matrix

$$U = [u_1 \ldots u_d] \in R^{n \times d}$$

as follows:

$$g \approx UU^\mathsf{T} g = \sum_{i=1}^d u_i u_i^\mathsf{T} g \approx \sum_{i=1}^d u_i \frac{\varphi(x + \zeta u_i) - \varphi(x)}{\zeta},$$

where $\zeta > 0$ takes sufficiently small values. Note that taking $d \geq n$ implies

$$UU^\mathsf{T} = I_n,$$

which yields the exact equality $g = UU^\mathsf{T}g$. This results in the following iterative method,

$$x_{t+1} = x_t - \alpha_t \sum_{i=1}^{d} \frac{\varphi(x_t + \zeta_t u_i^{(t)}) - \varphi(x_t)}{\zeta_t} u_i^{(t)},$$

which was studied, for instance, in [15] (see also [12] and references therein), where an analysis of structured finite difference algorithms was given without smoothness and convexity assumptions. This was done using a smooth approximation of the target function as a surrogate of the target. Moreover, it was proved in [15] that the surrogate of the gradient built with structured directions is an estimation of the gradient of a suitable target smoothing.

Note that the orthogonal matrix U_t is formed using quasirandom data and it must be different for each iteration.

In the case of nonlinear least squares setting (2), similar techniques can be used with account for the structure of the gradient:

$$\operatorname{grad} \varphi(x) = J^\mathsf{T}(x)f(x) \in R^n,$$

where

$$J(x) = \frac{\partial f}{\partial x} \in R^{m \times n}$$

is the Jacobian matrix of f at x. Therefore, the gradient method takes the form

$$x_{t+1} = x_t - \alpha_t J^\mathsf{T}(x_t)f(x_t), \qquad t = 0, 1, \dots, \qquad \alpha_t > 0,$$

and the derivatives involved can be approximated as

$$J^\mathsf{T}(x)f(x) \approx UU^\mathsf{T}J^\mathsf{T}(x)f(x) = \sum_{i=1}^{d} u_i u_i^\mathsf{T} J^\mathsf{T}(x)f(x)$$

$$= \sum_{i=1}^{d} u_i \big(J(x)u_i\big)^\mathsf{T} f(x) \approx \sum_{i=1}^{d} u_i \left(\frac{f(x+\zeta u_i)-f(x)}{\zeta}\right)^\mathsf{T} f(x).$$

The resulting derivative-free iterative method takes the form

$$x_{t+1} = x_t + \alpha_t U_t s_t,$$

where

$$s_t = [\sigma_1^{(t)} \dots \sigma_d^{(t)}]^\mathsf{T},$$

and

$$\sigma_i^{(t)} = -\left(\frac{f(x_t + \zeta_t u_i^{(t)}) - f(x_t)}{\zeta_t}\right)^\mathsf{T} f(x_t).$$

4   General Description of the Proposed Method

For the construction of better descent directions, let us apply the techniques similar to that introduced in [11]. To simplify the notations, below we omit the iteration index t. As above, let the next approximation x_+ to x_* be

$$x_+ = x + \alpha U s,$$

where the stepsize parameter α_t satisfies

$$0 < \alpha < 2,$$

and $U \in R^{n \times d}$ is an orthogonal matrix. Since the convergence of simple gradient methods can be very slow, we propose another choice of $s \in R^d$.

Let us introduce the rectangular matrix

$$F = \left[\frac{f(x + \zeta u_1) - f(x)}{\zeta} \mid \ldots \mid \frac{f(x + \zeta u_d) - f(x)}{\zeta} \right] \in R^{n \times d}, \qquad (4)$$

and consider the identity

$$\|f(x + \alpha U s)\| \equiv \|(f + \alpha F s) + \alpha \left(\frac{f(x + \alpha U s) - f}{\alpha} - F s \right) \|,$$

yielding the following residual norm estimate:

$$\|f(x + \alpha U s)\| \leq \|f + \alpha F s\| + \alpha \left\| \frac{f(x + \alpha U s) - f}{\alpha} - F s \right\|. \qquad (5)$$

Similar to the Levenberg-Marquardt type techniques, let us determine the vector s as a solution of the following linear least squares problem:

$$s = \arg \min_{s \in R^d} \left(\|f + F s\|^2 + \xi \|s\|^2 \right),$$

where $0 < \xi \ll \|F\|^2$ is a small regularizing parameter. Clearly, s satisfies linear equation

$$(F^\mathsf{T} F + \xi I) s = -F^\mathsf{T} f$$

and is equal to

$$s = -(F^\mathsf{T} F + \xi I)^{-1} F^\mathsf{T} f.$$

With this choice of s, for the first term in (4), the following upper bound holds:

$$\begin{aligned}
\|f + \alpha F s\|^2 &= \|f\|^2 + 2\alpha f^\mathsf{T} F s + \alpha^2 s^\mathsf{T} F^\mathsf{T} F s \\
&< \|f\|^2 + 2\alpha f^\mathsf{T} F s + \alpha^2 s^\mathsf{T} (F^\mathsf{T} F + \xi I) s \\
&= \|f\|^2 - \alpha(2 - \alpha) f^\mathsf{T} F (F^\mathsf{T} F + \xi I)^{-1} F^\mathsf{T} f.
\end{aligned}$$

Finally, note that (for a sufficiently small α) the second term in (4) is nothing but the norm of the difference between two different approximations of the same vector JUs. Therefore, subjecting the norm of the this term to a convenient upper bound would be a natural condition. Thus, introducing the quantity $\theta \in (0, 1)$ as

$$\theta^2 = \frac{f^\mathsf{T} F (F^\mathsf{T} F + \xi I)^{-1} F^\mathsf{T} f}{f^\mathsf{T} f},$$

we require the following *convergence assumption* to hold:

$$\left\| \frac{f(x + \alpha U s) - f}{\alpha} - F s \right\| \leq \left(1 - \frac{\alpha}{2} \right) \theta^2 \|f\|. \qquad (6)$$

Using the latter with (4) one has the following residual norm upper bound:

$$\frac{\|f(x + \alpha Us)\|}{\|f\|} \leq \sqrt{1 - \alpha(2 - \alpha)\theta^2} + \frac{\alpha(2 - \alpha)\theta^2}{2}.$$

The latter estimate can be simplified if one uses the inequality

$$\sqrt{1 - \eta} + \frac{\eta}{2} \leq \sqrt{1 - \frac{\eta^2}{4}},$$

which holds for any $0 \leq \eta \leq 1$ with $\eta = \alpha(2 - \alpha)\theta^2$:

$$\frac{\|f(x + \alpha Us)\|}{\|f\|} \leq \sqrt{1 - \left(\frac{\alpha(2 - \alpha)\theta^2}{2}\right)^2}. \tag{7}$$

Thus, (6) presents a sufficient condition for the descent of the residual norm, that is, $\varphi(x_+) < \varphi(x)$. Similar approach was earlier presented in [8–10] and adjusted to the derivative-free framework in [11].

Note that (7) can conveniently be used as the criteria for acceptance of the stepsize $\alpha \in \{1, 1/2, 1/4, \ldots\}$ in the construction of a backtracking Armijo type procedure [1] needed for practical implementation of the nonlinear solver.

5   Implementation Issues

Further we consider the actual computation formulae designed to improve the numerical stability and reduce computational expenses.

5.1   Nonlinear Krylov Subspaces and Bases Orthogonalization

Let $k \geq 1$ and $l \geq 0$ be fixed integers such that

$$d = k + l \leq n.$$

Here k stand for the dimension of Krylov subspace used, and l is the maximum number of the previous search directions

$$p_{t-1} = U_{t-1}s_{t-1} = (x_t - x_{t-1})/\alpha_{t-1}, \ldots.$$

Introduce the rectangular matrices $K_t \in R^{m \times n}$ serving as a kind of variable preconditioner, see Sect. 5.2 below. We further omit the iteration index t; for instance, the notation $p_{k-i} = p_{t+k-i}$ will be used.

Following [11], we apply Arnoldi-type orthogonalization procedure for the construction of bases of subspaces containing search directions:

$$v_1 \chi_{1,0} = f,$$

$$\tilde{u}_i = \begin{cases} K^\mathsf{T} v_i, & i \leq k; \\ u_i = p_{k-i}, & i > k, \end{cases}$$

$$\kappa_{i,i} u_i = \tilde{u}_i - \sum_{j=1}^{i-1} u_j \kappa_{j,i},$$

$$v_{i+1} \chi_{i+1,i} = \frac{f(x + \zeta u_i) - f(x)}{\zeta} - \sum_{j=1}^{i} v_j \chi_{j,i},$$

$$\text{where} \quad i = 1, \ldots, k + l,$$

and the coefficients $\chi_{1,0} = \|f\|$, $\kappa_{1,1} = \|u_1\|$,

$$\kappa_{j,i} = u_j^\mathsf{T} \tilde{u}_i, \quad j = 1, \ldots, i-1, \qquad \kappa_{i,i} = \left\| \tilde{u}_i - \sum_{j=1}^{i-1} v_j \kappa_{j,i} \right\|$$

$$\chi_{j,i} = v_j^\mathsf{T} \frac{f(x + \zeta u_i) - f(x)}{\zeta}, \quad j = 1, \ldots, i,$$

$$\chi_{i+1,i} = \left\| \frac{f(x + \zeta u_i) - f(x)}{\zeta} - \sum_{j=1}^{i} v_j \chi_{j,i} \right\|$$

are determined in a standard manner to satisfy the orthonormality conditions $v_i^\mathsf{T} v_j = \delta_{i-j}$ and $u_i^\mathsf{T} u_j = \delta_{i-j}$. A similar procedure was proposed in [11] as a generalization of the ones presented in [8–10], but the orthonormality of u_i was not provided there. An equivalent form of the above recurrences is the matrix factorization

$$F = VH, \qquad V^\mathsf{T} V = I_{k+l+1}, \tag{8}$$

where F was defined in (4) and

$$V = [v_1 \mid \ldots \mid v_{k+l+1}] \in R^{m \times (k+l+1)}, \qquad H \in R^{(k+l+1) \times (k+l)};$$

the latter is the upper Hessenberg matrix as usually arises in Arnoldi type processes. Recall that the search direction is determined as

$$p = Us, \quad s \in R^{k+l}, \tag{9}$$

where s was specified in preceding Section. In terms of decomposition (8), one has, with account for

$$f = Ve_1 \|f\|, \qquad e_1 = [1 \ 0 \ \ldots \ 0]^\mathsf{T} \in R^{k+l+1},$$

the following representations:

$$s = -(H^\mathsf{T} H + \xi I)^{-1} H^\mathsf{T} e_1 \|f\|, \tag{10}$$

$$\theta^2 = e_1^\mathsf{T} H (H^\mathsf{T} H + \xi I)^{-1} H^\mathsf{T} e_1 \tag{11}$$

used for numerical implementation.

5.2   Efficient Quasirandom Rectangular Preconditioner

As the preconditioner we consider a full column rank matrix $K_t \in R^{m \times n}$ satisfying

$$K_t^\mathsf{T} K_t \approx I_n. \tag{12}$$

Clearly, the forming of K_t and multiplying it by a vector $q = K_t v$ must be as cheap as possible. Here we will consider preconditionings having a potential for a quite efficient implementation on modern high-performance computers. In the proposed designs, the so called logistic sequence (see, e.g., [17] and references cited therein) is used, defined as

$$\xi_0 = 0.4, \qquad \xi_k = 1 - 2\xi_{k-1}^2, \qquad k = 1, 2, \ldots. \tag{13}$$

The idea of randomized preconditionings was already studied in [14] (see also the references cited therein), though in rather different context.

Here we consider K_t taken as the leading $m \times n$ submatrix of $H D_t H$, where D_t is a diagonal matrix with quasirandom entries generated by the logistic sequence, and H is the Hadamard matrix of the order $2^b \geq \max(m, n)$, defined recursively as

$$H_2 = \begin{bmatrix} 1 & 1 \\ 1 & -1 \end{bmatrix}, \qquad H_{2k} = \begin{bmatrix} H_k & H_k \\ H_k & -H_k \end{bmatrix}, \qquad k = 2, 4, 8, \ldots, 2^{b-1}.$$

Clearly, such matrix K_t is formed implicitly in the factorized form, and the multiplication of K_t^T by a vector requires only $O(m \log m)$ additions and $O(m)$ multiplications.

5.3   Description of Computational Algorithm

The above described preconditioned subspace descent algorithm can be summarized as follows. Note that indicating $f(x)$ as an input means the availability of a computational module for the evaluation of the vector $f(x)$ for any given x.

Algorithm 1.

Key notations:

$$V_t = [v_1 | \ldots | v_{i_{max}+1}] \in R^{m \times (i_{max}+1)}, \qquad U_t = [u_1 | \ldots | u_{i_{max}}] \in R^{n \times i_{max}},$$

$$H_i = \begin{bmatrix} \chi_{1,1} & \chi_{1,2} & \chi_{1,3} & \cdots & \chi_{1,i} \\ \chi_{2,1} & \chi_{2,2} & \chi_{2,3} & \cdots & \chi_{2,i} \\ 0 & \chi_{3,2} & \chi_{3,3} & \cdots & \chi_{3,i} \\ \cdots & \cdots & \cdots & \cdots & \cdots \\ 0 & \cdots & 0 & \chi_{i,i-1} & \chi_{i,i} \\ 0 & \cdots & 0 & 0 & \chi_{i+1,i} \end{bmatrix} \in R^{(i+1) \times i}, \qquad h_i = \begin{bmatrix} \chi_{1,1} \\ \chi_{1,2} \\ \chi_{1,3} \\ \cdots \\ \chi_{1,i} \end{bmatrix} \in R^i;$$

Input: $f(x) : R^n \to R^m$, $x_0 \in R^n$;

Initialization:

01     $d = k + l \leq n$,

02     $\eta = 10^{-12}$,

03     $\zeta = 10^{-8}$,

04      $\varepsilon = 10^{-10}$,
05      $\tau_{\min} = 10^{-10}$,
06      $t_{\max} = 1000$,
07      $l_{\max} = 30$,
08      $f_0 = f(x_0)$,
09      $\rho_0 = f_0^\mathsf{T} f_0$;

Iterations:

for $t = 0, 1, \ldots, t_{\max} - 1$:
10   generate new quasirandom $K_t \in R^{m \times n}$
11          $v_1 := f_t / \sqrt{\rho_t}$
12          $w := v_1$
13          $i_{\max} := k + \min(l, t)$
14   **for** $i = 1, \ldots, i_{\max}$:
15   **if** $(i \leq k)$ **then**
16                  $u_i := K_t^\mathsf{T} w$
17   **end if**
18   **for** $j = 1, \ldots, i - 1$:
19                  $\kappa_{j,i} = u_j^\mathsf{T} u_i$
20                  $u_i := u_i - u_j \kappa_{j,i}$
21   **end for**
22              $\kappa_{i,i} = \sqrt{u_i^\mathsf{T} u_i}$
23              $u_i := u_i / \kappa_{i,i}$
24              $w := (f(x_t + \zeta u_i) - f(x_t)) / \zeta$
25   **for** $j = 1, \ldots, i$:
26                  $\chi_{j,i} = v_j^\mathsf{T} w$
27                  $w := w - v_j \chi_{j,i}$
28   **end for**
29              $\chi_{i+1,i} = \sqrt{w^\mathsf{T} w}$
30              $w := w / \chi_{i+1,i}$
31              $v_{i+1} = w$
32   **end for**
33          $L_t L_t^\mathsf{T} = H_{i_{\max}}^\mathsf{T} H_{i_{\max}} + \eta \, \mathrm{trace}(H_{i_{\max}}^\mathsf{T} H_{i_{\max}}) I$
34          $z_t := (L_t)^{-1} h_{i_{\max}}$
35          $\theta_t := z_t^\mathsf{T} z_t$
36          $z_t := (L_t)^{-\mathsf{T}} z_t \rho_t$
37          $p_t = -U_t z_t$
38          $u_{k+1+(t \bmod l)} := p_t$
39          $\alpha^{(0)} = 1$
40   **for** $l = 0, 1, \ldots, l_{\max} - 1$:
41              $x_t^{(l)} = x_t + \alpha^{(l)} p_t$
42              $f_t^{(l)} = f(x_t^{(l)})$
43              $\rho_t^{(l)} = (f_t^{(l)})^\mathsf{T} f_t^{(l)}$
44              $\tau = \alpha^{(l)} (2 - \alpha^{(l)}) \theta_t$
43   **if** $(\tau < \tau_{\min})$ **return** x_t
44   **if** $(\rho_t^{(l)} / \rho_t > 1 - (\tau/2)^2)$ **then**

45                              $\alpha^{(l+1)} = \alpha^{(l)}/2$
46                              $x_t^{(l+1)} = x_t + \alpha^{(l+1)} p_t$
47   **else**
48   **go to** NEXT
49   **end if**
50   **end for**
51   NEXT: $x_{t+1} = x_t^{(l)}$, $f_{t+1} = f_t^{(l)}$, $\rho_{t+1} = \rho_t^{(l)}$;
52   **if** $(\rho_{t+1} < \varepsilon^2 \rho_0)$ **or** $(\rho_{t+1} \geq \rho_t)$ **return** x_{t+1}
end for

Comments to Algorithm 1 are as follows.

- Line 10: a new quasirandom diagonal matrix D_t is generated to form the preconditioner K_t as explained in Sect. 5.2.
- Lines 11 to 32: generation of two orthonormalized systems of basis vectors as determined in Sect. 5.1.
- Lines 33 to 35: the evaluation of quantity θ_t can be explained as follows. To simplify the notation, let us drop the indices $i_{\max}$, and (l). Then, by $H^\mathsf{T} H + \delta I = LL^\mathsf{T}$ and $h = H^\mathsf{T} e_1$, it holds

$$\theta_t = z_t^\mathsf{T} z_t = h^\mathsf{T} L_t^{-\mathsf{T}} L_t^{-1} h = e_1^\mathsf{T} H (H^\mathsf{T} H + \delta I)^{-1} H^\mathsf{T} e_1, \qquad \frac{\tau}{2} = \frac{\alpha(2-\alpha)}{2} \theta_t.$$

Comparing these equalities with (7) and (11) (note that $\theta_t = \theta^2$) gives exactly the backtracking condition $\rho_t^{(l)}/\rho_t > 1 - (\tau/2)^2$ used in Algorithm 1 for the refinement of stepsize α.
- Lines 36 to 37: the evaluation of new direction p_t according to its definition in (10).
- Line 38: adding the current direction to the basis for the subsequent orthogonalization.
- Lines 39 to 50: evaluate the stepsize α by performing the backtracking based on the upper bound (7).

6   Test Problems and Numerical Results

Below some results of application of Algorithm 1 to several nonlinear test problems taken from the literature are presented. It must be stressed that in the vicinity of the solution, the derivatives of the residual function f are not only discontinuous, but even not bounded.

For the test runs, one core of Pentium(R) Dual-Core CPU E6600 3.06 GHz, 3.25 Gbytes RAM desktop PC was used. We will consider sufficiently large subspace dimensions with $k \geq l$ and $k + l \leq n$. For the nonzero residual problems, the iterations typically terminate by the condition $\tau < \tau_{\min} = 10^{-10}$, see the corresponding line in Algorithm 1.

6.1   Generalized Dennis-Woods Function

Following [7], consider the problem (1) with

$$\varphi(x) = \frac{1}{2} \max \left(\|x - c\|^2, \|x + c\|^2 \right) - \frac{1}{2}\|c\|^2, \quad c_j = (-1)^{n-j+1}, \quad j = 1, \ldots, n.$$

In order to reformulate it as a NLS problem, one can observe that

$$\varphi(x) = \frac{1}{2} \left(x^{\mathsf{T}} x + 2|c^{\mathsf{T}} x| \right).$$

One of the numerous ways to construct $f(x)$ such that $\varphi = \|f\|^2/2$ is to set $m = n + 1$ and define

$$f_i = |x_i|, \quad 1 \leq i \leq n, \qquad f_{n+1} = \sqrt{2} \left| \sum_{j=1}^{n} (-1)^{n-j+1} x_j \right|^{1/2}.$$

The optimum value is $\|f(x_*)\|/2 = 0$ at the optimum point $x_* = 0$. The results obtained for $n = 100$ and starting point $x_0 = [1, 1/2, \ldots, 1/n]^{\mathsf{T}}$ are presented in Table 1. Note that the termination criterion was $\alpha < 2^{-30}$.

Table 1. Generalized Dennis-Woods test: comparing various subspace dimensions

$d = k + l$	#iter	#fun.eval.	$\varphi(x_{\text{iter}}) - \varphi(x_*)$	$\|x\|_\infty$
75+23	8	670	1.5E−05	1.3E−03
50+50	17	1046	1.0E−05	1.3E−03
51+47	22	1427	6.3E−08	8.4E−05
50+1	17	972	3.4E−05	8.2E−03
26+25	29	1217	7.9E−06	9.9E−04

6.2   Nesterov Nonsmooth Function

This test function was used in [13] and defined as

$$\varphi(x) = |x_1 - 1| + \sum_{i=2}^{n} |1 + x_i - 2x_{i-1}|.$$

Its NLS reformulation can be taken as

$$f_1 = \sqrt{2|x_1 - 1|}, \qquad f_i = \sqrt{2|1 + x_i - 2x_{i-1}|}, \qquad i = 2, \ldots, n.$$

The optimum value is $f^{\mathsf{T}} f/2 = 0$ at $x_* = [1 \ldots 1]^{\mathsf{T}}$ and the starting point is $x_0 = 0$. The results are given in Table 2 for n equal to $16, 32, 64, 128, 256, 512$. The subspace sizes were defined by $k = n/2 + 1$ and $l = n - k - 2$.

Apparently setting $x_0 = 0$ makes the problem too easy, so in Table 3 we present the results for $x_0 = [1, 1/2, \ldots, 1/n]^\mathsf{T}$. In this case, the convergence history demonstrates the behavior typical for linear conjugate gradients, when a certain residual norm decrease at initial steps is followed by a near stagnation phase and concluded with fast superlinear decrease at the final stage.

Remark 1. The direct comparison of the data in Table 3 for $n = 256$ with the corresponding results in Table 3 of [13] shows > 100 times reduction in the number of function evaluation provided by Algorithm 1.

Table 2. Nesterov Nonsmooth test: comparing problems of various size ($x_0 = 0$)

n	#iter	#fun.eval.	$\varphi(x_{\text{iter}}) - \varphi(x_*)$	$\|x\|_\infty$
16	3	65	1.0E−08	1.0E+00
32	3	89	1.3E−08	1.0E+00
64	3	137	2.0E−08	1.0E+00
128	3	233	2.8E−08	1.0E+00
256	3	425	4.0E−08	1.0E+00
512	3	809	5.6E−08	1.0E+00

Table 3. Nesterov Nonsmooth test: comparing problems of various size ($x_0 \neq 0$)

n	#iter	#fun.eval.	$\varphi(x_{\text{iter}}) - \varphi(x_*)$	$\|x\|_\infty$
16	30	487	6.1E−06	1.0E+00
32	38	1172	2.0E−07	1.0E+00
64	36	1870	6.3E−06	1.0E+00
128	180	21719	1.6E−07	1.0E+00
256	128	24800	2.4E−05	1.0E+00
512	323	133433	1.1E−06	1.0E+00

6.3   Rando e.a. Nonsmooth Function

Test function

$$\varphi(x) = \sum_{i=1}^{n} |x_i - i + 1|$$

was used in [15]. Its NLS reformulation can be taken as

$$f_i = \sqrt{2|x_i - i + 1|}, \qquad i = 1, \ldots, n.$$

The optimum value is $f^\mathsf{T} f/2 = 0$ at $x_* = [0, 1 \ldots n-1]^\mathsf{T}$ and the starting point is set to $x_0 = 0$. The results are given in Table 4 for $n = 50$ (as used in [15]) and different choices of k and l. Note that with $d = k + l \approx n$ the problem seems to be not too hard-to-solve for such moderate size. Therefore, we also present data for the larger size $n = 500$ in Table 5.

From the results obtained for this problem, it is clear that the best performance of the iterative solver (with respect to both solution cost and quality of the result) is observed for the subspace dimension $d = k + l$ slightly smaller than the size of the problem $\min(m, n)$.

Table 4. Rando e.a. test: comparing various subspace dimensions ($n = 50$)

$d = k + l$	#iter	#fun.eval.	$\varphi(x_{\mathrm{iter}}) - \varphi(x_*)$	$\|x\|_\infty$
26+22	61	2931	2.9E−08	49.E+00
25+25	37	1692	8.3E−07	49.E+00
30+20	26	1168	6.5E−07	49.E+00
40+10	21	1123	1.5E−07	49.E+00
20+10	>1000	>32142	2.1E+00	49.E+00

Table 5. Rando e.a. test: comparing various subspace dimensions ($n = 500$)

$d = k + l$	#iter	#fun.eval.	$\varphi(x_{\mathrm{iter}}) - \varphi(x_*)$	$\|x\|_\infty$
251+247	334	136673	2.6E−07	499.E+00
250+250	253	95413	9.4E−05	499.E+00
300+200	204	82186	1.4E−04	499.E+00
400+100	104	47108	2.2E−05	499.E+00
200+100	>1000	>295951	1.4E+03	504.E+00

7   Concluding Remarks

In the present paper, a nonlinear least squares solver is presented which is based on derivative-free computations and is formally applicable to all types of least squares problems, even with nonsmooth residual function. Key feature of the algorithm is the use of quasirandom rectangular preconditioners for the construction of approximate Krylov subspaces containing descent directions. The proposed algorithmic implementation of the method is justified by theoretical considerations related to the residual norm reduction. The results of numerical testing on several hard-to-solve problems have confirmed the efficiency and robustness of the derivative-free Preconditioned Subspace Descent method. The main contribution of the paper is that an additional orthogonalization stage was introduced, which placed the resulting method into the class of ZO solvers with structured quasirandom gradients.

Acknowledgement. The author thanks the anonymous referees for insightful comments and suggestions which allow to significantly improve the exposition of the paper.

References

1. Armijo, L.: Minimization of functions having Lipschitz continuous first partial derivatives. Pac. J. Math. **16**(1), 1–3 (1966)
2. Brent, R.P.: Algorithms for matrix multiplication. (Report No. STAN-CS-70-157), 58p. Department of Computer Science, Stanford Univ. CA (1970)
3. Brown, P.N.: A local convergence theory for combined inexact-Newton/finite-difference projection methods. SIAM J. Numer. Anal. **24**(2), 407–434 (1987)
4. Brown, P.N., Saad, Y.: Hybrid Krylov methods for nonlinear systems of equations. SIAM J. Sci. Stat. Comput. **11**(3), 450–481 (1990)
5. Brown, P.N., Saad, Y.: Convergence theory of nonlinear Newton-Krylov algorithms. SIAM J. Optim. **4**(2), 297–330 (1994)
6. Conn, A.R., Scheinberg, K., Vicente, L.N.: Introduction to Derivative Free Optimization. MPS-SIAM Series on Optimization. SIAM, Philadelphia (2009)
7. Dennis Jr., J.E., Woods, D.J.: Optimization on microcomputers: the Nelder-Mead simplex algorithm. In: Wouk, A. (ed.) New Computing Environments: Microcomputers in Large-Scale Scientific Computing, pp. 116–122. SIAM, Philadelphia (1987)
8. Kaporin, I.E.: The use of preconditioned Krylov subspaces in conjugate gradient type methods for the solution of nonlinear least square problems (Russian). Vestnik Mosk. Univ., Ser. 15 (Computational Math. and Cybernetics), vol. 3, pp. 26–31 (1995)
9. Kaporin, I.E., Axelsson, O.: On a class of nonlinear equation solvers based on the residual norm reduction over a sequence of affine subspaces. SIAM J. Sci. Comput. **16**(1), 228–249 (1994)
10. Kaporin, I.: Preconditioned subspace descent method for nonlinear systems of equations. Open Comput. Sci. **10**(1), 71–81 (2020)
11. Kaporin, I.: A derivative-free nonlinear least squares solver. In: Olenev, N.N., Evtushenko, Y.G., Jaćimović, M., Khachay, M., Malkova, V. (eds.) OPTIMA 2021. LNCS, vol. 13078, pp. 217–230. Springer, Cham (2021). https://doi.org/10.1007/978-3-030-91059-4_16
12. Kozak, D., Molinari, C., Rosasco, L., Tenorio, L., Villa, S.: Zeroth order optimization with orthogonal random directions. Math. Program. 1–41 (2022)
13. Nesterov, Y., Spokoiny, V.: Random gradient-free minimization of convex functions. Found. Comput. Math. **17**(2), 527–566 (2017)
14. Pan, V.Y., Qian, G.: Randomized preprocessing of homogeneous linear systems of equations. Linear Algebra Appl. **432**, 3272–3318 (2010)
15. Rando, M., Molinari, C., Villa, S., Rosasco, L.: Stochastic zeroth order descent with structured directions. arXiv preprint arXiv:2206.05124 (2022)

16. Strassen, V.: Gaussian elimination is not optimal. Numer. Math. **13**, 354–356 (1969)
17. Yu, L., Barbot, J.P., Zheng, G., Sun, H.: Compressive sensing with chaotic sequence. IEEE Sig. Process. Lett. **17**(8), 731–734 (2010)

Stochastic Adversarial Noise in the "Black Box" Optimization Problem

Aleksandr Lobanov[1,2,3]([✉])(iD)

[1] Moscow Institute of Physics and Technology, Dolgoprudny, Russia
[2] ISP RAS Research Center for Trusted Artificial Intelligence, Moscow, Russia
[3] Moscow Aviation Institute, Moscow, Russia
lobanov.av@mipt.ru

Abstract. This paper is devoted to the study of the solution of a stochastic convex black box optimization problem. Where the black box problem means that the gradient-free oracle only returns the value of objective function, not its gradient. We consider non-smooth and smooth setting of the solution to the black box problem under adversarial stochastic noise. For two techniques creating gradient-free methods: smoothing schemes via L_1 and L_2 randomizations, we find the maximum allowable level of adversarial stochastic noise that guarantees convergence. Finally, we analyze the convergence behavior of the algorithms under the condition of a large value of noise level.

Keywords: Gradient-free methods · Black-box problem · Adversarial stochastic noise · Smooth and non-smooth setting

1 Intoduction

The study of optimization problems, in which only a limited part of information is available, namely, the value of the objective function, in recent years is relevant and in demand. Such problems are usually classified as zero-order optimization problems [12] or black-box problems [4]. Where the latter intuitively understands that the black box is some process that has only two features (input, output), and about which nothing is known. The essential difference of this class of optimization problems is that the oracle returns only the value of the objective function and not its gradient or higher order derivatives, which are popular in numerical methods [33]. This kind of oracle is commonly referred to as a zero-order/gradient-free oracle [36], which acts as a black box. One of several leading directions for solving black-box optimization problems are zero-order numerical methods, which are based on first-order methods, approximating the gradient via finite-difference models. The first such method that gave rise to the field of research is the Kiefer-Wolfowitz method [25], which was proposed in 1952. Since

The research was supported by Russian Science Foundation (project No. 21-71- 30005), https://rscf.ru/en/project/21-71-30005/.

then, many gradient-free algorithms have been developed for applied problems in machine learning [30,31], distributed learning [3,37,42], federated learning [27,32], black-box attack to deep neural networks [11], online [5,21,38] and standard [13,22,39] optimization. In particular, gradient-free methods are actively used in the hyperparameters tuning of the deep learning model [16,24,26], as well as to solve the classical problem of adversarial multi-armed bandit [6,10,18].

In modern works [19] authors try to develop optimal gradient-free algorithms according to three criteria at once: iteration complexity (the number of iterations performed successively guaranteeing convergence), oracle complexity (the total number of gradient-free oracle calls guaranteeing convergence), and the maximum admissible level of adversarial noise guaranteeing convergence. While the first two criteria seem understandable, the maximum level of adversarial noise can raise questions. However, in practice it is very common that the gradient-free oracle returns an inaccurate value of the objective function at the requested point. In other words, the gradient-free oracle outputs the value of the objective function with some adversarial noise. A simple example is rounding error (machine accuracy). There are also problems in which the computational complexity clearly depends on the level of adversarial noise, i.e., the greater the level of adversarial noise, the better the algorithm works in terms of computational complexity. This is why it is necessary to consider this criterion when creating a gradient-free optimization algorithm.

In this paper we focus on solving a stochastic convex black-box optimization problem. We consider two black-box problem settings: smooth and non-smooth settings. We assume that the gradient-free oracle corrupted by an adversarial stochastic noise. Using smoothing schemes via L_1 and L_2 randomization, we derive the maximum permissible level of adversarial stochastic noise for each setting. And we also provide convergence results for the three-criteria optimal algorithm under adversarial stochastic noise, and provide discussions of how the algorithm would converge if the noise exceeded the optimal estimate.

1.1   Related Works

Gradient Approximation.   In many works [1,2,8,14,15,17,20,23,27,29,34, 41] have developed methods, using various techniques to create algorithms via gradient approximation. For example, in [8] a full gradient approximation instead of an exact gradient was used. Also in the smooth case, instead of an exact gradient, the some works use coordinate-wise randomization [23,41] and random search randomization [15,17]. It is worth noting that these approximations assume that the gradient of the function is available. In [2,34], the authors developed a gradient-free algorithm using a kernel approximation, which takes into account the increased smoothness of the objective function. For the non-smooth case [14,20] describes a smoothing technique for creating gradient-free algorithms using L_2 randomization. And the paper [1] derived a better theoretical estimate of the variance of the gradient approximation, using L_1 randomization instead of L_2 randomization. In turn, the work [27] generalized the results of L_1 randomization to the non-smooth case and compared two smoothing schemes: L_1 and L_2

randomization, showing that in practice the clear advantage of L_1 randomization is not observed. In this paper, using smoothing schemes via L_1 and L_2 randomizations, we derive optimal estimates of the level of noise, in which the optimality of oracle and iterative complexity are not degraded.

Adversarial Noise.   There are many works [3,7,9,14,27,35,40] that study optimization problems under adversarial noise. For example, the works [7,9] provided an optimal algorithm in terms of oracle complexity, but not optimal in terms of maximum allowable adversarial noise. Other works [35,40] have proposed algorithms that are optimal in terms of the maximum permissible level of adversarial noise, but are not optimal according to the criterion of oracle complexity. A gradient-free algorithm, which is optimal according to two criteria: oracle complexity and the maximum permissible level of adversarial noise, is proposed in [14]. Whereas work [27] provided an optimal gradient-free algorithm for all three criteria: iteration complexity, oracle complexity, and the maximum allowable level of adversarial noise. However, these works considered the concept of adversarial deterministic noise, whereas for adversarial stochastic noise (see., e.g. [3]) the optimal bound for solving the black box optimization problem in smooth and non-smooth settings has not been obtained. In this paper, we will solve the black box problem in two settings with adversarial stochastic noise.

1.2   Paper Organization

The structure of this paper is as follows. In Sect. 2, we describe the problem statement, as well as the basic assumptions and notations. We present the main result in Sect. 3. In Sect. 4, we discuss the theoretical results obtained. While Sect. 5 concludes this paper.

2   Setting and Assumptions

We study a standard stochastic convex black-box optimization problem:

$$f^* = \min_{x \in Q} \left\{ f(x) := \mathbb{E}_\xi \left[f(x, \xi) \right] \right\}, \tag{1}$$

where $Q \subseteq \mathbb{R}^d$ is a convex and compact set, $f : Q \to \mathbb{R}$ is convex function. Since problem (1) is a general problem formulation, we introduce standard assumptions and definitions that will narrow down the class of problems under consideration.

Definition 1 (Gradient-free oracle). *Gradient-free oracle returns a function value $f(x, \xi)$ at the requested point x with some adversarial stochastic noise, i.e. for all $x \in Q$*

$$f_\delta(x, \xi) := f(x, \xi) + \delta.$$

Assumption 1 (Lipschitz continuity of objective function). *The function $f(x, \xi)$ is an M-Lipschitz continuous function in the l_p-norm, i.e. for all $x, y \in Q$ we have*

$$|f(y, \xi) - f(x, \xi)| \leq M(\xi) \|y - x\|_p.$$

Moreover, there is a positive constant M, which is defined in the following way: $\mathbb{E}\left[M^2(\xi)\right] \leq M^2$. *In particular, for $p = 2$ we use the notation M_2 for the Lipschitz constant.*

Assumption 2 (Convexity on the set Q_γ). *Let $\gamma > 0$ a small number to be defined later and $Q_\gamma := Q + B_p^d(\gamma)$, then the function f is convex on the set Q_γ.*

The following assumption we need to solve problem (1) in a smooth setting.

Assumption 3 (Smoothness of function). *The function f is smooth, that is, differentiable on Q and such that for all $x, y \in Q$ with $L > 0$ we have*

$$\|\nabla f(y) - \nabla f(x)\|_q \leq L\|y - x\|_p.$$

Next, we introduce the assumption about adversarial noise.

Assumption 4 (Adversarial noise). *It holds, that the random variables δ_1 and δ_2 are independent from $e \in S_p^d(1)$ as well as $\mathbb{E}\left[\delta_1^2\right] \leq \Delta^2$ and $\mathbb{E}\left[\delta_2^2\right] \leq \Delta^2$.*

Our Assumption 1 is necessary for theoretical proofs in each setting: smooth and non-smooth. This Assumption 1 is common in literature (see e.g. [20,28]). Assumption 2 is standard for works using smoothing technique (see e.g. [27,35]). Assumption 3 was introduced only for smooth tuning, and is also often found in the literature (e.g., in previous works such as [2,5]). Whereas Definition 1 is similar to [19], only using adversarial stochastic noise instead of adversarial deterministic noise. Finally, Assumption 4 is the same as in the previous work [1].

Notation

We use $\langle x, y \rangle := \sum_{i=1}^d x_i y_i$ to denote standard inner product of $x, y \in \mathbb{R}^d$, where x_i and y_i are the i-th component of x and y respectively. We denote l_p-norms (for $p \geq 1$) in $\mathbb{R}^d$ as $\|x\|_p := \left(\sum_{i=1}^d |x_i|^p\right)^{1/p}$. Particularly for l_2-norm in $\mathbb{R}^d$ it follows $\|x\|_2 := \sqrt{\langle x, x \rangle}$. We denote l_p-ball as $B_p^d(r) := \left\{x \in \mathbb{R}^d : \|x\|_p \leq r\right\}$ and l_p-sphere as $S_p^d(r) := \left\{x \in \mathbb{R}^d : \|x\|_p = r\right\}$. Operator $\mathbb{E}[\cdot]$ denotes mathematical expectation. To denote the distance between the initial point x^0 and the solution of the initial problem x_* we introduce $R := \tilde{\mathcal{O}}\left(\|x^0 - x_*\|_p\right)$, where we notation $\tilde{\mathcal{O}}(\cdot)$ to hide logarithmic factors.

3   Main Result

In this section, we build our narrative on solving the black-box optimization problem (1) in a non-smooth setting. We will discuss the smooth setting as a special case of the non-smooth setting (presence of Assumption 3) at the end of the section in Remark 1. This section is organized as follows: in Subsect. 3.1 we introduce the smooth approximation of a non-smooth function and its properties, the gradient approximation by L_1 randomization and its properties, i.e. we describe the smoothing scheme via L_1 randomization. In Subsect. 3.2, we do the same and describe the smoothing scheme via L_2 randomization. And in Subsect. 3.3 we present maximum allowed level of adversarial stochastic noise. So we begin by describing the smoothing technique via L_1 randomization.

3.1   Smoothing Scheme via L_1 Randomization

Since problem (1) is non-smooth, we introduce the following smooth approximation of the non-smooth function:

$$f_\gamma(x) := \mathbb{E}_{\tilde{e}}\left[f(x + \gamma\tilde{e})\right], \tag{2}$$

where $\gamma > 0$ is a smoothing parameter, $\tilde{e}$ is a random vector uniformly distributed on $B_1^d(1)$. Here $f(x) := \mathbb{E}\left[f(x, \xi)\right]$. The following lemma provides the connection between the smoothed and the original function.

Lemma 1. *Let Assumptions 1, 2 hold, then for all $x \in Q$ we have*

$$f(x) \leq f_\gamma(x) \leq f(x) + \frac{2}{\sqrt{d}}\gamma M_2.$$

Proof. For the first inequality we use the convexity of the function $f(x)$

$$f_\gamma(x) = \mathbb{E}_{\tilde{e}}\left[f(x + \gamma\tilde{e})\right] \geq \mathbb{E}_{\tilde{e}}\left[f(x) + \langle \nabla f(x), \gamma\tilde{e}\rangle\right] = \mathbb{E}_{\tilde{e}}\left[f(x)\right] = f(x).$$

For the second inequality, applying Lemma 1 of [1] = ①, we have

$$|f_\gamma(x) - f(x)| = |\mathbb{E}_{\tilde{e}}\left[f(x + \gamma\tilde{e})\right] - f(x)| \leq \mathbb{E}_{\tilde{e}}\left[|f(x + \gamma\tilde{e}) - f(x)|\right]$$

$$\leq \gamma M_2 \mathbb{E}_{\tilde{e}}\left[\|\tilde{e}\|_2\right] \overset{①}{\leq} \frac{2}{\sqrt{d}}\gamma M_2,$$

using the fact that f is M_2-Lipschitz function.

$\square$

The following lemmas confirm that the Lipschitz continuity property holds and provide the Lipschitz constant of gradient for the smoothed function.

Lemma 2. *Let Assumptions 1, 2 hold, then for $f_\gamma(x)$ from (2) we have*

$$|f_\gamma(y) - f_\gamma(x)| \leq M\|y - x\|_p, \quad \forall x, y \in Q.$$

Proof. Using M-Lipschitz continuity of function f we obtain

$$|f_\gamma(y) - f_\gamma(x)| \leq \mathbb{E}_{\tilde{e}}\left[|f(y + \gamma\tilde{e}) - f(x + \gamma\tilde{e})|\right] \leq M\|y - x\|_p.$$

$\square$

Lemma 3 (Lemma 1, [27]). *Let Assumptions 1, 2 hold, then $f_\gamma(x)$ has $L_{f_\gamma} = \frac{dM}{\gamma}$-Lipschitz gradient*

$$\|\nabla f_\gamma(y) - \nabla f_\gamma(x)\|_q \leq L_{f_\gamma}\|y - x\|_p, \quad \forall x, y \in Q.$$

The gradient of $f_\gamma(x, \xi)$ can be estimated by the following approximation:

$$\nabla f_\gamma(x, \xi, e) = \frac{d}{2\gamma}\left(f_{\delta_1}(x + \gamma e, \xi) - f_{\delta_2}(x - \gamma e, \xi)\right)\text{sign}(e), \tag{3}$$

where $f_\delta(x, \xi)$ is gradient-free oracle from Definition 1, e is a random vector uniformly distributed on $S_1^d(\gamma)$. The following lemma provides properties of the gradient $\nabla f_\gamma(x, \xi, e)$.

Lemma 4 (Lemma 4 [1]**).** *Gradient* $\nabla f_\gamma(x, \xi, e)$ *has bounded variance (second moment) for all* $x \in Q$

$$\mathbb{E}_{\xi, e}\left[\|\nabla f_\gamma(x, \xi, e)\|_q^2\right] \leq \kappa(p, d)\left(M_2^2 + \frac{d^2\Delta^2}{12(1 + \sqrt{2})^2\gamma^2}\right),$$

where $1/p + 1/q = 1$ *and*

$$\kappa(p, d) = \kappa(p, d) = 48(1 + \sqrt{2})^2 d^{2 - \frac{2}{p}}.$$

3.2   Smoothing Scheme via L_2 Randomization

Since problem (1) is non-smooth, we introduce the following smooth approximation of the non-smooth function:

$$\tilde{f}_\gamma(x) := \mathbb{E}_{\tilde{e}}\left[f(x + \gamma\tilde{e})\right], \tag{4}$$

where $\gamma > 0$ is a smoothing parameter, $\tilde{e}$ is a random vector uniformly distributed on $B_2^d(1)$. Here $f(x) := \mathbb{E}\left[f(x, \xi)\right]$. The following lemma provides the connection between the smoothed and the original function.

Lemma 5. *Let Assumptions 1, 2 hold, then for all* $x \in Q$ *we have*

$$f(x) \leq \tilde{f}_\gamma(x) \leq f(x) + \gamma M_2.$$

Proof. For the first inequality we use the convexity of the function $f(x)$

$$\tilde{f}_\gamma(x) = \mathbb{E}_{\tilde{e}}\left[f(x + \gamma\tilde{e})\right] \geq \mathbb{E}_{\tilde{e}}\left[f(x) + \langle\nabla f(x), \gamma\tilde{e}\rangle)\right] = \mathbb{E}_{\tilde{e}}\left[f(x)\right] = f(x).$$

For the second inequality we have

$$|\tilde{f}_\gamma(x) - f(x)| = |\mathbb{E}_{\tilde{e}}\left[f(x + \gamma\tilde{e})\right] - f(x)| \leq \mathbb{E}_{\tilde{e}}\left[|f(x + \gamma\tilde{e}) - f(x)|\right]$$
$$\leq \gamma M_2\mathbb{E}_{\tilde{e}}\left[\|\tilde{e}\|_2\right] \leq \gamma M_2,$$

using the fact that f is M_2-Lipschitz function.

$\square$

The following lemmas confirm that the Lipschitz continuity property holds and provide the Lipschitz constant of gradient for the smoothed function.

Lemma 6. *Let Assumptions 1, 2 hold, then for* $\tilde{f}_\gamma(x)$ *from* (2) *we have*

$$|\tilde{f}_\gamma(y) - \tilde{f}_\gamma(x)| \leq M\|y - x\|_p, \quad \forall x, y \in Q.$$

Proof. Using M-Lipschitz continuity of function f we obtain

$$|\tilde{f}_\gamma(y) - \tilde{f}_\gamma(x)| \leq \mathbb{E}_{\tilde{e}}\left[|f(y + \gamma\tilde{e}) - f(x + \gamma\tilde{e})|\right] \leq M\|y - x\|_p.$$

$\square$

Lemma 7 (Theorem 1, [19]). *Let Assumptions 1, 2 hold, then $\tilde{f}_\gamma(x)$ has $L_{\tilde{f}_\gamma} = \frac{\sqrt{d}M}{\gamma}$-Lipschitz gradient*

$$\|\nabla \tilde{f}_\gamma(y) - \nabla \tilde{f}_\gamma(x)\|_q \leq L_{\tilde{f}_\gamma} \|y - x\|_p, \quad \forall x, y \in Q.$$

The gradient of $\tilde{f}_\gamma(x, \xi)$ can be estimated by the following approximation:

$$\nabla \tilde{f}_\gamma(x, \xi, e) = \frac{d}{2\gamma} \left(f_{\delta_1}(x + \gamma e, \xi) - f_{\delta_2}(x - \gamma e, \xi) \right) e, \tag{5}$$

where $f_\delta(x, \xi)$ is gradient-free oracle from Definition 1, e is a random vector uniformly distributed on $S_2^d(\gamma)$. The following lemma provides properties of the gradient $\nabla \tilde{f}_\gamma(x, \xi, e)$.

Lemma 8 ([27,38]). *Gradient $\nabla \tilde{f}_\gamma(x, \xi, e)$ has bounded variance (second moment) for all $x \in Q$*

$$\mathbb{E}_{\xi, e} \left[\|\nabla \tilde{f}_\gamma(x, \xi, e)\|_q^2 \right] \leq \kappa(p, d) \left(dM_2^2 + \frac{d^2 \Delta^2}{\sqrt{2}\gamma^2} \right),$$

where $1/p + 1/q = 1$ and

$$\kappa(p, d) = \sqrt{2} \min\{q, \ln d\} d^{1 - \frac{2}{p}}.$$

3.3   Maximum Level of Adversarial Stochastic Noise

In this subsection, we present our main result, namely the optimal bounds in terms of the maximum allowable level of the adversarial stochastic noise for smoothing techniques discussed in Subsects. 3.1 and 3.2. Next, we will consider case when $\Delta > 0$, i.e., there is adversarial noise. Then, before writing down main theorem, let us show that the gradient approximations (3) and (5) are unbiased.

- Gradient approximation (3) is unbiased:

$$\mathbb{E}_{e, \xi} \left[\nabla f_\gamma(x, \xi, e) \right] = \mathbb{E}_{e, \xi} \left[\frac{d}{2\gamma} \left(f_{\delta_1}(x + \gamma e, \xi) - f_{\delta_2}(x - \gamma e, \xi) \right) \text{sign}(e) \right]$$

$$= \mathbb{E}_{e, \xi} \left[\frac{d}{2\gamma} \left(f(x + \gamma e, \xi) + \delta_1 - f(x - \gamma e, \xi) - \delta_2 \right) \text{sign}(e) \right]$$

$$\overset{②}{=} \mathbb{E}_{e, \xi} \left[\frac{d}{2\gamma} \left(f(x + \gamma e, \xi) - f(x - \gamma e, \xi) \right) \text{sign}(e) \right]$$

$$= \mathbb{E}_e \left[\frac{d}{2\gamma} \left(f(x + \gamma e) - f(x - \gamma e) \right) \text{sign}(e) \right]$$

$$\overset{③}{=} \nabla f_\gamma(x),$$

where ② = we assume that Assumption 4 is satisfied, ③ = Lemma 1 [1].

- Gradient approximation (5) is unbiased:

$$\mathbb{E}_{e,\xi}\left[\nabla\tilde{f}_\gamma(x,\xi,e)\right] = \mathbb{E}_{e,\xi}\left[\frac{d}{2\gamma}\left(f_{\delta_1}(x+\gamma e,\xi) - f_{\delta_2}(x-\gamma e,\xi)\right)e\right]$$

$$= \mathbb{E}_{e,\xi}\left[\frac{d}{2\gamma}\left(f(x+\gamma e,\xi) + \delta_1 - f(x-\gamma e,\xi) - \delta_2\right)e\right]$$

$$\overset{②}{=} \mathbb{E}_{e,\xi}\left[\frac{d}{2\gamma}\left(f(x+\gamma e,\xi) - f(x-\gamma e,\xi)\right)e\right]$$

$$= \mathbb{E}_e\left[\frac{d}{2\gamma}\left(f(x+\gamma e) - f(x-\gamma e)\right)e\right]$$

$$\overset{③}{=} \nabla\tilde{f}_\gamma(x),$$

where ② = we assume that Assumption 4 is satisfied, ③ = Theorem 2.2 [20].

Since the adversarial noise does not accumulate in the bias (since at $\Delta > 0$ the gradient approximation is unbiased), the maximum allowable level of adversarial stochastic noise will only accumulate in the variance. Then the following Theorem 1 presents the optimal estimates for adversarial noise.

Theorem 1. *Let Assumptions 1, 2, 4 be satisfied, then the algorithm $A(L,\sigma^2)$ obtained by applying smoothing schemes (see Subsects. 3.1 and 3.2) based on the first order method*

1. for Smoothing scheme via L_1 randomization has level of adversarial noise

$$\Delta \lesssim \frac{\varepsilon}{\sqrt{d}};$$

2. for Smoothing scheme via L_2 randomization has level of adversarial noise

$$\Delta \lesssim \frac{\varepsilon}{\sqrt{d}},$$

where ε is accuracy solution to problem (1), $\mathbb{E}[f(x_N)] - f^* \leq \varepsilon$.

Proof. Since the adversarial noise accumulates only in the variance, in order to guarantee convergence (without losing in the oracle complexity estimates) it is necessary to guarantee that the following inequality is satisfied:

- for Smoothing scheme via L_1 randomization from Lemma 4

$$M_2^2 \geq \frac{d^2\Delta^2}{12(1+\sqrt{2})^2\gamma^2}$$

Then we have, using the fact that $\gamma = \frac{\sqrt{d}\varepsilon}{2M_2}$

$$\Delta \leq \sqrt{\frac{12(1+\sqrt{2})^2M_2^2\gamma^2}{d^2}} = \frac{2\sqrt{3}(1+\sqrt{2})M_2\sqrt{d}\varepsilon}{dM_2} \simeq \frac{\varepsilon}{\sqrt{d}}.$$

- for Smoothing scheme via L_2 randomization from Lemma 8

$$dM_2^2 \geq \frac{d^2\Delta^2}{\sqrt{2}\gamma^2}$$

Then we have, using the fact that $\gamma = \frac{\varepsilon}{2M_2}$

$$\Delta \leq \sqrt{\frac{\sqrt{2}dM_2^2\gamma^2}{d^2}} = \frac{2^{1/4}M_2\varepsilon}{\sqrt{d}M_2} \simeq \frac{\varepsilon}{\sqrt{d}}.$$

□

The results of Theorem 1 show that the maximum allowable level of adversarial stochastic noise is the same for the two smoothing schemes. Moreover, this estimation guarantees convergence without losing in other criteria, i.e. if we take as Algorithm $\mathbf{A}(L, \sigma^2)$ the accelerated batched first-order method and apply one of the two smoothing techniques, we will create an optimal algorithm for three criteria at once: oracle complexity, number of successive iterations, and the maximum permissible level of adversarial stochastic noise. An explanation of the choice of the smoothing parameter value can be found in Corollary 1 [27].

Remark 1 (Smoothing setting). Since Assumption 3 is satisfied in the smooth statement of problem (1), all statements above hold except Lemmas 1 and 5: $f(x) \leq \tilde{f}_\gamma(x) \leq f(x) + \frac{2}{d}\gamma^2 L^2$ (in Lemma 1) and $f(x) \leq \tilde{f}_\gamma(x) \leq f(x) + \gamma^2 L^2$ (in Lemma 5). Thus, we can conclude that if the Assumptions 1–4 are satisfied, then based on the accelerated first-order batched algorithm $\mathbf{A}(\sigma^2)$, and using any gradient approximation (L_1 or L_2 randomization) it will allow to create an optimal algorithm according to the three criteria, where the maximum allowed level of adversarial stochastic noise is $\Delta \lesssim \sqrt{\frac{\varepsilon}{d}}$.

4   Discussion

The essential difference between the stochastic adversarial noise considered in this paper and the deterministic adversarial noise is that in our case the adversarial noise does not accumulate into a bias, while in the deterministic noise concept the opposite is true. Precisely because this concept behaves less adversely, it is possible to solve the problem with a large value of adversarial noise without losing convergence, unlike deterministic adversarial noise, which has a maximum allowable noise level equal to $\mathcal{O}\left(\varepsilon^2 d^{-1/2}\right)$. In order to achieve optimality on the three criteria, it is necessary to rely on an accelerated (for optimal estimation of oracle complexity $\sim \mathcal{O}\left(d\varepsilon^{-2}\right)$ for case $p = 2$) batched (for optimal estimation of iterative complexity $\sim \mathcal{O}\left(d^{1/4}\varepsilon^{-2}\right)$) first-order method. And also using the concept of adversarial stochastic noise, by guaranteeing the prevalence of variance (e.g. from (8): $dM_2^2 \geq \frac{d^2\Delta^2}{\sqrt{2}\gamma^2}$), the Theorem 1 guarantees optimal convergence in three criteria. However, if the evaluation of the second moment is dominated by stochastic adversarial noise (i.e. $dM_2^2 < \frac{d^2\Delta^2}{\sqrt{2}\gamma^2}$), then the convergence of the algorithm will worsen (estimation of oracle complexity will be $\sim \varepsilon^{-4}$), since the second moment will be $\kappa(p,d)d^2\Delta^2\varepsilon^{-2}$, considering that $\gamma \sim \varepsilon$. Such a low estimate of oracle complexity corresponds to oracle complexity when a gradient approximation look like $\frac{d}{\gamma}f(x+\gamma e)e$ (e.g. for L_2 randomization), i.e. when gradient approximation requires only one gradient-free oracle calls.

5   Conclusion

In this paper, we studied stochastic convex black box optimization problems in two settings: non-smooth and smooth settings. For two smoothing techniques (with L_1 and L_2 randomization), we obtained the maximum allowable levels of adversarial stochastic noise. We also showed in this paper that for using any smoothing scheme via L_1 or L_2 randomization, as well as solving the black box problem in a non-smooth or smooth setting, the maximum value of adversarial noise is the same. Finally, we analyzed the convergence rate of the algorithm, provided that the noise level is large.

References

1. Akhavan, A., Chzhen, E., Pontil, M., Tsybakov, A.: A gradient estimator via l1-randomization for online zero-order optimization with two point feedback. Adv. Neural. Inf. Process. Syst. **35**, 7685–7696 (2022)
2. Akhavan, A., Pontil, M., Tsybakov, A.: Exploiting higher order smoothness in derivative-free optimization and continuous bandits. Adv. Neural. Inf. Process. Syst. **33**, 9017–9027 (2020)
3. Akhavan, A., Pontil, M., Tsybakov, A.: Distributed zero-order optimization under adversarial noise. Adv. Neural. Inf. Process. Syst. **34**, 10209–10220 (2021)
4. Audet, C., Hare, W.: Derivative-free and blackbox optimization (2017)
5. Bach, F., Perchet, V.: Highly-smooth zero-th order online optimization. In: Conference on Learning Theory, pp. 257–283. PMLR (2016)
6. Bartlett, P., Dani, V., Hayes, T., Kakade, S., Rakhlin, A., Tewari, A.: High-probability regret bounds for bandit online linear optimization. In: Proceedings of the 21st Annual Conference on Learning Theory-COLT 2008, pp. 335–342. Omnipress (2008)
7. Bayandina, A.S., Gasnikov, A.V., Lagunovskaya, A.A.: Gradient-free two-point methods for solving stochastic nonsmooth convex optimization problems with small non-random noises. Autom. Remote. Control. **79**, 1399–1408 (2018)
8. Berahas, A.S., Cao, L., Choromanski, K., Scheinberg, K.: A theoretical and empirical comparison of gradient approximations in derivative-free optimization. Found. Comput. Math. **22**(2), 507–560 (2022)
9. Beznosikov, A., Sadiev, A., Gasnikov, A.: Gradient-free methods with inexact oracle for convex-concave stochastic saddle-point problem. In: International Conference on Mathematical Optimization Theory and Operations Research, pp. 105–119. Springer (2020)
10. Bubeck, S., Cesa-Bianchi, N., et al.: Regret analysis of stochastic and nonstochastic multi-armed bandit problems. Found. Trends Mach. Learn. **5**(1), 1–122 (2012)

11. Chen, P.Y., Zhang, H., Sharma, Y., Yi, J., Hsieh, C.J.: Zoo: Zeroth order optimization based black-box attacks to deep neural networks without training substitute models. In: Proceedings of the 10th ACM Workshop on Artificial Intelligence and Security, pp. 15–26 (2017)
12. Conn, A.R., Scheinberg, K., Vicente, L.N.: Introduction to derivative-free optimization. SIAM (2009)
13. Duchi, J.C., Jordan, M.I., Wainwright, M.J., Wibisono, A.: Optimal rates for zero-order convex optimization: the power of two function evaluations. IEEE Trans. Inf. Theory **61**(5), 2788–2806 (2015)
14. Dvinskikh, D., Tominin, V., Tominin, I., Gasnikov, A.: Noisy zeroth-order optimization for non-smooth saddle point problems. In: International Conference on Mathematical Optimization Theory and Operations Research, pp. 18–33. Springer, Cham (2022). https://doi.org/10.1007/978-3-031-09607-5_2
15. Dvurechensky, P., Gorbunov, E., Gasnikov, A.: An accelerated directional derivative method for smooth stochastic convex optimization. Eur. J. Oper. Res. **290**(2), 601–621 (2021)
16. Elsken, T., Metzen, J.H., Hutter, F.: Neural architecture search: a survey. J. Mach. Learn. Res. **20**(1), 1997–2017 (2019)
17. Ermoliev, Y.: Stochastic programming methods (1976)
18. Flaxman, A.D., Kalai, A.T., McMahan, H.B.: Online convex optimization in the bandit setting: gradient descent without a gradient. arXiv preprint cs/0408007 (2004)
19. Gasnikov, A., Dvinskikh, D., Dvurechensky, P., Gorbunov, E., Beznosikov, A., Lobanov, A.: Randomized gradient-free methods in convex optimization. arXiv preprint arXiv:2211.13566 (2022)
20. Gasnikov, A., et al.: The power of first-order smooth optimization for black-box non-smooth problems. In: International Conference on Machine Learning, pp. 7241–7265. PMLR (2022)
21. Gasnikov, A.V., Krymova, E.A., Lagunovskaya, A.A., Usmanova, I.N., Fedorenko, F.A.: Stochastic online optimization. single-point and multi-point non-linear multi-armed bandits. convex and strongly-convex case. Autom. Remote Control **78**, 224–234 (2017)
22. Gorbunov, E., Dvurechensky, P., Gasnikov, A.: An accelerated method for derivative-free smooth stochastic convex optimization. arXiv preprint arXiv:1802.09022 (2018)
23. Hanzely, F., Kovalev, D., Richtarik, P.: Variance reduced coordinate descent with acceleration: New method with a surprising application to finite-sum problems. In: International Conference on Machine Learning, pp. 4039–4048. PMLR (2020)
24. Hazan, E., Klivans, A., Yuan, Y.: Hyperparameter optimization: a spectral approach. arXiv preprint arXiv:1706.00764 (2017)
25. Kiefer, J., Wolfowitz, J.: Stochastic estimation of the maximum of a regression function. The Annals of Mathematical Statistics, pp. 462–466 (1952)
26. Li, L., Jamieson, K., DeSalvo, G., Rostamizadeh, A., Talwalkar, A.: Hyperband: a novel bandit-based approach to hyperparameter optimization. J. Mach. Learn. Res. **18**(1), 6765–6816 (2017)
27. Lobanov, A., Alashqar, B., Dvinskikh, D., Gasnikov, A.: Gradient-free federated learning methods with l_1 and l_2-randomization for non-smooth convex stochastic optimization problems. arXiv preprint arXiv:2211.10783 (2022)
28. Lobanov, A., Anikin, A., Gasnikov, A., Gornov, A., Chukanov, S.: Zero-order stochastic conditional gradient sliding method for non-smooth convex optimization. arXiv preprint arXiv:2303.02778 (2023)

29. Nemirovskij, A.S., Yudin, D.B.: Problem complexity and method efficiency in optimization (1983)
30. Papernot, N., McDaniel, P., Goodfellow, I.: Transferability in machine learning: from phenomena to black-box attacks using adversarial samples. arXiv preprint arXiv:1605.07277 (2016)
31. Papernot, N., McDaniel, P., Goodfellow, I., Jha, S., Celik, Z.B., Swami, A.: Practical black-box attacks against machine learning. In: Proceedings of the 2017 ACM on Asia Conference on Computer and Communications Security, pp. 506–519 (2017)
32. Patel, K.K., Saha, A., Wang, L., Srebro, N.: Distributed online and bandit convex optimization. In: OPT 2022: Optimization for Machine Learning (NeurIPS 2022 Workshop) (2022)
33. Polyak, B.T.: Introduction to optimization. optimization software. Inc., Publications Division, New York 1, 32 (1987)
34. Polyak, B.T., Tsybakov, A.B.: Optimal order of accuracy of search algorithms in stochastic optimization. Problemy Peredachi Informatsii **26**(2), 45–53 (1990)
35. Risteski, A., Li, Y.: Algorithms and matching lower bounds for approximately-convex optimization. Advances in Neural Information Processing Systems 29 (2016)
36. Rosenbrock, H.: An automatic method for finding the greatest or least value of a function. Comput. J. **3**(3), 175–184 (1960)
37. Scaman, K., Bach, F., Bubeck, S., Lee, Y.T., Massoulié, L.: Optimal convergence rates for convex distributed optimization in networks. J. Mach. Learn. Res. **20**, 1–31 (2019)
38. Shamir, O.: An optimal algorithm for bandit and zero-order convex optimization with two-point feedback. J. Mach. Learn. Res. **18**(1), 1703–1713 (2017)
39. Shibaev, I., Dvurechensky, P., Gasnikov, A.: Zeroth-order methods for noisy hölder-gradient functions. Optim. Lett. **16**(7), 2123–2143 (2022)
40. Vasin, A., Gasnikov, A., Spokoiny, V.: Stopping rules for accelerated gradient methods with additive noise in gradient. Technical report, Berlin: Weierstraß-Institut für Angewandte Analysis und Stochastik (2021)
41. Vaswani, S., Bach, F., Schmidt, M.: Fast and faster convergence of sgd for over-parameterized models and an accelerated perceptron. In: The 22nd International Conference on Artificial Intelligence and Statistics, pp. 1195–1204. PMLR (2019)
42. Yu, Z., Ho, D.W., Yuan, D.: Distributed randomized gradient-free mirror descent algorithm for constrained optimization. IEEE Trans. Autom. Control **67**(2), 957–964 (2021)

Accelerated Zero-Order SGD Method for Solving the Black Box Optimization Problem Under "Overparametrization" Condition

Aleksandr Lobanov[1,2(✉)] and Alexander Gasnikov[1,2,3]

[1] Moscow Institute of Physics and Technology, Dolgoprudny, Russia
{gasnikov.av,lobanov.av}@mipt.ru
[2] ISP RAS Research Center for Trusted Artificial Intelligence, Moscow, Russia
[3] Institute for Information Transmission Problems RAS, Moscow, Russia

Abstract. This paper is devoted to solving a convex stochastic optimization problem in a overparameterization setup for the case where the original gradient computation is not available, but an objective function value can be computed. For this class of problems we provide a novel gradient-free algorithm, whose creation approach is based on applying a gradient approximation with l_2 randomization instead of a gradient oracle in the biased Accelerated SGD algorithm, which generalizes the convergence results of the AC-SA algorithm to the case where the gradient oracle returns a noisy (inexact) objective function value. We also perform a detailed analysis to find the maximum admissible level of adversarial noise at which we can guarantee to achieve the desired accuracy. We verify the theoretical results of convergence using a model example.

Keywords: Black-Box Optimization · Overparametrization · Accelerated Zero-Order SGD Method · Biased Gradient Oracle

1 Introduction

The black-box optimization problem [4,14,41] usually arises when we have few information about the objective function. Such problems can appear when the objective function is non-smooth [25] (i.e. no gradient computation is available) or, for instance, when the function is smooth [2] (may even have a higher order of smoothness [5]), but the process of computing the derivatives is too expensive in contrast to computing the value of the objective function. Moreover, this problem is often solved under conditions of privacy, when some data cannot be disseminated due to confidentiality. Then a gradient-free oracle [46] (or in other words zero-order oracle) acts as some "black box", which returns only

The research was supported by Russian Science Foundation (project No. 21-71- 30005), https://rscf.ru/en/project/21-71-30005/.

objective function value $f(x)$ at requested point x with some bounded adversarial noise $\delta(x) \leq \Delta$ (where Δ is level noise). Where the latter means that the gradient-free oracle can give an inexact value of the objective function. As practice shows [9], the lower the level of adversarial noise, the more expensive it is to call a gradient-free oracle, so it is important to understand how large the level of adversarial noise can be, at which a "good" convergence rate is still guaranteed (by "good" convergence rate is meant the convergence of the algorithm when $\Delta = 0$). This concept of a gradient-free oracle is common in the literature and has the interpretation of a adversarial attack on the black-box model [12]. The black box problem is currently actively researched in the optimization [17,50] and machine learning [39,40] community, since this problem has applications in the following areas: federated learning [34,42], distributed learning [36,47] and deep learning [22]. A particular need for solving such problems arises in the following applications: model hyperparameters tuning [20,28], reinforcement learning [13,38], multi-armed bandits [6,11,49], and many others.

There are various approaches to solving the black-box problem, but the most common in a theoretically proof-of-concept sense is to create gradient-free optimization algorithms based on the state of the art first-order methods and using various randomized gradient approximations [23]. The most common first-order optimization methods in machine learning models are momentum SGD, Adam, and others. But note that these algorithms are variants of Stochastic Gradient Descent (SGD) [10,45], which use stochastic gradient estimates. By adding a procedure for batching stochastic gradient estimates, it is possible to obtain algorithms that are easily parallelized on several computers. It is data distribution (parallelization) that significantly reduces the computational costs that certainly arise in a huge number of modern machine learning models. Also, with the addition of acceleration one can still achieve improvement in terms of estimates on the number of successive iterations. Thus, for many optimization problems, the state of the art first-order algorithms are accelerated batched variants of SGD.

In this paper, we focus on solving a stochastic convex black-box optimization problem in a smooth setting with an overparameterization condition. Where the latter means that the model has many more parameters than the data available. We can summarize our contributions as follows:

- We derive the convergence rate for a biased Accelerated Stochastic Gradient Descent, which covers smooth convex stochastic optimization problems under the overparameterization setup.
- We provide a novel Accelerated Zero-Order Stochastic Gradient Descent Method (AZO-SGD) for solving the black-box problem in a smooth setting under the overparameterization condition. We analyze the robustness of AZO-SGD algorithm to adversarial noise, providing an estimate for the maximum admissible level of adversarial noise at which the desired accuracy can still be achieved. We show that our algorithm is optimal on oracle calls in the class of gradient-free algorithms.
- We show the convergence of the Accelerated Zero-Order Stochastic Gradient Descent Method proposed in this paper using a model example of finite sums

in which the number of summands is less than the number of variables (the overparameterization condition).

1.1 Related Works

Adversarial Noise. Finding the maximum admissible noise level at which one can still guarantee convergence to the desired accuracy ε is an important issue for the black-box optimization problem. Special attention to this question was allocated by the works [19,31,33,34]. For example, in [19] the authors found the maximum admissible level of *adversarial deterministic noise* for a non-smooth convex black-box optimization problem $\sim \mathcal{O}\left(\varepsilon^2 d^{-1/2}\right)$, moreover in [34] it is shown that this estimate will be the same for l_1 and l_2 randomization. In [31], the authors showed that by assuming a strong convexity, the estimate maximum level of *adversarial deterministic noise* can be improved to $\sim \mathcal{O}\left(\mu^{1/2}\varepsilon^{3/2}d^{-1/2}\right)$ in non-smooth setting. And in [33] the authors were able to show that this estimation in a non-smooth one can be also improved to $\sim \mathcal{O}\left(\varepsilon d^{-1/2}\right)$ by using *adversarial stochastic noise*, since this concept of *adversarial stochastic noise* does not accumulate in the bias, but accumulates only in the variance. In addition, this paper shows that if the function is smooth, the estimate of the maximum level of adversarial stochastic noise can be improved to $\sim \mathcal{O}\left(\varepsilon^{1/2}d^{-1/2}\right)$. In this paper, we will use a gradient approximation via l_2 randomization to create a novel gradient-free algorithm, and we will find the maximum admissible level of deterministic noise in a smooth setting under overparameterization condition.

SGD Type Algorithms. Many works [1,24,26,32,37,43,56–58] study the Stochastic Gradient Descent and its variant in different setups. For example, in [32] the authors proposed an accelerated method of stochastic gradient descent, AC-SA. Later in [56] authors proposed optimal algorithms for federated learning architecture, which is based on AC-SA (Single-Machine Accelerated SGD and Mini-Batch Accelerated SGD) method. In [26] proposed an clipped accelerated SGD method for heavy-tailed optimization problems based on the accelerated SGD method: Stochastic Similar Triangle Method (SSTM) [24]. In [1], the authors studied the biased SGD method in the Polak-Lojasiewicz [37,43] setup. It is worth noting that in [58] it was shown that for problems satisfying the Polak-Lojasiewicz condition the non-accelerated SGD algorithm will be optimal. In [57], the study of the AC-SA (accelerated SGD) algorithm was continued already in the overparameterization setup. We in this paper generalize the analysis of the AC-SA algorithm from [57], to create a biased accelerated SGD algorithm in the overparameterization setting. A biased first-order algorithm is necessary because using l_2 randomization produces a bias in the case $\delta(x) > 0$. Therefore, based on the biased batched accelerated stochastic gradient descent and using l_2 randomization, we create a new gradient-free optimization algorithm to solve a convex stochastic black-box optimization problem under overparameterization setup.

Gradient Noise Assumptions. Recently there is a trend in works [8, 27,35,44,48,51,52,57] of relaxed stochastic gradient variance restriction condition. Very many works (e.g. see [27,44]) use the standard assumption:

$\mathbb{E}\left[\|\nabla f(x,\xi)\|^2\right] \leq \sigma^2$. However, already in the works [8,35,52,59] used in the analysis of the algorithm a more relaxed assumption of weak growth: $\mathbb{E}\left[\|\nabla f(x,\xi)\|^2\right] \leq M\|\nabla f(x)\|^2 + \sigma^2$. The following work [48] have set the constants so that the strong growth condition assumption is satisfied: $\mathbb{E}\left[\|\nabla f(x,\xi)\|^2\right] \leq M\|\nabla f(x)\|^2$. In [51,57] the condition satisfying the over-parameterized set: $\mathbb{E}\left[\|\nabla f(x^*,\xi)\|^2\right] \leq \sigma_*^2$. In this paper, we will also assume uniform smoothness of the function over ξ as well as the overparameterized condition, since our approach to creating gradient-free algorithms is based on the Accelerated Batched Stochastic Gradient Descent [57].

1.2   Notations

We use $\langle x,y \rangle := \sum_{i=1}^d x_i y_i$ to denote standard inner product of $x,y \in \mathbb{R}^d$, where x_i and y_i are the i-th component of x and y respectively. We denote Euclidean norm (l_2-norm) in $\mathbb{R}^d$ as $\|x\| = \|x\|_2 := \sqrt{\langle x,x \rangle}$. We use the following notation $B_2^d(r) := \left\{x \in \mathbb{R}^d : \|x\| \leq r\right\}$ to denote Euclidean ball (l_2-ball) and $S_2^d(r) := \left\{x \in \mathbb{R}^d : \|x\| = r\right\}$ to denote Euclidean sphere. Operator $\mathbb{E}[\cdot]$ denotes full mathematical expectation. We notation $\tilde{O}(\cdot)$ to hide logarithmic factors.

1.3   Paper Ogranization

This paper has the following structure. In Sect. 1 we introduce this paper and also provide related works. In Sect. 2 we formulate the problem statement. While in Sect. 3 we provide an accelerated SGD algorithm with a biased gradient oracle in the reparameterization setup. Section 4 presents the main result of this paper. We confirm the theoretical results via a model example in Sect. 5. While Sect. 6 concludes this paper. Detailed proofs are presented in the supplementary materials (Appendix)[4].

2   Technical Preliminaries

We study a standard stochastic convex optimization problem:

$$f^* = \min_{x \in \mathbb{R}^d} \left\{f(x) := \mathbb{E}\left[f(x,\xi)\right]\right\}, \tag{1}$$

where $f : \mathbb{R}^d \to \mathbb{R}$ is smooth convex function that we want to minimize over $\mathbb{R}^d$. This problem statement is a general smooth convex stochastic optimization problem, so to define the class of the problem considered in this paper we will introduce some assumptions on the objective function and on the gradient oracle.

---

[4] The full version of this article, which includes the Appendix can be found by the article title in the arXiv at the following link: https://arxiv.org/abs/2307.12725.

2.1 Assumptions on the Objective Function

In all proofs we assume convexity and smoothness of the function $f(x, \xi)$.

Assumption 1. *For almost every* ξ, $f(x, \xi)$ *is non-negative, convex w.r.t. x, i.e.*

$$\forall x, y, \xi \quad f(y, \xi) \geq f(x, \xi) + \langle \nabla f(x, \xi), y - x \rangle .$$

Assumption 2. *For almost every* ξ, $f(x, \xi)$ *is non-negative, L-smooth w.r.t. x, i.e.*

$$\forall x, y, \xi \quad f(y, \xi) \leq f(x, \xi) + \langle \nabla f(x, \xi), y - x \rangle + \frac{L}{2} \|y - x\|^2 .$$

Assumptions 1 and 2 are common in the literature (see, e.g., [21,55]). However, it is worth noting that these assumptions require uniform convexity and smoothness over $\xi \sim \mathcal{D}$ [53]. This is an essential difference from the standard assumptions, which define a narrower class of the objective function.

Assumption 3. *The function $f(x)$ is a convex and has the minimum value $f^* = \min_x f(x)$, which is attained at a point x^* with $\|x^*\| \leq R$.*

We explicitly introduce the problem solution f^* in Assumption 3, since our approach implies that the convergence rate depends on the solution f^* (see, e.g., [15]), i.e., our analysis will show an improvement in convergence at $f^* \to 0$.

2.2 Assumptions on the Gradient Oracle

In our analysis we consider the case when we obtain an inexact gradient value when calling the oracle. Therefore we first define a biased gradient oracle.

Definition 1 (Gradient Oracle). *A map* $\mathbf{g} : \mathbb{R}^d \times \mathcal{D} \to \mathbb{R}^d$ *s.t.*

$$\mathbf{g}(x, \xi) = \nabla f(x, \xi) + \mathbf{b}(x),$$

where $\mathbf{b} : \mathbb{R}^d \to \mathbb{R}^d$ *such that* $\forall x \in \mathbb{R}^d$: $\|\mathbf{b}(x)\|^2 \leq \zeta^2$.

Next, we assume that gradient noise is bounded as follows.

Assumption 4. *There exists $\sigma_*^2 \geq 0$ such that $\forall x \in \mathbb{R}^d$*

$$\mathbb{E}\left[\|\nabla f(x^*, \xi) - \nabla f(x^*)\|^2\right] \leq \sigma_*^2.$$

Assumption 4 is a common assumption for overparameterized optimization problems, since in this setup in many problems [3,7,30] f^* can be expected to be small. We introduced Assumptions 1–4 because in this paper we based on the results of [57], in which it was proved that the convergence rate of AC-SA method (from [32]) has the following form in overparameterization setup of problem (1):

$$\mathbb{E}\left[f(x_N^{ag}) - f^*\right] \leq c \cdot \left(\frac{LR^2}{N^2} + \frac{LR^2}{BN} + \sqrt{\frac{LR^2 f^*}{BN}}\right), \tag{2}$$

where N is a iteration number, B is a batch size and $\sigma_*^2 \leq 2Lf^*$ (it's proven [57]).

3   Accelerated SGD with Biased Gradient

Our approach to create a gradient-free algorithm (for a overparameterized setup) implies the use of l_2 randomization (see Sect. 4), based on the AC-SA algorithm from [57]. However, the standard concept of a gradient-free oracle implies the presence of adversarial noise, which can accumulate in both variance and bias. Therefore, it is important to investigate the adversarial noise for the question of maximum admissible noise level, when the desired accuracy can be guaranteed. As can be seen, the result (2) obtained in [57] does not account for the bias in the gradient oracle (see Definition 1 in the case when $\|\mathbf{b}(x)\| = 0$). Consequently, in Theorem 1 we provide a novel biased AC-SA algorithm that is robust to the overparameterized setup and accounts for bias in gradient oracle.

Theorem 1 (Convergence of Biased AC-SA). *Let f satisfy Assumptions 1–3 and gradient oracle from Definition 1 satisfy Assumption 4, then Biased AC-SA algorithm guarantees the convergence with a universal constant c*

$$\mathbb{E}\left[f(x_N^{ag}) - f^*\right] \le c \cdot \left(\frac{LR^2}{N^2} + \frac{LR^2}{BN} + \frac{\sigma_* R}{\sqrt{BN}} + \zeta R + \frac{\zeta^2}{2L}N\right).$$

The results of Theorem 1 show the convergence of the AC-SA algorithm, considering the bias $\mathbf{b}(x)$ in the gradient oracle. It is not difficult to see that if we consider the case without bias ($\zeta = 0$), the convergence result of Theorem 1 will fully correspond to the result (2). The last two terms in Theorem 1 are standard for the accelerated algorithm (see, for example, [18,54]). There are several ways to obtain these results: using the (δ, L)-oracle technique [16], modifying Assumptions 1, 2, or performing sequential reasoning with current assumptions. The proof of Theorem 1 can be found in supplementary materials (Appendix B).

4   Main Result

In this section, we present the main result of this work, namely a gradient-free algorithm for solving a convex smooth stochastic black-box optimization problem in an overparameterized setup. We further narrow the problem class (1) considered in this section to the black box problem, that is, when the calculation of the gradient oracle is not available for some reason. Unfortunately, we cannot apply the AC-SA algorithm or even the biased AC-SA algorithm to solving this problem class. Therefore, there is a need to create an algorithm that only requires calculations of function values. Such algorithms are usually called *gradient-free*, since the efficiency of this class of algorithms is determined by three criteria: the maximum admissible level of adversarial noise Δ, iterative complexity N, and in particular the total number of calls to the *gradient-free oracle* T. Our approach in creating gradient-free algorithm based on the biased AC-SA algorithm. Instead of the gradient oracle (see Definition 1) we use the gradient approximation:

$$\mathbf{g}(x, \xi, e) = \frac{d}{2\tau} \left(f_\delta(x + \tau e, \xi) - f_\delta(x - \tau e, \xi)\right) e, \tag{3}$$

where e is a vector uniformly distributed on unit sphere $S_2^d(1)$, τ is a smoothing parameter and $f_\delta(x, \xi) = f(x, \xi) + \delta(x)$ $(|\delta(x)| \leq \Delta)$ is a *gradient-free* oracle. Thus Algorithm 1 presents a novel gradient-free method, namely Accelerated Zero-Order Stochastic Gradient Descent (AZO-SGD) Method for solving the black-box optimization problem (1) under the overparameterization condition.

Algorithm 1. Accelerated Zero-Order Stochastic Gradient Descent (AZO-SGD)

Input: Start point $x_0^{ag} = x_0 \in \mathbb{R}^d$, maximum number of iterations $N \in \mathbb{Z}_+$.
Let stepsize $\eta_k > 0$, parameters $\beta_k, \tau > 0$, batch size $B \in \mathbb{Z}_+$.

1: **for** $k = 0, ..., N - 1$ **do**
2: $\quad$ $\beta_k = 1 + \frac{k}{6}$ and $\eta_k = \eta(k+1)$ for $\gamma = \min\left\{\frac{1}{12L}, \frac{B}{24L(N+1)}, \sqrt{\frac{BR^2}{Lf^*N^3}}\right\}$
3: $\quad$ $x_k^{md} = \beta_k^{-1}x_k + (1 - \beta_k^{-1})x_k^{ag}$
4: $\quad$ Sample $\{e_1, ..., e_B\}$ and $\{\xi_1, ..., \xi_B\}$ independently
5: $\quad$ Define $\mathbf{g}_k = \frac{1}{B}\sum_{i=1}^{B} \mathbf{g}(x_k^{md}, \xi_i, e_i)$ using (3)
6: $\quad$ $\tilde{x}_{k+1} = x_k - \eta_k \mathbf{g}_k$
7: $\quad$ $x_{k+1} = \min\left\{1, \frac{R}{\|\tilde{x}_{k+1}\|}\right\}\tilde{x}_{k+1}$
8: $\quad$ $x_{k+1}^{ag} = \beta_k^{-1}x_{k+1} + (1 - \beta_k^{-1})x_{k+1}^{ag}$
9: **end for**

Output: x_N^{ag}.

Next, we provide Theorem 2, in which we show the convergence results for Accelerated Zero-Order Stochastic Gradient Descent (AZO-SGD) Method.

Theorem 2. *Let f satisfy Assumptions 1–3 and gradient approximation (3) with parameter $\tau \leq \frac{\varepsilon}{LR}$ satisfy Assumption 4, then Accelerated Zero-Order Stochastic Gradient Descent (AZO-SGD) Method (see Algorithm 1) achieves ε-accuracy: $\mathbb{E}\left[f(x_N^{ag}) - f^*\right] \leq \varepsilon$ after*

$$N = \mathcal{O}\left(\sqrt{\frac{LR^2}{\varepsilon}}\right), \quad T = \max\left\{\mathcal{O}\left(\frac{LR^2}{\varepsilon}\right), \mathcal{O}\left(\frac{d\sigma_*^2R^2}{\varepsilon^2}\right)\right\}$$

number of iterations, total number of gradient-free oracle calls and at

$$\Delta \leq \frac{\varepsilon^2}{dLR^2}$$

the maximum admissible level of adversarial noise.

The result of Theorem 2 shows the effective iterative complexity N, since an accelerated method was taken as the base (biased AC-SA, see Theorem 1). It is also worth noting that the batch size $B = \max\left\{\mathcal{O}\left(\sqrt{\frac{LR^2}{\varepsilon}}\right), \mathcal{O}\left(\frac{d\sigma_*^2R}{L^{1/2}\varepsilon^{3/2}}\right)\right\}$ can change with time, i.e., it directly depends on σ_*^2, which leads to an optimal estimate of the number of gradient-free oracle calls T. Finally, one of the main results of this paper is the estimation for the maximum admissible noise level Δ. This estimation is inferior to the other estimations $\sim \mathcal{O}\left(\varepsilon^2 d^{-1/2}\right)$ in the non-smooth setting and $\sim \mathcal{O}\left(\varepsilon^{3/2} d^{-1/2}\right)$ in smooth setting, but this is expected, since Algorithm 1 is working in a different setup, namely in overparameterization condition. It's also noted that we don't guarantee that this evaluation is unimproved, but only states that at $\Delta \leq \mathcal{O}\left(\varepsilon^2 d^{-1}\right)$ there will be convergence to ε-accuracy. We present an open question for future research: improving the estimate to the maximum admissible noise level, as well as finding upper bounds on noise level beyond which convergence in the overparameterized setup cannot be guaranteed. The proof of Theorem 1 can be found in supplementary materials (Appendix C).

Remark 1 (General case). It is worth noting that this paper, and in particular Theorem 1 and Theorem 2, focus on the Euclidean case. However, using the work of [29] (namely, Algorithm 2) as a basis, and similarly generalizing the convergence results (Corollary 1, [29]) to the case with a biased gradient oracle (see Definition 1), we can obtain a gradient-free algorithm for a more general class of problems (L_p-norm and presence of constraints). In this case, the parameters (number of iterations N, the maximum admissible level of adversarial noise Δ, smoothing parameter τ) of the gradient-free algorithm will be the same as in Theorem 2, except for the total number of oracle calls: let given $1/p + 1/q = 1$

$$T = N \cdot B = \max \left\{ \mathcal{O}\left(\frac{LR^2}{\varepsilon}\right), \mathcal{O}\left(\frac{\min\{q, \ln d\} d^{2-\frac{2}{p}} \sigma_*^2 R^2}{\varepsilon^2}\right) \right\}.$$

This generalization allows one to solve problem (1) in a broader setting, for instance, by imposing a constraint. In particular, by solving the problem on a simplex ($p = 1$, $q = \infty$) we can achieve a reduction of the total number of calls to the zero-order oracle T by $\ln(d)$ compared to the Euclidean case.

5 Experiments

In this section, we will use a simple example to verify the theoretical results, namely to show the convergence of the proposed algorithm Accelerated Zero-Order Stochastic Gradient Descent Method (AZO-SGD, see Algorithm 1). The optimization problem (1) is as follows:

$$\min_{x \in \mathbb{R}^d} f(x) := \frac{1}{m} \sum_{i=1}^{m} (l_i(x))^2, \tag{4}$$

where $l(x) = Ax - b$ is system of m linear equations under overparameterization ($d > m$), $A \in \mathbb{R}^{m \times d}$, $x, b \in \mathbb{R}^d$. Problem (4) is a convex stochastic optimization problem (1), also known as the Empirical Risk Minimization problem, where $\xi = i$ is one of m linear equations.

In Fig. 1 we see the convergence of the proposed gradient-free algorithm. We can also conclude that as the batch size increases, the number of iterations

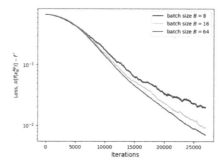

Fig. 1. Convergence of the Accelerated Zero-Order Stochastic Gradient Descent Method and the effect of parameter B (batch size) on the iteration complexity.

required to achieve the desired accuracy decreases. This effect occurs especially in applications with huge-scale machine learning models. We optimize $f(x)$ (4) with parameters: $d = 256$ (dimensional of problem), $m = 128$ (number of linear equations), $\tau = 0.001$ (smoothing parameter), $\eta = 0.0001$ (step size), $B = \{8, 16, 64\}$ (batch size). We also understand machine inaccuracy by noise level.

6   Conclusion

The optimization problem in the overparameterization setup has not yet been sufficiently studied. In this paper, we proposed a novel gradient-free algorithm: Accelerated Zero-Order Stochastic Gradient Descent Method for solving the smooth convex stochastic black-box optimization problem in the overparameterization setup. Our approach in creating the gradient-free Algorithm 1 was based on accelerated stochastic gradient descent. However, since there is an accumulation of adversarial noise in l_2 randomization, the result of [57] was generalized to the case of a biased gradient oracle. We also showed that the proposed gradient-free algorithm (AZO-SGD) is optimal in terms of iteration and oracle complexities. In addition, we obtained the first estimate, as far as we know, of the level of adversarial noise in the overparameterization setup, thereby opening up many potentially interesting future research questions in this setup.

The authors are grateful to Daniil Vostrikov.

References

1. Ajalloeian, A., Stich, S.U.: On the convergence of sgd with biased gradients. arXiv preprint arXiv:2008.00051 (2020)
2. Akhavan, A., Chzhen, E., Pontil, M., Tsybakov, A.: A gradient estimator via l1-randomization for online zero-order optimization with two point feedback. Adv. Neural. Inf. Process. Syst. **35**, 7685–7696 (2022)
3. Allen-Zhu, Z., Li, Y., Liang, Y.: Learning and generalization in overparameterized neural networks, going beyond two layers. Advances in neural information processing systems 32 (2019)

4. Audet, C., Hare, W.: Derivative-free and blackbox optimization (2017)
5. Bach, F., Perchet, V.: Highly-smooth zero-th order online optimization. In: Conference on Learning Theory, pp. 257–283. PMLR (2016)
6. Bartlett, P., Dani, V., Hayes, T., Kakade, S., Rakhlin, A., Tewari, A.: High-probability regret bounds for bandit online linear optimization. In: Proceedings of the 21st Annual Conference on Learning Theory-COLT 2008, pp. 335–342. Omnipress (2008)
7. Belkin, M., Hsu, D., Ma, S., Mandal, S.: Reconciling modern machine-learning practice and the classical bias-variance trade-off. Proc. Natl. Acad. Sci. **116**(32), 15849–15854 (2019)
8. Bertsekas, D., Tsitsiklis, J.N.: Neuro-dynamic programming. Athena Scientific (1996)
9. Bogolubsky, L., et al.: Learning supervised pagerank with gradient-based and gradient-free optimization methods. Advances in neural information processing systems 29 (2016)
10. Bottou, L., Curtis, F.E., Nocedal, J.: Optimization methods for large-scale machine learning. SIAM Rev. **60**(2), 223–311 (2018)
11. Bubeck, S., Cesa-Bianchi, N., et al.: Regret analysis of stochastic and nonstochastic multi-armed bandit problems. Found. Trends Mach. Learn. **5**(1), 1–122 (2012)
12. Chen, P.Y., Zhang, H., Sharma, Y., Yi, J., Hsieh, C.J.: Zoo: Zeroth order optimization based black-box attacks to deep neural networks without training substitute models. In: Proceedings of the 10th ACM Workshop on Artificial Intelligence and Security, pp. 15–26 (2017)
13. Choromanski, K., Rowland, M., Sindhwani, V., Turner, R., Weller, A.: Structured evolution with compact architectures for scalable policy optimization. In: International Conference on Machine Learning, pp. 970–978. PMLR (2018)
14. Conn, A.R., Scheinberg, K., Vicente, L.N.: Introduction to derivative-free optimization. SIAM (2009)
15. Cotter, A., Shamir, O., Srebro, N., Sridharan, K.: Better mini-batch algorithms via accelerated gradient methods. Advances in neural information processing systems 24 (2011)
16. Devolder, O.: Exactness, inexactness and stochasticity in first-order methods for large-scale convex optimization. Ph.D. thesis, CORE UCLouvain Louvain-la-Neuve, Belgium (2013)
17. Duchi, J.C., Jordan, M.I., Wainwright, M.J., Wibisono, A.: Optimal rates for zero-order convex optimization: the power of two function evaluations. IEEE Trans. Inf. Theory **61**(5), 2788–2806 (2015)
18. Dvinskikh, D., Gasnikov, A.: Decentralized and parallel primal and dual accelerated methods for stochastic convex programming problems. J. Inverse Ill-posed Problems **29**(3), 385–405 (2021)
19. Dvinskikh, D., Tominin, V., Tominin, I., Gasnikov, A.: Noisy zeroth-order optimization for non-smooth saddle point problems. In: International Conference on Mathematical Optimization Theory and Operations Research, pp. 18–33. Springer, Cham (2022). https://doi.org/10.1007/978-3-031-09607-5_2
20. Elsken, T., Metzen, J.H., Hutter, F.: Neural architecture search: a survey. J. Mach. Learn. Res. **20**(1), 1997–2017 (2019)
21. Fatkhullin, I., Etesami, J., He, N., Kiyavash, N.: Sharp analysis of stochastic optimization under global kurdyka-lojasiewicz inequality. Adv. Neural. Inf. Process. Syst. **35**, 15836–15848 (2022)

22. Gao, J., Lanchantin, J., Soffa, M.L., Qi, Y.: Black-box generation of adversarial text sequences to evade deep learning classifiers. In: 2018 IEEE Security and Privacy Workshops (SPW), pp. 50–56. IEEE (2018)
23. Gasnikov, A., Dvinskikh, D., Dvurechensky, P., Gorbunov, E., Beznosikov, A., Lobanov, A.: Randomized gradient-free methods in convex optimization. arXiv preprint arXiv:2211.13566 (2022)
24. Gasnikov, A., Nesterov, Y.: Universal fast gradient method for stochastic composit optimization problems. arXiv preprint arXiv:1604.05275 (2016)
25. Gasnikov, A., et al.: The power of first-order smooth optimization for black-box non-smooth problems. In: International Conference on Machine Learning, pp. 7241–7265. PMLR (2022)
26. Gorbunov, E., Danilova, M., Gasnikov, A.: Stochastic optimization with heavy-tailed noise via accelerated gradient clipping. Adv. Neural. Inf. Process. Syst. **33**, 15042–15053 (2020)
27. Hazan, E., Kale, S.: Beyond the regret minimization barrier: optimal algorithms for stochastic strongly-convex optimization. J. Mach. Learn. Res. **15**(1), 2489–2512 (2014)
28. Hazan, E., Klivans, A., Yuan, Y.: Hyperparameter optimization: a spectral approach. arXiv preprint arXiv:1706.00764 (2017)
29. Ilandarideva, S., Juditsky, A., Lan, G., Li, T.: Accelerated stochastic approximation with state-dependent noise. arXiv preprint arXiv:2307.01497 (2023)
30. Jacot, A., Gabriel, F., Hongler, C.: Neural tangent kernel: Convergence and generalization in neural networks. Advances in neural information processing systems 31 (2018)
31. Kornilov, N., Gasnikov, A., Dvurechensky, P., Dvinskikh, D.: Gradient free methods for non-smooth convex optimization with heavy tails on convex compact. arXiv preprint arXiv:2304.02442 (2023)
32. Lan, G.: An optimal method for stochastic composite optimization. Math. Program. **133**(1–2), 365–397 (2012)
33. Lobanov, A.: Stochastic adversarial noise in the "black box" optimization problem. arXiv preprint arXiv:2304.07861 (2023)
34. Lobanov, A., Alashqar, B., Dvinskikh, D., Gasnikov, A.: Gradient-free federated learning methods with l_1 and l_2-randomization for non-smooth convex stochastic optimization problems. arXiv preprint arXiv:2211.10783 (2022)
35. Lobanov, A., Gasnikov, A., Stonyakin, F.: Highly smoothness zero-order methods for solving optimization problems under pl condition. arXiv preprint arXiv:2305.15828 (2023)
36. Lobanov, A., Konin, G., Gasnikov, A., Kovalev, D.: Non-smooth setting of stochastic decentralized convex optimization problem over time-varying graphs. arXiv preprint arXiv:2307.00392 (2023)
37. Lojasiewicz, S.: Une propriété topologique des sous-ensembles analytiques réels. Les équations aux dérivées partielles **117**, 87–89 (1963)
38. Mania, H., Guy, A., Recht, B.: Simple random search of static linear policies is competitive for reinforcement learning. Advances in Neural Information Processing Systems 31 (2018)
39. Papernot, N., McDaniel, P., Goodfellow, I.: Transferability in machine learning: from phenomena to black-box attacks using adversarial samples. arXiv preprint arXiv:1605.07277 (2016)
40. Papernot, N., McDaniel, P., Goodfellow, I., Jha, S., Celik, Z.B., Swami, A.: Practical black-box attacks against machine learning. In: Proceedings of the 2017 ACM on Asia Conference on Computer and Communications Security, pp. 506–519 (2017)

41. Pardalos, P.M., Rasskazova, V., Vrahatis, M.N., et al.: Black Box Optimization, Machine Learning, and No-Free Lunch Theorems. Springer (2021)
42. Patel, K.K., Saha, A., Wang, L., Srebro, N.: Distributed online and bandit convex optimization. In: OPT 2022: Optimization for Machine Learning (NeurIPS 2022 Workshop) (2022)
43. Polyak, B.T.: Gradient methods for the minimisation of functionals. USSR Comput. Math. Math. Phys. **3**(4), 864–878 (1963)
44. Rakhlin, A., Shamir, O., Sridharan, K.: Making gradient descent optimal for strongly convex stochastic optimization. In: Proceedings of the 29th International Conference on International Conference on Machine Learning, pp. 1571–1578 (2012)
45. Robbins, H., Monro, S.: A stochastic approximation method. In: The Annals of Mathematical Statistics, pp. 400–407 (1951)
46. Rosenbrock, H.: An automatic method for finding the greatest or least value of a function. Comput. J. **3**(3), 175–184 (1960)
47. Scaman, K., Bach, F., Bubeck, S., Lee, Y.T., Massoulié, L.: Optimal convergence rates for convex distributed optimization in networks. J. Mach. Learn. Res. **20**, 1–31 (2019)
48. Schmidt, M., Roux, N.L.: Fast convergence of stochastic gradient descent under a strong growth condition. arXiv preprint arXiv:1308.6370 (2013)
49. Shamir, O.: An optimal algorithm for bandit and zero-order convex optimization with two-point feedback. J. Mach. Learn. Res. **18**(1), 1703–1713 (2017)
50. Shibaev, I., Dvurechensky, P., Gasnikov, A.: Zeroth-order methods for noisy hölder-gradient functions. Optim. Lett. **16**(7), 2123–2143 (2022)
51. Srebro, N., Sridharan, K., Tewari, A.: Optimistic rates for learning with a smooth loss. arXiv preprint arXiv:1009.3896 (2010)
52. Stich, S.U.: Unified optimal analysis of the (stochastic) gradient method. arXiv preprint arXiv:1907.04232 (2019)
53. Tran, T.H., Scheinberg, K., Nguyen, L.M.: Nesterov accelerated shuffling gradient method for convex optimization. In: International Conference on Machine Learning, pp. 21703–21732. PMLR (2022)
54. Vasin, A., Gasnikov, A., Dvurechensky, P., Spokoiny, V.: Accelerated gradient methods with absolute and relative noise in the gradient. Optimization Methods and Software, pp. 1–50 (2023)
55. Vaswani, S., Bach, F., Schmidt, M.: Fast and faster convergence of sgd for overparameterized models and an accelerated perceptron. In: The 22nd International Conference on Artificial Intelligence and Statistics, pp. 1195–1204. PMLR (2019)
56. Woodworth, B.E., Bullins, B., Shamir, O., Srebro, N.: The min-max complexity of distributed stochastic convex optimization with intermittent communication. In: Conference on Learning Theory, pp. 4386–4437. PMLR (2021)
57. Woodworth, B.E., Srebro, N.: An even more optimal stochastic optimization algorithm: minibatching and interpolation learning. Adv. Neural. Inf. Process. Syst. **34**, 7333–7345 (2021)
58. Yue, P., Fang, C., Lin, Z.: On the lower bound of minimizing polyak-{\L} ojasiewicz functions. arXiv preprint arXiv:2212.13551 (2022)
59. Zhang, Y., Chen, C., Shi, N., Sun, R., Luo, Z.Q.: Adam can converge without any modification on update rules. Adv. Neural. Inf. Process. Syst. **35**, 28386–28399 (2022)

Algorithms for Euclidean-Regularised Optimal Transport

Dmitry A. Pasechnyuk[1,2,3,4(✉)] [iD], Michael Persiianov[2,4] [iD],
Pavel Dvurechensky[5] [iD], and Alexander Gasnikov[2,4,6] [iD]

[1] Mohamed bin Zayed University of Artificial Intelligence, Abu Dhabi, UAE
dmitry.vilensky@mbzuai.ac.ae
[2] Moscow Institute of Physics and Technology, Dolgoprudny, Russia
persiianov.mi@phystech.edu, gasnikov.av@mipt.ru
[3] ISP RAS Research Center for Trusted Artificial Intelligence, Moscow, Russia
[4] Institute for Information Transmission Problems RAS, Moscow, Russia
[5] Weierstrass Institute for Applied Analysis and Stochastics, Berlin, Germany
pavel.dvurechensky@wias-berlin.de
[6] Caucasus Mathematic Center of Adygh State University, Maikop, Russia

Abstract. This paper addresses the Optimal Transport problem, which
is regularized by the square of Euclidean ℓ_2-norm. It offers theoretical
guarantees regarding the iteration complexities of the Sinkhorn–Knopp
algorithm, Accelerated Gradient Descent, Accelerated Alternating Minimisation, and Coordinate Linear Variance Reduction algorithms. Furthermore, the paper compares the practical efficiency of these methods
and their counterparts when applied to the entropy-regularized Optimal
Transport problem. This comparison is conducted through numerical
experiments carried out on the MNIST dataset.

Keywords: Optimal transport · Euclidean regularisation · Sinkhorn
algorithm · Primal-dual algorithm · Alternating optimisation

1 Introduction

Optimal Transport (OT) problem has a long history [9,15], has been extensively
studied [17,20] and piques interest in the modern statistical learning community [2,10]. This paper focuses on the discrete OT problem statement and the
numerical optimisation methods applied to it. Formally, the original problem to
solve is:

$$\min_{\substack{X\mathbf{1}_m=a \\ X^\top \mathbf{1}_n=b \\ x_{ij}\geq 0}} \langle C, X \rangle, \tag{1}$$

The research was supported by Russian Science Foundation (project No. 23-11-00229),
https://rscf.ru/en/project/23-11-00229/, and by the grant of support for leading scientific schools NSh775.2022.1.1.
D. A. Pasechnyuk and M. Persiianov—Equal contribution.

where $a \in \mathcal{S}_n$ and $b \in \mathcal{S}_m$ are the source and destination distributions (measures), the unit simplex $\mathcal{S}_d \equiv \{x \in \mathbb{R}_+^d \mid \sum_{i=1}^d x_i = 1\}$, $X \in \mathbb{R}_+^{n \times m}$ is a transportation plan such that x_{ij} is the mass to transport from the i-th source to the j-th destination, and $C \in \mathbb{R}_+^{n \times m}$ is the cost of the transportation matrix.

An algorithm applied to the OT problem must derive an ε-optimal transportation plan, denoted by X_ε and defined as one that meets the following condition:

$$\langle C, X_\varepsilon \rangle - \varepsilon \leq \langle C, X^* \rangle \equiv \min_{\substack{X\mathbf{1}_m = a \\ X^\top \mathbf{1}_n = b \\ x_{ij} \geq 0}} \langle C, X \rangle,$$

and strictly adheres to constraints $X_\varepsilon \mathbf{1}_m = a$, $X_\varepsilon^\top \mathbf{1}_n = b$, and $X_\varepsilon \in \mathbb{R}_+^{n \times m}$. To obtain such a solution, we consider the Euclidean-regularised OT problem:

$$\min_{\substack{X\mathbf{1}_m = a \\ X^\top \mathbf{1}_n = b \\ x_{ij} \geq 0}} \{f(X) \equiv \langle C, X \rangle + \tfrac{\gamma}{2}\|X\|_2^2\}, \tag{2}$$

where $\|X\|_2^2 \equiv \sum_{i=1,j=1}^{n,m} x_{ij}^2$, and apply convex optimisation methods to solve it. It is noteworthy that if $\gamma \propto \varepsilon$, then the ε-optimum of this optimisation problem is a ($\propto \varepsilon$)-optimal transportation plan for the original problem (1). Unlike (1), problem statement (2) allows one to leverage convex optimisation tools like duality and acceleration.

Contribution. We provide the first arithmetic complexity bounds for Euclidean-regularised OT. The results of this paper are summarised in Table 1 below. Each cell contains an estimate of the number of arithmetic operations number needed for an Algorithm in the leftmost column to achieve target accuracy ε for problem (1) with given n, m (we assume without loss of generality that $n > m$), and C in the worst case. Constant factors are omitted, and ε is assumed to be sufficiently small. The arithmetic complexities for original algorithms applied to entropy-regularised OT [4] are known and are presented in the right column. The left column contains the estimates obtained in this paper.

Table 1. Theoretical guarantees on the arithmetic complexity of methods considered in this paper, compared with those of analogous methods for entropy-regularised problems.

#a.o.	Euclidean-reg. OT	entropy-reg. OT
Sinkhorn, Alg. 1	$\dfrac{n^{7/2}\|C\|_\infty^2}{\varepsilon^2}$, Thm 2	$\dfrac{n^2\|C\|_\infty^2 \log n}{\varepsilon^2}$, [6]
APDAGD, Alg. 3	$\dfrac{n^3\|C\|_\infty}{\varepsilon}$, Thm 4	$\dfrac{n^{5/2}\|C\|_\infty \sqrt{\log n}}{\varepsilon}$, [6]
AAM, Alg. 5	$\dfrac{n^3\|C\|_\infty}{\varepsilon}$, Thm 6	$\dfrac{n^{5/2}\|C\|_\infty \sqrt{\log n}}{\varepsilon}$, [8]
CLVR, Alg. 6 (rand.)	$\dfrac{n^3\|C\|_\infty}{\varepsilon}$ (on avg.), Thm 8	—

The organisation of this paper is as follows. Section 2 provides a short literature review, highlighting the works that underpin the proofs presented in this paper and tracing the history of applying quadratic regularisation in OT. Section 3 encompasses all the theoretical results of this paper. Subsections 3.2, 3.3, 3.4, and 3.5 delve into the details of the Sinkhorn, Accelerated Gradient, Alternating Minimisation, and Coordinate Linear Variance Reduction algorithms, respectively. Finally, Sect. 4 contains results of numerical experiments that compare the practical performance of the proposed algorithms and their counterparts applied to entropy-regularised OT.

2   Background

The Sinkhorn–Knopp algorithm [4,18] stands out as the most widely-known method to solve the OT problem. The works [1,6] justify its worst-case arithmetic complexity in terms of ε and n. Our analysis of the arithmetic complexity of the Sinkhorn–Knopp algorithm applied to Euclidean-regularised OT draws from the framework outlined in [6] as well. As an alternative to the Sinkhorn–Knopp algorithm, the works [6,12] show that accelerated gradient descent applied to entropy-regularised OT problem improves iteration complexity with respect to ε. on the other hand, acceleration can be applied directly to the Sinkhorn–Knopp algorithm by viewing it as an alternating minimisation procedure, as proposed in [8]. Both approaches yield similar iteration complexities and require only minor adjustments in proofs for applying to Euclidean-regularised OT.

The standard approach for effectively applying convex optimisation methods to the OT is entropy regularisation [4]. Recently, there has been a growing interest in Euclidean regularisation [7,11,14]. A practically valuable property of Euclidean-regularised OT is the sparsity of the optimal plan [3], which holds significance in various applications, such as image colour transfer. Additionally, algorithms used for Euclidean-regularised OT are anticipated to be more computationally stable and more robust for small regularisation parameter. For instance, the Sinkhorn–Knopp algorithm for entropy-regularised OT requires computing the exponent with γ in the denominator. Besides, none of the aforementioned papers that study Euclidean regularisation provide arithmetic complexity estimates for particular algorithms applied to Euclidean-regularised OT.

3   Theoretical Guarantees for Various Approaches

3.1   Common Reasoning

We have two discrete probability measures, $a \in \mathcal{S}_n$ and $b \in \mathcal{S}_m$ from the unit simplex, such that $a^\top \mathbf{1}_n = 1, b^\top \mathbf{1}_m = 1$, along with the cost matrix $C \in \mathbb{R}_+^{n \times m}$. Our objective is to find the transport plan $X \in \mathbb{R}_+^{n \times m}$ determined by optimisation problem (2), which represents the Euclidean-regularised version of the classical problem (1).

The problems under consideration are in the generalised linear form and allow for the use of convex duality to eliminate linear constraints. Let us consider the

Lagrange saddle-point problem $\max_{\lambda \in \mathbb{R}^n, \mu \in \mathbb{R}^m} \min_{X \in \mathbb{R}_+^{n \times m}} \mathcal{L}(X, \lambda, \mu)$, where the Lagrangian function is defined as follows:

$$\mathcal{L}(X, \lambda, \mu) \equiv \langle C, X \rangle + \tfrac{\gamma}{2} \|X\|_2^2 + \lambda^\top (X \mathbf{1}_m - a) + \mu^\top (X^\top \mathbf{1}_n - b).$$

The first-order optimality condition for this problem implies

$$\tfrac{\partial \mathcal{L}(X, \lambda, \mu)}{\partial x_{ij}} = 0 = c_{ij} + \gamma x_{ij} + \lambda_i + \mu_j,$$

yielding the following closed-form expression for the optimal transport plan $X(\lambda, \mu) = \left[-C - \lambda \mathbf{1}_m^\top - \mathbf{1}_n \mu^\top \right]_+ / \gamma$, given the dual multipliers λ and μ, where $[x]_+ \equiv \max\{0, x\}$. Upon substituting $X(\lambda, \mu)$ into the formula for $\mathcal{L}$, we derive the following dual problem:

$$\max_{\lambda \in \mathbb{R}^n, \mu \in \mathbb{R}^m} \left\{ \varphi(\lambda, \mu) \equiv -\tfrac{1}{2\gamma} \sum_{j=1}^m \left\| \left[-C_j - \lambda - \mu_j \mathbf{1}_n \right]_+ \right\|_2^2 - \lambda^\top a - \mu^\top b \right\}, \quad (3)$$

where C_j is the j-th row of matrix C.

3.2   The Sinkhorn–Knopp Algorithm

Following the reasoning of [4] regarding the justification of the Sinkhorn–Knopp algorithm for the entropy-regularised OT problem, we come to an analogous Sinkhorn–Knopp method for the Euclidean-regularised OT problem.

The first-order optimality conditions for the dual problem (3) with respect to λ and μ are, respectively,

$$\begin{cases} f_i(\lambda_i) - \gamma a_i = 0, \ i = 1, ..., n \\ g_j(\mu_j) - \gamma b_j = 0, \ j = 1, ..., m, \end{cases} \quad (4)$$

$$f_i(\lambda) = \sum_{j=1}^m [-c_{ij} - \lambda - \mu_j]_+, \quad g_j(\mu) = \sum_{i=1}^n [-c_{ij} - \lambda_i - \mu]_+.$$

Let us denote the i-th order statistic of the elements of the vector x as $x_{(i)}$, and choose l as the largest index j such that $f_i(-(C_i^\top + \mu)_{(j)}) \le \gamma a_i$, and k as the largest index i such that $g_j(-(C_j + \lambda)_{(i)}) \le \gamma b_j$), respectively [14]. Then, by holding μ and λ constant, the explicit solutions of (4) are

$$\begin{cases} \lambda_i = -\left(\gamma a_i + \sum_{j=1}^l (C_i^\top + \mu)_{(j)} \right) / l, \ i = 1, ..., n, \\ \mu_j = -\left(\gamma b_j + \sum_{i=1}^k (C_j + \lambda)_{(i)} \right) / k, \ j = 1, ..., m. \end{cases} \quad (5)$$

The alternating updates of λ and μ according to the formulas above yield the Sinkhorn–Knopp algorithm applied to Euclidean-regularised OT. Its pseudocode is listed in Algorithm 1. The following proposition estimates the algorithmic complexity of each iteration of Algorithm 1.

Proposition 1. *One iteration of Algorithm 1 requires $\mathcal{O}((n + m)^2)$ amortised arithmetic operations per iteration (only +, -, * and $\le$; $\mathcal{O}(n + m)$ /; no built-in functions calculations).*

Algorithm 1. Euclidean Sinkhorn–Knopp

Input: $a, b, C, \gamma, \varepsilon, K$
1: **for** $k = 0, 1, \ldots, K$ **do**
2:     **if** k is even **then**
3:         Iterate over $(C_i^\top + \mu)$ and choose l
4:         $\lambda_i = -\left(\gamma a_i + \sum_{j=1}^{l}(C_i^\top + \mu)_{(j)}\right) / l, \ i = 1, \ldots, n$
5:     **else**
6:         Iterate over $(C_j + \lambda)$ and choose k
7:         $\mu_j = -\left(\gamma b_j + \sum_{i=1}^{k}(C_j + \lambda)_{(i)}\right) / k, \ j = 1, \ldots, m$
8:     **end if**
9:     $x_{ij} := [-c_{ij} - \lambda_i - \mu_j]_+/\gamma, \ i = 1, \ldots, n, \ j = 1, \ldots, m$
10:    **if** $\sum_{i=1}^{n}|\sum_{j=1}^{m} x_{ij} - a_i| + \sum_{j=1}^{m}|\sum_{i=1}^{n} x_{ij} - b_j| \leq \varepsilon$ **then**
11:        **break**
12:    **end if**
13: **end for**

Following Lemmas 1, 2 and Theorem 1 correspond to Lemmas 1, 2 and Theorem 1 from [6], but the proofs are significantly different from that of their analogues due to the use of specific properties of Euclidean regularisation.

Lemma 1. *For* $R = \|C\|_\infty + \frac{\gamma}{\min\{n,m\}}(1 - \max_{\substack{i=1,\ldots,n \\ j=1,\ldots,m}} \{a_i, b_j\})$, *it holds that*

$$\max_{j=1,\ldots,m} \mu_j - \min_{j=1,\ldots,m} \mu_j \leq R, \quad \max_{i=1,\ldots,n} \lambda_i - \min_{i=1,\ldots,n} \lambda_i \leq R,$$
$$\max_{j=1,\ldots,m} \mu_j^* - \min_{j=1,\ldots,m} \mu_j^* \leq R, \quad \max_{i=1,\ldots,n} \lambda_i^* - \min_{i=1,\ldots,n} \lambda_i^* \leq R.$$

Proof. Firstly, thanks to the form of updates (5), we can guarantee the non-positivity of dual variables. Indeed, initial values of μ and λ are zero, so non-positive. Then, for all $j = 1, \ldots, m$,

$$\tfrac{n-1}{\gamma}\mu_j + b_j = \tfrac{1}{\gamma}\sum_{i=1}^{n}(-c_{ij} - \lambda_i - \mu_j) \leq \tfrac{1}{\gamma}\sum_{i=1}^{n}[-c_{ij} - \lambda_i - \mu_j]_+ = X^\top \mathbf{1}_n = b_j,$$

that implies $\mu_j \leq 0$. Similarly, one can prove $\lambda_i \leq 0$ for all $i = 1, \ldots, n$.

Further, let's relate dual variables with corresponding marginal distributions of X. Here we consider only μ, assuming that we just updated it. Similar reasoning can be applied to just updated λ as well, that gives the right column of statements from Lemma.

$$-\mu_j - \|C\|_\infty - \tfrac{1}{n}\mathbf{1}_n^\top \lambda \leq \tfrac{\gamma}{n}[X^\top \mathbf{1}_n]_i = \tfrac{\gamma}{n}b_j \leq \tfrac{\gamma}{n}$$
$$-\mu_j - \tfrac{1}{n}\mathbf{1}_n^\top \lambda \geq \tfrac{\gamma}{n}[X^\top \mathbf{1}_n]_i = \tfrac{\gamma}{n}b_j, \quad \forall j = 1, \ldots, m.$$

This implies

$$\mu_j \geq -\|C\|_\infty - \tfrac{1}{n}(\mathbf{1}_n^\top \lambda + \gamma), \quad \mu_j \leq -\tfrac{1}{n}(\mathbf{1}_n^\top \lambda + \gamma b_j), \quad \forall j = 1, \ldots, m.$$

Finally,

$$\max_{j=1,\ldots,m} \mu_j - \min_{j=1,\ldots,m} \mu_j \leq -\tfrac{1}{n}\left(\mathbf{1}_n^\top \lambda + \gamma \max_{j=1,\ldots,m} b_j\right) + \|C\|_\infty + \tfrac{1}{n}\left(\mathbf{1}_n^\top \lambda + \gamma\right)$$
$$= \|C\|_\infty + \tfrac{\gamma}{n}\left(1 - \max_{j=1,\ldots,m} b_j\right).$$

Reasoning for μ^* and λ^* is similar, since the gradient of objective in (3) vanishes, so $X^\top 1_n = b$ and $X 1_m = a$, correspondingly.

Lemma 2. *For λ, μ, and X taken from each iteration of Algorithm 1 it holds that*

$$\varphi(\lambda^*, \mu^*) - \varphi(\lambda, \mu) \le 4R\sqrt{n+m}(\|X 1_m - a\|_2 + \|X^\top 1_n - b\|_2).$$

Proof. Due to concavity of φ, we have

$$\varphi(\lambda^*, \mu^*) \le \varphi(\lambda, \mu) + \langle \nabla\varphi(\lambda, \mu), (\lambda^*, \mu^*) - (\lambda, \mu)\rangle.$$

Then, by Hölder inequality and Lemma 1,

$$\varphi(\lambda^*, \mu^*) - \varphi(\lambda, \mu) \le \sqrt{n+m}\|\nabla\varphi(\lambda, \mu)\|_2\|(\lambda^*, \mu^*) - (\lambda, \mu)\|_\infty$$
$$\le 4R\sqrt{n+m}\|\nabla\varphi(\lambda, \mu)\|_2 \le 4R\sqrt{n+m}(\|X 1_m - a\|_2 + \|X^\top 1_n - b\|_2).$$

Theorem 1. *To obtain ε solution of problem (2), its sufficient to perform $2 + \frac{8\max\{n,m\}^{3/2}R}{\gamma\varepsilon}$ iterations of Algorithm 1.*

Proof. Below, λ_+ and μ_+ will denote values of λ and μ after the current iteration, and λ_{+k} and μ_{+k} denote values of λ and μ after k iterations. Let current update relate to λ. Denoting $S = -C - 1_n\mu^\top - \lambda 1_m^\top$ and $\delta = \lambda - \lambda_+$, we have

$$\varphi(\lambda_+, \mu_+) - \varphi(\lambda, \mu) = \frac{1}{2\gamma}\sum_{i,j=0,0}^{n,m}(\max\{0, S_{ij} + \delta_i\}^2 - \max\{0, S_{ij}\}^2) + \delta^\top a$$
$$\ge \frac{1}{2\gamma}\sum_{S_{ij}>0, \delta_i<0}(\max\{0, S_{ij} + \delta_i\}^2 - S_{ij}^2) + \delta^\top a$$
$$\ge \delta^\top(a + [\delta]_- - 2\gamma X 1_m) \ge \|\delta\|_2^2 + \delta^\top(a - 2\gamma X 1_m)$$
$$\ge \delta^\top(a - X 1_m) \ge \frac{\gamma}{n}\|a - X 1_m\|_2^2,$$

due to $\lambda_i - [\lambda_+]_i = \frac{\gamma}{l}a_i - \frac{1}{l}\sum_{j=1}^{l}(-C_i^\top - \mu - \lambda_i)_{(j)} \ge \frac{\gamma}{l}a - \frac{\gamma}{l}X 1_m$ and for small enough γ. Then, by Lemma 2, we have

$$\varphi(\lambda_+, \mu_+) - \varphi(\lambda, \mu) \ge \max\left\{\frac{\gamma}{16n^2}\frac{[\varphi(\lambda^*, \mu^*) - \varphi(\lambda, \mu)]^2}{R^2}, \frac{\gamma}{n}\varepsilon^2\right\}, \qquad (6)$$

which implies, similarly to §2.1.5 from [16], that

$$k \le 1 + \frac{16n^2R^2}{\gamma}\frac{1}{[\varphi(\lambda^*, \mu^*) - \varphi(\lambda_+, \mu_+)]} - \frac{16n^2R^2}{\gamma}\frac{1}{[\varphi(\lambda^*, \mu^*) - \varphi(\lambda, \mu)]}. \qquad (7)$$

In the other case of (6), we have

$$[\varphi(\lambda^*, \mu^*) - \varphi(\lambda_{+k}, \mu_{+k})] \le [\varphi(\lambda^*, \mu^*) - \varphi(\lambda, \mu)] - \frac{k\gamma\varepsilon^2}{n}. \qquad (8)$$

To combine bounds on k from (7) and (8), we take minimum of their sum over all options for current objective function value

$$k \le \min_{0 \le s \le [\varphi(\lambda^*, \mu^*) - \varphi(\lambda, \mu)]}\left\{2 + \frac{16n^2R^2}{\gamma s} - \frac{16n^2R^2}{\gamma}\frac{1}{[\varphi(\lambda^*, \mu^*) - \varphi(\lambda, \mu)]} + \frac{sn}{\gamma\varepsilon^2}\right\}$$
$$= \begin{cases} 2 + \frac{n}{\gamma}\left(\frac{8\sqrt{n}R}{\varepsilon} - \frac{16nR^2}{[\varphi(\lambda^*, \mu^*) - \varphi(\lambda, \mu)]}\right) & [\varphi(\lambda^*, \mu^*) - \varphi(\lambda, \mu)] \ge 4\varepsilon\sqrt{n}R^2, \\ 2 + \frac{n}{\gamma}\frac{[\varphi(\lambda^*, \mu^*) - \varphi(\lambda, \mu)]}{\varepsilon^2} & [\varphi(\lambda^*, \mu^*) - \varphi(\lambda, \mu)] < 4\varepsilon\sqrt{n}R^2, \end{cases}$$

which implies the statement of Theorem.

We have not set R and γ in the bound above. By Lemma 1, $R \leq \|C\|_\infty + \frac{\gamma}{n}$, so $k \leq 2 + \frac{8n^{3/2}\|C\|_\infty}{\gamma\varepsilon} + \frac{8n^{1/2}}{\varepsilon}$, and one can take $\gamma = \varepsilon/2$, such that solving regularised problem with accuracy $\varepsilon/4$ will give $(\varepsilon/2)$-solution of original problem. Besides, by Lemma 7 from [1] we have

$$\langle C, X \rangle \leq \langle C, X^* \rangle + \tfrac{\gamma}{2}\|X\|_2^2 + 2(\|a - X\mathbf{1}_m\|_1 + \|b - X^\top\mathbf{1}_n\|_1)\|C\|_\infty,$$

so one should set target accuracy to $\varepsilon/(4\|C\|_\infty)$. This proves the following result.

Theorem 2. *Number of iterations of Algorithm 1, sufficient for Algorithm 2 to return ε-optimal transport plan X such that $X\mathbf{1}_m = a$, $X^\top\mathbf{1}_n = b$, is*

$$\mathcal{O}\left(\frac{(n+m)^{3/2}\|C\|_\infty^2}{\varepsilon^2}\right).$$

Algorithm 2. Approximate OT by Algorithm 1

Input: a, b, C, ε
1: Find X' for given $C, a, b, \gamma = \varepsilon/2$, with accuracy $\varepsilon/(4\|C\|_\infty)$ using Algorithm 1
2: Find projection X of X' onto the feasible set using Algorithm 2 [1]

Note that correction $a' = (1 - \varepsilon/8)\,(a + \mathbf{1}_n\varepsilon/(n(8 - \varepsilon)))$ of target marginal distributions a and b, which is required for original Sinkhorn–Knopp algorithm [6], is not necessary in Algorithms 2 and 4, since formula for R from Lemma 1 makes sense even if $a_i = 0$ and $b_j = 0$ for some i and j.

3.3 Adaptive Accelerated Gradient Descent

To apply accelerated gradient method to the problem (2), let us consider it as problem of convex optimisation with linear constrains:

$$\min_{\substack{A[X]=B \\ x_{ij} \geq 0}} f(X), \tag{9}$$

where operator $A : \mathbb{R}^{n \times m} \to \mathbb{R}^{n+m}$ is defined by $A[X] = (X\mathbf{1}_m, X^\top\mathbf{1}_n)$, $B = (a, b) \in \mathbb{R}_+^{n+m}$, f is defined in (2), and corresponding dual problem is equivalent to (3). The following theorem gives iteration complexity for primal-dual Algorithm 3, which will be further applied to obtain the solution for problem (2). Note, that for given operator A it holds that

$$\|A\|_{2,2} \equiv \sup_{\|X\|_2=1} \|A[X]\|_2 = \sqrt{n + m}. \tag{10}$$

Theorem 3 (Theorem 3 [6]). *Assume that optimal dual multipliers satisfy $\|(\lambda^*, \mu^*)\|_2 \leq R_2$. Then, Algorithm 3 generates sequence of approximate solutions for primal and dual problems (9) and (3), which satisfy*

$$f(X_k) - f(X^*) \leq f(X_k) - \varphi(\lambda_k, \mu_k) \leq \frac{16\|A\|_{2,2}^2 R^2}{\gamma k^2},$$

$$\|A[X_k] - B\|_2 \leq \frac{16\|A\|_{2,2}^2 R}{\gamma k^2}, \quad \|X_k - X^*\|_2 \leq \frac{8\|A\|_{2,2} R}{\gamma k}.$$

Algorithm 3. Adaptive Primal-Dual Accelerated Gradient Descent

---

Input: $a, b, \varphi(\cdot), \nabla\varphi(\cdot), L_0, \varepsilon, K$

1: $\beta_0 = 0$

2: $\lambda_0 = \lambda_0' = \tilde{\lambda}_0 = 0$, $\mu_0 = \mu_0' = \tilde{\mu}_0 = 0$

3: **for** $k = 0, 1, \ldots, K$ **do**

4:    $i = 0$

5:    **repeat**

6:       $L_{k+1} = 2^{i-1} \cdot L_k$, $i = i + 1$

7:       Solve $L_{k+1}\alpha_{k+1}^2 - \alpha_{k+1} + \beta_k = 0$ with respect to α_{k+1}

8:       $\beta_{k+1} = \beta_k + \alpha_{k+1}$

9:       $\tilde{\lambda}_k = \lambda_k + \frac{\alpha_{k+1}}{\beta_{k+1}}(\lambda_k' - \lambda_k)$, $\tilde{\mu}_k = \mu_k + \frac{\alpha_{k+1}}{\beta_{k+1}}(\mu_k' - \mu_k)$

10:      $\lambda_{k+1}' = \lambda_k + \alpha_{k+1}\nabla_\lambda\varphi(\tilde{\lambda}_k, \tilde{\mu}_k)$, $\mu_{k+1}' = \mu_k + \alpha_{k+1}\nabla_\mu\varphi(\tilde{\lambda}_k, \tilde{\mu}_k)$

11:      $\lambda_{k+1} = \lambda_k + \frac{\alpha_{k+1}}{\beta_{k+1}}(\lambda_{k+1}' - \lambda_k)$, $\mu_{k+1} = \mu_k + \frac{\alpha_{k+1}}{\beta_{k+1}}(\mu_{k+1}' - \mu_k)$

12:    **until** $\varphi(\lambda_{k+1}, \mu_{k+1}) \geq \varphi(\tilde{\lambda}_k, \tilde{\mu}_k) + \langle\nabla_\lambda\varphi(\tilde{\lambda}_k, \tilde{\mu}_k), \lambda_{k+1} - \tilde{\lambda}_k\rangle +$
     $\langle\nabla_\mu\varphi(\tilde{\lambda}_k, \tilde{\mu}_k), \mu_{k+1} - \tilde{\mu}_k\rangle + \frac{L_{k+1}}{2}(\|\lambda_{k+1} - \tilde{\lambda}_k\|_2^2 + \|\mu_{k+1} - \tilde{\mu}_k\|_2^2)$

13:    $X_{k+1} = X_k + \frac{\alpha_{k+1}}{\beta_{k+1}}(X(\lambda_{k+1}, \mu_{k+1}) - X_k)$

14:    **if** $f(X_{k+1}) - \varphi(\lambda_{k+1}, \mu_{k+1}) \leq \varepsilon$, $\|X_{k+1}\mathbf{1}_m - a\|_2^2 + \|X_k^\top\mathbf{1}_n - b\|_2^2 \leq \varepsilon^2$ **then**

15:       **break**

16:    **end if**

17: **end for**

Following the proof scheme chosen in [6], we estimate the error of solution X for the original problem (1):

$$\langle C, X \rangle = \langle C, X^* \rangle + \langle C, X_{\text{reg.}}^* - X^* \rangle + \langle C, X_k - X_{\text{reg.}}^* \rangle + \langle C, X - X_k \rangle$$
$$\leq \langle C, X^* \rangle + \langle C, X_{\text{reg.}}^* - X^* \rangle + \langle C, X - X_k \rangle + f(X_k) + \varphi(\lambda_k, \mu_k) + \gamma,$$

where $X_{\text{reg.}}^*$ is the exact solution of problem (2). By choosing $\gamma \leq \varepsilon/3$, obtaining X_k such that $f(X_k) - \varphi(\lambda_k, \mu_k) \leq \varepsilon/3$ by Algorithm 3 and making $\langle C, X - X_k \rangle \leq \varepsilon/3$, we guarantee arbitrarily good approximate solution X. Let us consider the latter condition in more details. By Lemma 7 [1] and Theorem 3 one has

$$\langle C, X - X_k \rangle \leq \|C\|_\infty \|X - X_k\|_1 \leq 2\|C\|_\infty(\|X_k\mathbf{1}_m - a\|_1 + \|X_k^\top\mathbf{1}_n - b\|_1)$$
$$\leq_1 2\sqrt{n+m}\|C\|_\infty\|A[X_k] - B\|_2 \leq \frac{32(n+m)^{3/2}\|C\|_\infty R}{\gamma k^2}$$
$$\leq_2 2\sqrt{n+m}\|C\|_\infty\|X_k - X_{\text{reg.}}^*\|_2 \leq \frac{16(n+m)\|C\|_\infty R}{\gamma k}.$$

To ensure the latter, it is sufficient to choose k such that

$$k = \mathcal{O}\left(\min\left\{\frac{n\|C\|_\infty R}{\varepsilon^2}, \frac{n^{3/4}\sqrt{\|C\|_\infty R}}{\varepsilon}\right\}\right). \tag{11}$$

On the other hand, $f(X_k) - \varphi(\lambda_k, \mu_k) \leq \varepsilon/3$ together with Theorem 3 imply

$$k = \mathcal{O}\left(\frac{\sqrt{n+m}R}{\varepsilon}\right),$$

which is majorated by (11) and does not contribute to iteration complexity. This proves, taking into account (10), Lemma 1, and that $R_2 \leq R\sqrt{n+m}$, the following result.

Theorem 4. *Number of iterations of Algorithm 3, sufficient for Algorithm 4 to return ε-optimal transport plan X such that $X\mathbf{1}_m = a, X^\top\mathbf{1}_n = b$, is*

$$\mathcal{O}\left(\min\left\{\frac{(n+m)^{3/2}\|C\|_\infty^2}{\varepsilon^2}, \frac{(n+m)\|C\|_\infty}{\varepsilon}\right\}\right).$$

Algorithm 4. Approximate OT by Algorithms 3, 5, or 6

Input: a, b, C, ε
 1: Find X' for given $C, a, b, \gamma = \varepsilon/3$, with accuracy $\varepsilon/3$ using Algorithms 3, 5, or 6
 2: Find projection X of X' onto the feasible set using Algorithm 2 [1]

3.4 Accelerated Alternating Minimisation

Note that Sinkhorn–Knopp algorithm is based on the simplest alternating optimisation scheme: dual function φ is explicitly optimised with respect to λ and μ alternately. Thus, if there is a way to accelerate some alternating optimisation algorithm, similar technique can be applied to Sinkhorn–Knopp algorithm. Moreover, iteration complexity will correspond to that of taken accelerated alternating optimisation method, while the arithmetic complexity of optimisation with respect to one variable will be the same as for Sinkhorn algorithm.

The following theorem gives iteration complexity for general primal-dual alternating minimisation Algorithm 5, which can be used similarly to Algorithm 3 to obtain the solution for problem (2). Note that b, which denotes the number of independent variables blocks in [8], can be set to $b = 2$ in our case, because $\|\nabla_\lambda\varphi(\lambda, \mu)\|_2 > \|\nabla_\mu\varphi(\lambda, \mu)\|_2$ implies $\|\nabla_\lambda\varphi(\lambda, \mu)\|_2^2 > \frac{1}{2}\|\nabla\varphi(\lambda, \mu)\|_2^2$. But since dimensionalities of λ and μ are different, one of the variables which has bigger dimensionality will be updated more often a priori.

Theorem 5 (Theorem 3 [8] for $b = 2$). *Assume that optimal dual multipliers satisfy $\|(\lambda^*, \mu^*)\|_2 \leq R_2$. Then, Algorithm 5 generates sequence of approximate solutions for primal and dual problems (9) and (3), which satisfy*

$$f(X_k) - f(X^*) \leq f(X_k) - \varphi(\lambda_k, \mu_k) \leq \frac{16\|A\|_{2,2}^2 R^2}{\gamma k^2},$$

$$\|A[X_k] - B\|_2 \leq \frac{16\|A\|_{2,2}^2 R}{\gamma k^2}, \quad \|X_k - X^*\|_2 \leq \frac{8\|A\|_{2,2} R}{\gamma k},$$

Instead of arg max operator taking place in the listing of general Algorithm 5 one should use formulas (5). The advantage of this approach consists in simplicity of obtaining the solution for these auxiliary problems. It is expected that while accelerated gradient descent considered before was making one gradient step

---

Algorithm 5. Primal-Dual Accelerated Alternating Minimisation

---

Input: $a, b, \varphi(\cdot), \nabla\varphi(\cdot), L_0, \varepsilon, K$

1: $\beta_0 = 0, \lambda_0 = \lambda'_0 = \widetilde{\lambda}_0 = 0, \mu_0 = \mu'_0 = \widetilde{\mu}_0 = 0$

2: **for** $k = 0, 1, \ldots, K$ **do**

3:    $i = 0$

4:    **repeat**

5:       $L_{k+1} = 2^{i-1} \cdot L_k, \, i = i + 1$

6:       $\alpha_{k+1} = \frac{1}{L_{k+1}} + \sqrt{\frac{1}{4L_{k+1}^2} + \frac{\alpha_k L_k}{L_{k+1}}}$

7:       $\widetilde{\lambda}_k = \lambda_k + \frac{1}{\alpha_{k+1}L_{k+1}}(\lambda'_k - \lambda_k), \, \widetilde{\mu}_k = \mu_k + \frac{1}{\alpha_{k+1}L_{k+1}}(\mu'_k - \mu_k)$

8:       **if** $\|\nabla_\lambda\varphi(\widetilde{\lambda}_{k+1}, \widetilde{\mu}_{k+1})\| > \|\nabla_\mu\varphi(\widetilde{\lambda}_{k+1}, \widetilde{\mu}_{k+1})\|$ **then**

9:         $\lambda_{k+1} = \arg\max_\lambda \varphi(\lambda, \widetilde{\mu}_k), \, \mu_{k+1} = \mu_k$

10:      **else**

11:         $\lambda_{k+1} = \lambda_k, \, \mu_{k+1} = \arg\max_\mu \varphi(\widetilde{\lambda}_k, \mu)$

12:      **end if**

13:    **until** $\varphi(\lambda_{k+1}, \mu_{k+1}) \geq \varphi(\widetilde{\lambda}_{k+1}, \widetilde{\mu}_{k+1}) + \frac{\|\nabla\varphi(\widetilde{\lambda}_{k+1}, \widetilde{\mu}_{k+1})\|_2^2}{2L_{k+1}}$

14:    $\lambda'_{k+1} = \lambda_k + \alpha_{k+1}\nabla_\lambda\varphi(\widetilde{\lambda}_k, \widetilde{\mu}_k), \, \mu'_{k+1} = \mu_k + \alpha_{k+1}\nabla_\mu\varphi(\widetilde{\lambda}_k, \widetilde{\mu}_k)$

15:    $X_{k+1} = \frac{1}{\alpha_{k+1}L_{k+1}}X(\widetilde{\lambda}_k, \widetilde{\mu}_k) + \frac{\alpha_k^2 L_k}{\alpha_{k+1}^2 L_{k+1}}X_k$

16:    **if** $f(X_{k+1}) - \varphi(\lambda_{k+1}, \mu_{k+1}) \leq \varepsilon, \|X_{k+1}\mathbf{1}_m - a\|_2^2 + \|X_{k+1}^\top\mathbf{1}_n - b\|_2^2 \leq \varepsilon^2$ **then**

17:      **break**

18:    **end if**

19: **end for**

---

at each iteration, this algorithm makes optimal step with respect to half of dual variables, so expected progress per iteration is bigger, while the number of iterations is the same up to small $\mathcal{O}(1)$ factor. Using the proof scheme similar to which is provided in Sect. 3.3 and the same problem pre- and post-processing Algorithm 4, one can guarantee, taking into account (10) and Lemma 1, that the following result holds.

Theorem 6. *Number of iterations of Algorithm 5, sufficient for Algorithm 4 to return ε-optimal transport plan X such that $X\mathbf{1}_m = a, X^\top\mathbf{1}_n = b$, is*

$$\mathcal{O}\left(\min\left\{\frac{(n+m)^{3/2}\|C\|_\infty^2}{\varepsilon^2}, \frac{(n+m)\|C\|_\infty}{\varepsilon}\right\}\right).$$

3.5 Coordinate Linear Variance Reduction

One can also consider problem (2) as generalised linear problem with strongly-convex regulariser and sparse constraints. By using the property that dual variables or problem (3) are separable into two groups (λ and μ), one can apply primal-dual incremental coordinate methods. One of the modern algorithms which is based on dual averaging and has implicit variance reduction effect was proposed in [19]. The following theorem presents simplified form of iteration complexity estimate for Algorithm 6 adopted to our particular problem.

Algorithm 6. Coordinate Linear Variance Reduction

Input: $a, b, C, \gamma, \varepsilon, K, \alpha$

1:   $A_0 = a_0 = 1/(2\sqrt{n+m})$, $X_0 = \{1/(nm)\}$, $\lambda_0 = 0$, $\mu_0 = 0$, $z_{-1} = 0$, $q_{-1} = a_0 C$

2: **for** $k = 0, 1, \ldots, K$ **do**

3:      $X_{k+1} = [\alpha X_0 - q_{k-1}]_+ / (\alpha + \gamma A_{k+1})$

4:      Generate uniformly random $\xi_k \in [0, 1]$

5:      **if** $\xi_k < 0.5$ **then**

6:         $\lambda_{k+1} = \lambda_k + 2\gamma a_k (X_{k+1}\mathbf{1}_m - a)$, $\mu_{k+1} = \mu_k$

7:      **else**

8:         $\lambda_{k+1} = \lambda_k$, $\mu_{k+1} = \mu_k + 2\gamma a_k (X_{k+1}^\top \mathbf{1}_n - b)$

9:      **end if**

10:     $a_{k+1} = \frac{1}{4}\sqrt{\frac{1+\gamma A_k/\alpha}{n+m}}$, $A_{k+1} = A_k + a_k$

11:     **if** $\xi_k < 0.5$ **then**

12:        $z_k = z_{k-1} + (\lambda_k - \lambda_{k-1})\mathbf{1}_m^\top$

13:     **else**

14:        $z_k = z_{k-1} + \mathbf{1}_n(\mu_k - \mu_{k-1})^\top$

15:     **end if**

16:     $q_k = q_{k-1} + a_{k+1}(z_k + C) + 2a_k(z_k - z_{k-1})$

17:     $\widetilde{X}_{k+1} = \frac{1}{A_{k+1}}\sum_{i=1}^{k+1} a_i X_i$

18:     **if** $f(\widetilde{X}_{k+1}) - \varphi(\lambda_{k+1}, \mu_{k+1}) \le \varepsilon$, $\|\widetilde{X}_{k+1}\mathbf{1}_m - a\|_2^2 + \|\widetilde{X}_{k+1}^\top \mathbf{1}_n - b\|_2^2 \le \varepsilon^2$ **then**

19:        **break**

20:     **end if**

21: **end for**

Theorem 7 (Corollary 1 [19] for $b = 2$). *Assume that optimal dual multipliers satisfy $\|(\lambda^*, \mu^*)\|_2 \le R_2$. Then, Algorithm 6 generates sequence of approximate solutions for primal and dual problems (9) and (3), which satisfy*

$$\mathbb{E}[f(\widetilde{X}_k) - f(X^*)] = \mathcal{O}\left(\frac{\|A\|_{2,2}^2 R^2}{\gamma k^2}\right), \quad \mathbb{E}[\|A[\widetilde{X}_k] - B\|_2] = \mathcal{O}\left(\frac{\|A\|_{2,2}^2 R}{\gamma k^2}\right).$$

Taking into account (10) and Lemma 1, using the same reasoning as for Theorem 4, one has

Theorem 8. *Number of iterations of Algorithm 6, sufficient for Algorithm 4 to return expected ε-optimal transport plan X such that $X\mathbf{1}_m = a$, $X^\top \mathbf{1}_n = b$, is*

$$\mathcal{O}\left(\min\left\{\frac{(n+m)^{3/2}\|C\|_\infty^2}{\varepsilon^2}, \frac{(n+m)\|C\|_\infty}{\varepsilon}\right\}\right),$$

where "expected ε-optimal" means that $\mathbb{E}[\langle C, X\rangle] - \varepsilon \le \langle C, X^*\rangle$.

One can see that asymptotic of iteration complexity is the same as that of Algorithms 3 and 5. This allows to use the same pre- and post-processing Algorithm 4 to apply this algorithm to the OT problem. The advantage of this algorithm is the simplicity of iterations. It is expect that despite the same $\mathcal{O}(nm)$ arithmetic complexity of one iteration, constant of it in practice is significantly smaller than for accelerated methods considered before.

4   Numerical Experiments

All the optimisation algorithms described in previous section are implemented in Python 3 programming language. Reproduction package including source code of algorithms and experiments settings is hosted on GitHub[1]. We consider OT problem for the pair of images from MNIST dataset [5], where distributions are represented by vectorised pixel intensities and cost matrix contains pairwise Euclidean distances between pixels.

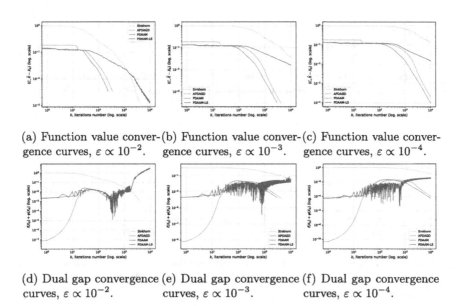

(a) Function value conver-(b) Function value conver-(c) Function value convergence curves, $\varepsilon \propto 10^{-2}$.   gence curves, $\varepsilon \propto 10^{-3}$.   gence curves, $\varepsilon \propto 10^{-4}$.

(d) Dual gap convergence (e) Dual gap convergence (f) Dual gap convergence curves, $\varepsilon \propto 10^{-2}$.   curves, $\varepsilon \propto 10^{-3}$.   curves, $\varepsilon \propto 10^{-4}$.

Fig. 1. Practical efficiency of Sinkhorn–Knopp, Adaptive Accelerated Gradient, and Accelerated Alternating methods applied to entropy-regularised OT problem on MNIST dataset.

Firstly, experiment on comparison of algorithms applied to entropy-regularised OT was carried out. Following algorithms were compared: Sinkhorn–Knopp algorithm (Sinkhorn) [6], Adaptive Primal-dual Accelerated Gradient Descent (APDAGD) [6], Primal-dual Accelerated Alternating Minimisation (PDAAM) [8] and its modification which uses one-dimensional optimisation to choose step size (PDAAM-LS). Results of the experiment are shown in Fig. 1. There are presented convergence curves of methods for two progress measures: function value for original problem (1) and dual gap for problem (3). The range of target accuracy value is $\varepsilon \in \{2 \cdot 10^{-2}, 1.85 \cdot 10^{-3}, 5 \cdot 10^{-4}\}$ (each target accuracy value requires separate experiment, because ε is a parameters of Algorithms 2 and 4 and affects the convergence from the beginning).

---

[1] Repository   is   available   at   https://github.com/MuXauJl11110/Euclidean-Regularised-Optimal-Transport.

All the plots show that PDAAM is leading algorithm, and performance of APDAGD is competitive with it. On the other hand, Sinkhorn–Knopp algorithm converges slowly, especially for small ε. PDAAM-LS demonstrates unstable behaviour in our experiment.

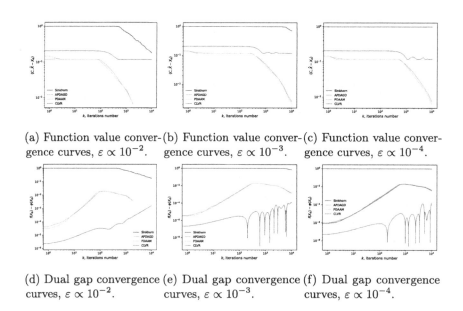

(a) Function value conver-(b) Function value conver-(c) Function value convergence curves, $\varepsilon \propto 10^{-2}$.   gence curves, $\varepsilon \propto 10^{-3}$.   gence curves, $\varepsilon \propto 10^{-4}$.

(d) Dual gap convergence (e) Dual gap convergence (f) Dual gap convergence curves, $\varepsilon \propto 10^{-2}$.   curves, $\varepsilon \propto 10^{-3}$.   curves, $\varepsilon \propto 10^{-4}$.

Fig. 2. Practical efficiency of Sinkhorn–Knopp, Adaptive Accelerated Gradient, and Accelerated Alternating methods applied to Euclidean-regularised OT problem on MNIST dataset.

Secondly, the same algorithms were compared while applied to Euclidean-regularised OT problem. Figure 2 shows convergence curves of methods, organisation of the plots is the same as above. One can see that ordering of the methods' performance remain the same as in the case of entropy-regularised OT. Specifically, the PDAAM algorithm convergence is faster than that of APDAGD and Sinkhorn. On the other hand, difference between PDAAM and APDAGD performance is less significant in the case of Euclidean-regularised OT (we conclude that progress of step which is optimal with respect to one of the dual variables is not much bigger than progress of the gradient step), and Sinkhorn algorithm performs significantly worse than in entropy-regularised OT and is not efficient in practice. CLVR did not displayed itself an efficient method in our experiment. Generally, convergence of all of the algorithms in the case of Euclidean regularisation is more prone to slowing down on the latter iterations.

The expected property of Euclidean-regularised OT that the optimal transport plan obtained with it is sparse is approved in our experiments. One can see the examples of transport plans in Fig. 3, the fraction of zero elements (which are $< 10^{-21}$) in them is around 99.5%.

5   Discussion

Euclidean regularisation for OT problems has been recently explored in several papers due to its practically valuable properties, such as robustness to small regularisation parameter and sparsity of the optimal transport plan. This paper provides a theoretical analysis of various algorithms that are applicable efficiently to Euclidean-regularised OT. We demonstrate and compare their practical performance. Our findings reveal that these desirable properties come at a cost. Namely, the slower convergence of all the algorithms and faster increase in arithmetic complexity as dimensionality grows.

Our plans involve considering different convex optimisation algorithms applied to Euclidean-regularised OT, focusing on splitting algorithms that are to be more computationally stable with small regularisation parameter [13]. Additionally, we aim to explore the application of Euclidean regularisation for the Wasserstein barycenter problem.

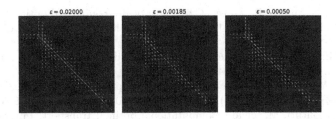

Fig. 3. Sparse optimal transport plans obtained by Adaptive Accelerated Gradient Descent applied to Euclidean-regularised OT problem on MNIST dataset, 99.5% of zero elements.

References

1. Altschuler, J., Niles-Weed, J., Rigollet, P.: Near-linear time approximation algorithms for optimal transport via Sinkhorn iteration. Adv. Neural Inf. Process. **30** (2017)
2. Arjovsky, M., Chintala, S., Bottou, L.: Wasserstein generative adversarial networks. In: International Conference on Machine Learning, pp. 214–223. PMLR (2017)
3. Blondel, M., Seguy, V., Rolet, A.: Smooth and sparse optimal transport. In: International Conference on Artificial Intelligence and Statistics, pp. 880–889. PMLR (2018)
4. Cuturi, M.: Sinkhorn distances: lightspeed computation of optimal transport. Adv. Neural Inf. Process. Syst. **26** (2013)
5. Deng, L.: The MNIST database of handwritten digit images for machine learning research. IEEE Signal Process. Mag. **29**(6), 141–142 (2012)
6. Dvurechensky, P., Gasnikov, A., Kroshnin, A.: Computational optimal transport: complexity by accelerated gradient descent is better than by Sinkhorn's algorithm. In: International Conference on Machine Learning, pp. 1367–1376. PMLR (2018)

7. Essid, M., Solomon, J.: Quadratically regularized optimal transport on graphs. SIAM J. Sci. Comput. **40**(4), A1961–A1986 (2018)

8. Guminov, S., Dvurechensky, P., Tupitsa, N., Gasnikov, A.: On a combination of alternating minimization and Nesterov's momentum. In: International Conference on Machine Learning, pp. 3886–3898. PMLR (2021)

9. Kantorovich, L.V.: On the translocation of masses. Dokl. Akad. Nauk. USSR (NS) **37**, 199–201 (1942)

10. Kolouri, S., Park, S.R., Thorpe, M., Slepcev, D., Rohde, G.K.: Optimal mass transport: signal processing and machine-learning applications. IEEE Signal Process. Mag. **34**(4), 43–59 (2017)

11. Li, L., Genevay, A., Yurochkin, M., Solomon, J.M.: Continuous regularized wasserstein barycenters. Adv. Neural. Inf. Process. Syst. **33**, 17755–17765 (2020)

12. Lin, T., Ho, N., Jordan, M.: On efficient optimal transport: an analysis of greedy and accelerated mirror descent algorithms. In: International Conference on Machine Learning, pp. 3982–3991. PMLR (2019)

13. Lindbäck, J., Wang, Z., Johansson, M.: Bringing regularized optimal transport to lightspeed: a splitting method adapted for GPUs. arXiv preprint arXiv:2305.18483 (2023)

14. Lorenz, D.A., Manns, P., Meyer, C.: Quadratically regularized optimal transport. Appl. Math. Optim. **83**(3), 1919–1949 (2019). https://doi.org/10.1007/s00245-019-09614-w

15. Monge, G.: Mémoire sur la théorie des déblais et des remblais. Mem. Math. Phys. Acad. Royale Sci., pp. 666–704 (1781)

16. Nesterov, Y.: Introductory Lectures on Convex Optimization: A Basic Course, vol. 87. Springer, New York (2003). https://doi.org/10.1007/978-1-4419-8853-9

17. Peyré, G., Cuturi, M., et al.: Computational optimal transport. In: Foundations and Trends® in Machine Learning (2019)

18. Sinkhorn, R.: Diagonal equivalence to matrices with prescribed row and column sums. Am. Math. Mon. **74**(4), 402–405 (1967)

19. Song, C., Lin, C.Y., Wright, S., Diakonikolas, J.: Coordinate linear variance reduction for generalized linear programming. Adv. Neural. Inf. Process. Syst. **35**, 22049–22063 (2022)

20. Villani, C., et al.: Optimal Transport: Old and New, vol. 338. Springer, Heidelberg (2009). https://doi.org/10.1007/978-3-540-71050-9

Real Acceleration of Communication Process in Distributed Algorithms with Compression

Svetlana Tkachenko[1], Artem Andreev[1], Aleksandr Beznosikov[1,2(✉)], and Alexander Gasnikov[1,2]

[1] Moscow Institute of Physics and Technology, Moscow, Russia
beznosikov.an@phystech.edu
[2] Institute for Information Transmission Problems, Moscow, Russia

Abstract. Modern applied optimization problems become more and more complex every day. Due to this fact, distributed algorithms that can speed up the process of solving an optimization problem through parallelization are of great importance. The main bottleneck of distributed algorithms is communications, which can slow down the method dramatically. One way to solve this issue is to use compression of transmitted information. In the current literature on theoretical distributed optimization, it is generally accepted that as much as we compress information, so much we reduce communication time. But in reality, the communication time depends not only on the size of the transmitted information, but also, for example, on the message startup time. In this paper, we study distributed optimization algorithms under the assumption of a more complex and closer-to-reality dependence of transmission time on compression. In particular, we describe the real speedup achieved by compression, analyze how much it makes sense to compress information, and present an adaptive way to select the power of compression depending on unknown or changing parameters of the communication process.

Keywords: distributed optimization · compression · acceleration

1 Introduction

Modern realities pose more and more complex optimization problems that need to be solved. For example, to improve the generalization of deployed models, machine learning engineers need to rely on training datasets of ever increasing sizes and on elaborate large-scale over-parametrized models [3]. Therefore, it is increasingly necessary to resort to the use of distributed approaches to solving the optimization problem. The essence of distributed optimization is to the process of streamlining the target function by using multiple computing resources that are scattered across different machines or servers. It enables optimization algorithms to run in parallel, which can greatly increase the speed and efficiency of finding the optimal solution. Therefore, distributed optimization is widely

The research was supported by Russian Science Foundation (project No. 23-11-00229).

used in various domains, including machine learning, data science, and operations research [20].

However, when utilizing parallel computation in a distributed optimization environment, a common challenge is the communication between the computational devices. Since the agents function independently, they must exchange information to harmonise their local solutions and revise the global solution. Meanwhile, communication time is a waste that prevents full parallelization. Therefore, to struggle for effective communication and to address the communication bottleneck issue is a key point in distributed optimization [9,13,19].

Employing compression of forwarded information is one of the viable solutions to decrease communication expenses [1,18]. It assists in reducing file size while preserving important information. With the use of effective compression algorithms, transmission time can be considerably reduced both in theory and in practice [10].

Several models describing the dependence of transmission time on message size can be found in literature. The most frequently utilized model in the theoretical optimization is $T = \beta s$, where T is the transmission time, β is the delay-size relationship, and s is the message size. Meanwhile, there is a more practical and widespread model that has stayed away from theoretical optimization. This model is $T = \beta s + \alpha$, where α is the server initialization time [8]. The simpler model indicate that transmission time can be reduced by a factor of n by transmitting n times less information. Nevertheless, the practical results contrast with the theoretical ones. In actuality, messaging involves initializing the channel, which refers to establishing a connection between the sender and the recipient. The second model accounts for this. This implies that there is minimal distinction when transmitting 1 or 2 bits, but once we send 100 Mb and 200 Mb, the variance is substantial. Therefore, we necessitate an accurate representation to characterise the communications.

Considering the issue of communication expenses and the proposed solution, the main questions of this study can be posed:

1. *Which model better describes the real world of messaging?*
2. *How does this change the theory of distributed optimization?*
3. *How can we determine the parameters of this model?*
4. *What is the most efficient method of calculating the parameters of this model in the event of frequent data updates?*

1.1   Contributions

More practical communication model: Instead of the classical delay versus size model of $T = \beta s$ (where β represents the relationship between delay and size), we have adopted a more realistic approach of $T = \beta s + \alpha$ (taking into account α – the server initialization time). When a worker sends a message to the server, the channel initialization time contributes significantly to the small size of message transmission. It is, therefore, essential to consider this factor.

Impact of model on communication complexities: We analyze how the more practical model from the previous paragraph affects the communication costs of modern distributed algorithms with compression. We consider state-of-the-art methods that have the best theoretical guarantees for convex and non-convex problems.

Estimate of α, β: In order to calculate the two coefficients α and β based on the real data on the dependence of delay on message volume, we assume that $\alpha = \alpha_{const} + \delta_\alpha$ and $\beta = \beta_{const} + \delta_\alpha$ where δ_α and δ_β follow independent normal distributions. α_{const} and β_{const} are the true values of these coefficients, $\delta\alpha$ and $\delta\beta$ are errors of measurement or calculation.

Using this information, it is possible to calculate the coefficients α_{const} and β_{const} through statistical techniques, such as least squares. Rather than storing the complete dataset of message size and delay time, we can update the summation of variables. This approach enables us to update just four variables without needing to recalculate the coefficients using the least squares method.

The estimation process aims to find the values of α and β that best fit the observed data, thus providing insight into the server initialization time and the delay-volume relationship.

2   Problem Statement

We consider the optimization problems of the form:

$$\min_{x \in \mathbb{R}^d} \left\{ f(x) := \frac{1}{n} \sum_{i=1}^{n} f_i(x) \right\},$$

where x is the optimization variable. For example, in the context of ML, $x \in \mathbb{R}^d$ contains the parameters of the statistical model to be trained, n – number of employees/devices and functions $f_i(x) : \mathbb{R}^d \to \mathbb{R}$ – model data loss x, stored on the device i.

2.1   Distributed Optimization with Compression

We give an illustration of the traditional use of distributed optimization, utilizing the gradient descent algorithm as an instance – see Algorithm 1.

As noted above to handle large data sets, compression is necessary. In Algorithm 1 this can be represented by the compression operator $\mathcal{C} : \mathbb{R}^d \to \mathbb{R}^d$. In particular, $\nabla f_i(x^k)$ in line 4 should be replaced by $\mathcal{C}(\nabla f_i(x^k))$ and, accordingly, in line 6 we aggregate the compressed gradients $\mathcal{C}(\nabla f_i(x^k))$:

$$x^{k+1} = x^k - \gamma_k \cdot \frac{1}{n} \sum_{i=1}^{n} \mathcal{C}(\nabla f_i(x^k)).$$

This approach is basic, but does not give the best convergence results [6,11]. Once can note that more advanced methods with compression use more tricky

---

Algorithm 1.

1: **Initialization:** choose $x^0 \in \mathbb{R}^d$ and stepsizes $\{\gamma_k\}_{k=0}^K$
2: **for** $k = 0, 1, \ldots, K$ **do**
3:     Server sends x^k to all n nodes
4:     Each i-th node, in parallel with the others, calculates the gradient of its corresponding function f_i:

$$\nabla f_i(x^k)$$

5:     All nodes send $\nabla f_i(x^k)$ to the server
6:     Server performs aggregation:

$$x^{k+1} = x^k - \gamma_k \cdot \frac{1}{n} \sum_{i=1}^{n} \nabla f_i(x^k)$$

7: **end for**

---

schemes, in particular, they are based on various variance reduction techniques, which prescribe to compress not the gradient itself, but the difference between the gradient and some reference value [4,5,7,10,14,15,17].

The theory of convergence of methods with compression is based on a formal definition of the properties of $\mathcal{C}$ operators. In particular, two classes of operators: unbiased and biased, are often distinguished in the literature.

Definition 1. $\mathcal{C}$ *is an unbiased compression with* $\zeta \geq 1$ *if* $\mathcal{C}$ *is unbiased* ($\mathbb{E}[\mathcal{C}(x)] = x$) *and* $\mathbb{E}\left[\|\mathcal{C}(x)\|_2^2\right] \leq \zeta \|x\|_2^2$ *for all* $x \in \mathbb{R}^d$.

Definition 2. $\mathcal{C}$ *is a biased compression with* $\delta \geq 1$ *if* $\mathbb{E}\left[\|\mathcal{C}(x) - x\|_2^2\right] \leq (1 - 1/\delta)\|x\|_2^2$ *for all* $x \in \mathbb{R}^d$.

Meanwhile, these definitions do not give a complete picture about compression operators. The definitions are interesting for proving convergence and obtaining iterative complexity of algorithms. But to obtain the communication cost in the amount of transmitted information, it is necessary to understand how much the operator reduces the transmitted information.

2.2   Degree of Compression

In this subsection, we estimate the degree of compression $\omega_{inf} = \frac{\text{len}(x)}{\text{len}(\mathcal{C}(x))}$, where $\text{len}(x)$ is the number of bits of information to send $x \in \mathbb{R}^d$. We consider different classical compression operators.

Definition 3. *For* $k \in [d] := \{1, \ldots, d\}$, *the unbiased random (aka Rand-k) sparsification operator is defined via*

$$\mathcal{C}(x) := \frac{d}{k} \sum_{i \in S} x_i e_i,$$

where $S \subseteq [d]$ *is the k-nice sampling; i.e., a subset of* $[d]$ *of cardinality k chosen uniformly at random, and* $e_1, \ldots, e_d$ *are the standard unit basis vectors in* $\mathbb{R}^d$.

Lemma 1. *For the unbiased random sparsification $w_{inf} = \frac{d}{k}$.*

Proof. Initial vector x contains d non-zero coordinates, and the compressed one contains k. Then $w_{inf} = \frac{d}{k}$. Here it is important to clarify that in the general case it is necessary to forward numbers of non-zero coordinates as well. But if the same random generator with the same seed is installed on the sending and receiving devices, it is possible to synchronize the randomness for free, and then there is no need to send additional information.

Definition 4 (see [2]). *Top-k sparsification operator is defined via*

$$C(x) := \sum_{i=d-k+1}^{d} x_{(i)} e_{(i)},$$

where coordinates are ordered by their magnitudes so that $|x_{(1)}| \leq |x_{(2)}| \leq \cdots \leq |x_{(d)}|$.

Top-k is a greedy version of unbiased random sparsification.

Lemma 2. *For Top-k sparsification $w_{inf} = \frac{d \cdot len(x)}{k \cdot len(x) + k \cdot \lceil \log_2 d \rceil}$.*

Proof. Similar to Unbiased random sparsification initial vector x contains d non-zero coordinates, and the compressed one contains k. But here, unlike random sparsification, we have to pass the numbers of selected non-zero coordinates. To encode the numbers from 1 to d, $\lceil \log_2 d \rceil$ bits are needed. The total number of transmitted bits is $k \cdot len(x) + k \cdot \lceil \log_2 d \rceil$. Then $w_{inf} = \frac{d \cdot len(x)}{k \cdot len(x) + k \cdot \lceil \log_2 d \rceil}$.

Definition 5 (see [12]). *Natural compression operator C_{nat} is defined as follows:*

$$C(x) = \begin{cases} sign(x) \cdot 2^{\lfloor \log_2 |x| \rfloor}, & with\ p(x), \\ sign(x) \cdot 2^{\lceil \log_2 |x| \rceil}, & with\ 1 - p(x), \end{cases}$$

where probability $p(x) := \frac{2^{\lceil \log_2 |x| \rceil} - |x|}{2^{\lfloor \log_2 |x| \rfloor}}$.

The essence of this compression is random rounding to the nearest power of two. In terms of computing on a computer with 32bit float type, this is simply equivalent to using only the sign bit and 8 bits from the exponent.

Lemma 3. *For Natural compression $w_{inf} = \frac{32}{9}$.*

Proof. The statement follows directly from the use of such compression with 32bit float. Instead of 32 bits we send 9.

Definition 6 (see [21]). *Rank-r Power compression introduced by [21] is a compressed-decompressed approach based on the low-rank approximate decomposition of the matrix $X \in \mathbb{R}^{n \times m}$ (transformed version of the original parcel vector x).*

Lemma 4. *For Rank-r PowerSGD compression* $\omega_{inf} = \frac{nm}{r(n+m)}$.

Proof. The product of matrices $PQ^T, P \in \mathbb{R}^{n \times r}, Q \in \mathbb{R}^{m \times r}$ approximates the matrix $X \in \mathbb{R}^{n \times m}$, Thus, instead of storing $n \cdot m$ numbers must be $r \cdot n + r \cdot m$. Then $\omega_{inf} = \frac{nm}{r(n+m)}$.

The results obtained above are summarized in Table 1.

Table 1. ω_{inf} for different compression operators.

Compression operator	ω_{inf}
Unbiased random sparsification	$\dfrac{d}{k}$
Top-k sparsification [2]	$\dfrac{d \cdot \text{len}(x)}{k \cdot \text{len}(x) + k \cdot \lceil \log_2 d \rceil}$
Natural compression [12]	$\dfrac{32}{9}$
Rank-r Power compression [21]	$\dfrac{nm}{r(n+m)}$

As previously stated, the estimation of communication time necessitates ω_{inf}. This coefficient serves as a reliable indicator of the effectiveness of the compression operator in each instance, particularly if one knows the optimal frequency of message compression (which is the objective of this research paper).

3   Main Part

3.1   Transmission Time Model and Convergence Complexities

We consider the following model of transmission time:

$$T(s) = \alpha + \beta \cdot s,$$

where s is the size of the packages, β represents the time to transmit one unit of the information, α represents the time to initialize the channel, which is the delay that occurs before any message transmission occurs. This delay may involve activities such as creating a connection, verifying the user's identity, or loading essential resources. As mentioned earlier, in papers on theoretical optimization and convergence estimates, it is assumed that $\alpha = 0$.

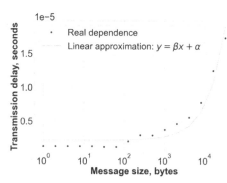

Fig. 1. Dependence of communication time on the size of the transmitted messages for MSU supercomputer "Lomonosov": blue dots – real values, orange line – approximation. (Color figure online)

But let us examine the plot presented in Fig. 1, which portrays the relationship between the transmission delay and message size in a live network. Using this plot we can see the effect of α on communication time.

Note that the theoretical results on the communication cost of distributed algorithms with compression depend on the parameter

$$\eta = \frac{T(\text{len}(\nabla f_i(x)))}{T(\text{len}(\nabla f_i(x))/w_{inf})}. \tag{1}$$

In particular, the best results for method with unbiased compression for convex [14] and non-convex [10] problems linear depends on $\left(\frac{1}{\eta} + \frac{\varsigma}{\eta\sqrt{n}}\right)$. The state-of-the-art results for biased compression [16,17] linear depends on $\left(\frac{1}{\eta} + \frac{\delta}{\eta}\right)$. It is easy to see that if $\alpha = 0$, the expression (1) gives $\eta = w_{inf}$. But if $\alpha \neq 0$, it is possible that $\eta = 1$, thus the impact of even a large w_{inf} can be almost completely canceled.

3.2 Division into Areas

Let us examine the original plot (Fig. 1) and divide it into three conditional ranges (Fig. 2). In the first range, the coefficient α is the most significant. This means that if the size of the message s falls within this range, then α greatly exceeds βs. In the second range, both coefficients α and β hold value, with α and βs being close in value. In the third range, β carries the most significance, with βs greatly exceeding α.

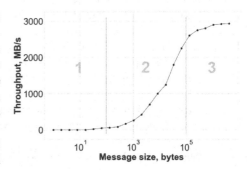

Fig. 2. Division into fields according to the importance of the summands α and βs

Let us examine how the transmission delay varies with changes in message size (see Fig. 3, Fig. 4, and Fig. 5). We determine the number of times the message size alters during compression, and subsequently how many times the transmission time changes:

- When transitioning between areas 3 to 3 and 3 to 2, the message is compressed by a factor of n, while communication time is reduced by $0,95 \cdot n$.
- When transitioning between areas 2 to 2 and 2 to 1, compression is approximately 40 times greater than the reduction in communication cost.
- When transitioning between areas 3 to 1, the time saved is insignificant compared to the compression size (with a compression of 5000 times, the transmission time is only reduced by approximately 100 times).

Conclusion: It is most feasible to travel from area 3 to 3 or 2, and it is not financially viable to travel from area 3 to 1.

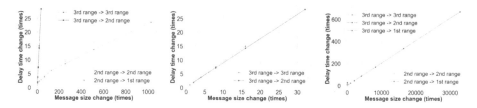

Fig. 3. Transitions from area 2 to areas 1, 2   **Fig. 4.** Transitions from area 3 to areas 2, 3   **Fig. 5.** Transitions from area 3 to area 1

Let us consider an example how compression of a message affects transmission time. We use the distributed system from Fig. 1. The size of an uncompressed message is 10^{22}. Figure 6 demonstrates that the variation in time is almost linear at first, but then the compression loses its effectiveness. Hence, it can be concluded that the compression of a message has an impact on the transmission time.

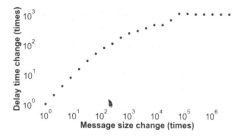

Fig. 6. Transmission delay reduction on message size reduction

Conclusion: High compression does not result in a significant time savings, thus extensively compressing a message is not an efficient approach.

3.3 A Way to Find α and β for an Unknown Network Parameters

Here is a method for determining α and β when the network parameters are unknown, when we can specify areas as in Fig. 2. We formalize the problem of finding or estimating α and β as follows.

Condition: It is possible to transmit messages of varying sizes ranging from 0 to P_{max}, which represents the maximum message size that we can send. It is imperative to consider technological constraints when evaluating the feasibility of message transmission. For instance, for each message, the values α and β are stochastic and have the laws: $\alpha(t) = \alpha_{const} + \delta\alpha$ and $\beta(t) = \beta_{const} + \delta\beta$. That is, $\alpha(t)$ and $\beta(t)$ vary among samples and consist of a constant value plus stochastic noise δ (which follows a normal distribution with mean 0 and variance σ^2, where $\sigma = \alpha_m \cdot \alpha_{const}$ or $\sigma = \beta_m \cdot \beta_{const}$ depending on the nature of δ).

Suggested solution: Let us apply the formulas of the method of least squares to recalculate α and β. During the operation of the main optimization algorithm we vary message sizes. Firstly, we calculate the delay at 2 points. Then, for each subsequent step, we select (determistically or randomly) the next point from within the interval $(0, P_{max})$ and calculate the delay. Here is an example of code that executes this algorithm:

Algorithm 2.

1: **Parameters:** largest message size P_{max};
2: **Initialization:** compute times y_1, y_2 for message sizes x_1, x_2 respectively ($y = \beta \cdot x + \alpha$). Set $s_x^2 = x_1 + x_2$, $s_y^2 = y_1 + y_2$, $s_{xy}^2 = y_1 \cdot x_1 + y_2 \cdot x_2$, $s_{xx}^2 = x_1 \cdot x_1 + x_2 \cdot x_2$;
3: **for** $k = 3, 4, \ldots$ **do**
4:     for new $x_k \in [0, P_{max}]$, compute y_k
5:     $s_x^k = s_x^{k-1} + x_k$
6:     $s_y^k = s_y^{k-1} + y_k$
7:     $s_{xy}^k = s_{xy}^{k-1} + x_k \cdot y_k$
8:     $s_{xx}^k = s_{xx}^{k-1} + x_k \cdot x_k$
9:     $\beta_k = \frac{k \cdot s_{xy}^k - s_x^k \cdot s_y^k}{k \cdot s_{xx}^k - (s_x^k)^2}$
10:     $\alpha_k = \frac{s_y^k - \beta \cdot s_x^k}{k}$
11: **end for**

The algorithm recalculates the α and β coefficients using the least squares formulas. It is worth pointing out that it is very expensive to recalculate the parametrs α and β using the least squares method and to store all data of message sizes and times of transmission $\{x_i, y_i\}$. But Algorithm 2 can works online. We need only 4 variables: the sum of message sizes s_x, the sum of delays s_y, and the sums needed for the least squares calculation s_{xy} and s_{xx}.

Proposition 1. *β_k, α_k from Algorithm 2 are unbiased estimations of β and α, namely $\mathbb{E}[\beta_k] = \beta_{const}$ and $\mathbb{E}[\alpha_k] = \alpha_{const}$.*

Proof. We start from β_k:

$$\mathbb{E}[\beta_k] = \mathbb{E}\left[\frac{k \cdot s_{xy}^k - s_x^k \cdot s_y^k}{k \cdot s_{xx}^k - (s_x^k)^2}\right] = \frac{k \cdot \mathbb{E}[s_{xy}^k] - s_x^k \cdot \mathbb{E}[s_y^k]}{k \cdot s_{xx}^k - (s_x^k)^2}. \tag{2}$$

Next, we estimate $\mathbb{E}[s_{xy}^k]$ and $s_x^k \cdot \mathbb{E}[s_y^k]$

$$\mathbb{E}[s_{xy}^k] = \mathbb{E}\left[\sum_{i=1}^{k}(x_i y_i)\right] = \left(\sum_{i=1}^{k}(x_i \mathbb{E}[y_i])\right) = \sum_{i=1}^{k} x_i(\beta_{const} x_i + \alpha_{const})$$

$$= \beta_{const}\sum_{i=1}^{k} x_i^2 + \alpha_{const}\sum_{i=1}^{k} x_i, \tag{3}$$

$$s_x^k \cdot \mathbb{E}[s_y^k] = \left(\sum_{i=1}^{k} x_i\right) \cdot \mathbb{E}\left[\sum_{i=1}^{k} y_i\right] = \left(\sum_{i=1}^{k} x_i\right) \cdot \sum_{i=1}^{k}(\beta_{const} x_i + \alpha_{const}) =$$

$$= \beta_{const}\left(\sum_{i=1}^{k} x_i\right)^2 + k \cdot \alpha_{const}\sum_{i=1}^{k} x_i. \tag{4}$$

Substituting (4) and (3) to (2), we get

$$\mathbb{E}[\beta_k] = \frac{k \cdot \left(\beta_{const} \cdot s_{xx}^k + \alpha_{const} \cdot s_x^k\right) - \beta_{const} \cdot (s_x^k)^2 - k \cdot \alpha_{const} \cdot s_x^k}{k \cdot s_{xx}^k - (s_x^k)^2} = \beta_{const}.$$

Finally, for $\mathbb{E}[\alpha_k]$ we obtain

$$\mathbb{E}[\alpha_k] = \frac{\mathbb{E}[s_y^k - \beta_k \cdot s_x^k]}{k} = \frac{\mathbb{E}[s_y^k] - \mathbb{E}[\beta_k] \cdot s_x^k}{k}$$

$$= \frac{\sum\limits_{i=1}^{k}(\beta_{const} x_i + \alpha_{const}) - \beta_{const} \cdot s_x^k}{k} = \frac{k \cdot \alpha_{const}}{k} = \alpha_{const}.$$

4   Conclusions

In this paper, we considered a realistic communication cost model $T(s) = \beta s + \alpha$, which takes into account α – the server initialization time. We tried to discuss how it affects to communication time complexities of algorithms. We also provided the algorithm for determining the coefficients α and β utilizing statistical techniques such as the least squares method, alongside estimated uncertainties related to this approach. Rather than storing the complete message size and delay time sets, it is viable to update some combinations of the variables.

One can note that the model considered in this paper can also be improved. For example, we can also include the time required for the communication operator counting. In some cases, this can be quite expensive, which slows down the computational process. Taking this time into account is an important detail for future research.

References

1. Alistarh, D., Grubic, D., Li, J., Tomioka, R., Vojnovic, M.: QSGD: communication-efficient SGD via gradient quantization and encoding. In: Advances in Neural Information Processing Systems, pp. 1709–1720 (2017)
2. Alistarh, D., Hoefler, T., Johansson, M., Khirirat, S., Konstantinov, N., Renggli, C.: The convergence of sparsified gradient methods (2018)
3. Arora, S., Cohen, N., Hazan, E.: On the optimization of deep networks: implicit acceleration by overparameterization. In: International Conference on Machine Learning, pp. 244–253. PMLR (2018)
4. Beznosikov, A., Gasnikov, A.: Compression and data similarity: combination of two techniques for communication-efficient solving of distributed variational inequalities. arXiv preprint arXiv:2206.09446 (2022)
5. Beznosikov, A., Gasnikov, A.: Similarity, compression and local steps: three pillars of efficient communications for distributed variational inequalities. arXiv preprint arXiv:2302.07615 (2023)
6. Beznosikov, A., Horváth, S., Richtárik, P., Safaryan, M.: On biased compression for distributed learning. arXiv preprint arXiv:2002.12410 (2020)

7. Beznosikov, A., Richtárik, P., Diskin, M., Ryabinin, M., Gasnikov, A.: Distributed methods with compressed communication for solving variational inequalities, with theoretical guarantees. Adv. Neural. Inf. Process. Syst. **35**, 14013–14029 (2022)
8. Chan, E., Heimlich, M., Purkayastha, A., Van De Geijn, R.: Collective communication: theory, practice, and experience. Concurr. Comput. Pract. Exp. **19**(13), 1749–1783 (2007)
9. Ghosh, A., Maity, R.K., Mazumdar, A., Ramchandran, K.: Communication efficient distributed approximate newton method. In: 2020 IEEE International Symposium on Information Theory (ISIT), pp. 2539–2544. IEEE (2020)
10. Gorbunov, E., Burlachenko, K.P., Li, Z., Richtárik, P.: Marina: faster non-convex distributed learning with compression. In: International Conference on Machine Learning, pp. 3788–3798. PMLR (2021)
11. Gorbunov, E., Hanzely, F., Richtárik, P.: A unified theory of SGD: variance reduction, sampling, quantization and coordinate descent. In: International Conference on Artificial Intelligence and Statistics, pp. 680–690. PMLR (2020)
12. Horvath, S., Ho, C.Y., Horvath, L., Sahu, A.N., Canini, M., Richtarik, P.: Natural compression for distributed deep learning (2022)
13. Konečnỳ, J., McMahan, H.B., Yu, F.X., Richtárik, P., Suresh, A.T., Bacon, D.: Federated learning: strategies for improving communication efficiency. arXiv preprint arXiv:1610.05492 (2016)
14. Li, Z., Kovalev, D., Qian, X., Richtárik, P.: Acceleration for compressed gradient descent in distributed and federated optimization. arXiv preprint arXiv:2002.11364 (2020)
15. Mishchenko, K., Gorbunov, E., Takáč, M., Richtárik, P.: Distributed learning with compressed gradient differences. arXiv preprint arXiv:1901.09269 (2019)
16. Qian, X., Richtárik, P., Zhang, T.: Error compensated distributed SGD can be accelerated. Adv. Neural. Inf. Process. Syst. **34**, 30401–30413 (2021)
17. Richtárik, P., Sokolov, I., Fatkhullin, I.: EF21: a new, simpler, theoretically better, and practically faster error feedback. arXiv preprint arXiv:2106.05203 (2021)
18. Seide, F., Fu, H., Droppo, J., Li, G., Yu, D.: 1-bit stochastic gradient descent and its application to data-parallel distributed training of speech DNNs. In: Fifteenth Annual Conference of the International Speech Communication Association (2014)
19. Smith, V., Forte, S., Chenxin, M., Takáč, M., Jordan, M.I., Jaggi, M.: Cocoa: a general framework for communication-efficient distributed optimization. J. Mach. Learn. Res. **18**, 230 (2018)
20. Verbraeken, J., Wolting, M., Katzy, J., Kloppenburg, J., Verbelen, T., Rellermeyer, J.S.: A survey on distributed machine learning. ACM Comput. Surv. **53**(2), 1–33 (2020)
21. Vogels, T., Karimireddy, S.P., Jaggi, M.: PowerSGD: practical low-rank gradient compression for distributed optimization. Adv. Neural Inf. Process. Syst. **32** (2019)

Two-Stage Algorithm for Bi-objective Black-Box Traffic Engineering

Alexander Yuskov[1]([✉]) [ID], Igor Kulachenko[2] [ID], Andrey Melnikov[2] [ID],
and Yury Kochetov[2] [ID]

[1] Novosibirsk State University, Novosibirsk, Russia
`a.yuskov@g.nsu.ru`
[2] Sobolev Institute of Mathematics of Siberian Branch of Russian Academy of
Sciences, Novosibirsk, Russia
`{ink,melnikov,jkochet}@math.nsc.ru`

Abstract. We have a directed graph describing a network and an origin-destination matrix for customer internet traffic demands. Our aim is to optimize the routing of the traffic by adjusting the weights of the graph links. Though the internal design of the routing protocol is unavailable, we have access to the simulator to model it. Given the link weights, the simulator provides the values for traffic flow on each link. If the flow on a link exceeds its capacity, this link is considered overloaded. The objectives of the problem are to minimize the total number of overloaded links and the distance from the initial weight vector. We have developed a scheme based on a novel integer linear programming model. It uses values of the traffic flow changes depending on the link weights modifications. In the two-stage approach, this scheme is used to provide the initial Pareto set approximation. The approach outperforms the state-of-the-art multi-objective evolutionary algorithms.

Keywords: Simulation-based approach · Mathematical programming · Constrained gray-box optimization · Evolutionary algorithm

1 Introduction and Relevant Literature

It is essential to manage network resources efficiently due to the ever-growing demand for internet traffic. It can be achieved by routing the traffic on the paths where the network bandwidth is available. Paths to route the traffic are selected by internet routing protocols (IRPs). We consider IRPs that are configured by the weights of the network links. Often, IRPs follow the shortest path routing strategy [2]. However, we do not take into account any assumptions about the protocol design. That is because IRPs used in practice can be hard to describe explicitly in the model. However, the traffic flows induced by a protocol can often be simulated within a computer program. Such problems, where we cannot evaluate the objectives and constraints explicitly, are called black-box optimization ones [5].

The problems dealing with the performance optimization of communication networks are called traffic engineering (TE) problems in the literature [15]. Techniques for performing intradomain traffic engineering include adjusting link weights (e.g., open shortest path first (OSPF) and intermediate system to intermediate system (IS-IS) protocols [2]), using Multiprotocol Label Switching (MPLS) to control routing paths [13], using centralized controllers like a Software Defined Network (SDN) [1], segment routing [7], and others. In most cases, the objectives of the TE problems are to minimize the maximum link load [25], delays in the network [6], or improve load balancing [8] and energy efficiency [3].

The link weights in the OSPF protocol are configured by network administrators. This can be done following some simple rules, for example by setting the weights inversely proportional to the link capacities [9]. There are also some metaheuristic approaches to the problem. For example, in the paper [16] authors developed a local search heuristic. Also, the paper [14] presents a genetic algorithm capable of dealing with the problem. The authors of the [17] develop an algorithm for reacting to the changed network demands.

Because of the multicriteria nature of TE, there are many papers dealing with the optimization of several objectives. For example, load balancing and energy efficiency are optimized in [3]. The authors propose a heuristic scheme and compare it to the optimal network performance. In [12], the nondominated sorting genetic algorithm (NSGA-II) is applied for the optimization of routing costs and load balancing costs. The authors also solve the problem using the exact approach and compare the performance of the heuristic with solutions on the true Pareto front. The paper [26] discusses the usage of a strength Pareto evolutionary algorithm (SPEA2) and NSGA-II for congestion and delay cost minimization. This paper presents an optimization framework to help network administrators choose an adequate configuration to meet the required demands. SPEA2 and NSGA-II appeared to be two of the most-known multi-objective evolutionary algorithms (MOEAs). A comparison of SPEA2 and NSGA-II for the weight setting TE problem with possible link failures is presented in [24]. The authors discovered that the NSGA-II algorithm demonstrates better overall results for large problems. A literature review on multi-objective evolutionary algorithms can be found in [21].

We consider the optimization problem that aims to minimize the total number of overloaded links in the network by adjusting the link weights. We call a link overloaded if the flow on it exceeds its capacity. The flows are calculated during the simulation of the network protocol, which assigns paths for requests based on the weighted lengths of the paths. In the problem, it is also preferable to not change the weights significantly since it could lead to unexpected behaviour when realized in practice. Hence, the second objective of the problem is to minimize the distance from the initial weight vector. After the optimization of an existing network, we should provide a quality of service for customers that is not worse than it was before. Thus, the link that was not overloaded before optimization should remain so. Furthermore, the total load on the network links should not increase. We know an analytic form of the constraints and the objective functions, which depend on the values of traffic flows obtained from a black-box function. Therefore, the problem can be considered to be a grey-box traffic engineering optimization problem. To the best of our knowledge, the TE problem in this formulation was not considered before in the literature.

We design a scheme based on an integer linear programming (ILP) model to solve the formulated problem. It uses the values of the traffic flow changes depending on the modifications of link weights. This scheme is then incorporated into the two-stage approach, where its results are provided to an evolutionary algorithm for further optimization. The main contribution of the paper is this two-stage approach using the ILP-based heuristic.

The rest of this paper is structured as follows. We first introduce the mathematical model of the problem in Sect. 2. The model-based approach is presented in Sect. 3. The computational experiments are discussed in Sect. 4. The last Sect. 5 concludes the paper.

2   Problem Formulation

We use the following notation to formulate the problem under consideration:

A is a set of arcs corresponding to the links in the network;

$\boldsymbol{w}^0 = (w_a^0)$, $a \in A$ is the initial vector of link weights;

c_a is the capacity of the link $a \in A$;

Our decision variables are the variable vector $\boldsymbol{w} = (w_a)$, $a \in A$, representing link weights. After setting their values from the box $W = \{\boldsymbol{w} \in \mathbb{Z}^{|A|} | \boldsymbol{w}^{\min} \leq \boldsymbol{w} \leq \boldsymbol{w}^{\max}\}$, we could simulate the network behaviour and compute the following characteristics of its workflow:

$l_a(\boldsymbol{w})$ is the total load of the link a;

$$o_a(\boldsymbol{w}) = \begin{cases} 1, & \text{if link } a \in A \text{ is overloaded,} \\ 0, & \text{otherwise;} \end{cases}$$

$$no_a(\boldsymbol{w}) = \begin{cases} 1, & \text{if } o_a(\boldsymbol{w}) = 1 \text{ and } o_a(\boldsymbol{w}^0) = 0, \\ 0, & \text{otherwise.} \end{cases}$$

Having the simulation results, we can compute the following characteristics of a solution:

- the total number of overloaded links $O(\boldsymbol{w}) = \sum_{a \in A} o_a(\boldsymbol{w})$;
- the distance between the initial and current vectors of weights $D(\boldsymbol{w}) = ||\boldsymbol{w} - \boldsymbol{w}^0||_{\ell_1}$;
- the total load on the links $L(\boldsymbol{w}) = \sum_{a \in A} l_a(\boldsymbol{w})$;
- the number of newly overloaded links $NO(\boldsymbol{w}) = \sum_{a \in A} no_a(\boldsymbol{w})$.

Now, we can formulate the problem as the following bi-objective black-box integer programming problem.

$$\min_{\boldsymbol{w} \in W} O(\boldsymbol{w}), \qquad (1)$$

$$\min_{\boldsymbol{w} \in W} D(\boldsymbol{w}), \qquad (2)$$

$$L(\boldsymbol{w}) \le L(\boldsymbol{w}^0), \qquad\qquad (3)$$

$$NO(\boldsymbol{w}) = 0, \qquad\qquad (4)$$

$$\boldsymbol{w} \in W. \qquad\qquad (5)$$

The formulae (1)–(5) state that the aim is to find link weight vectors that minimize the number of overloaded links and have the smallest distance from the initial weight vector $\boldsymbol{w}^0$. The solutions must have a total load on the links that is not worse than the initial one and must not induce newly overloaded links.

A weight vector $\boldsymbol{w} \in W$ is called *feasible* if it satisfies inequalities (3) and (4). Given two feasible weight vectors $\boldsymbol{w}^1, \boldsymbol{w}^2 \in W$, we say that $\boldsymbol{w}^1$ *dominates* $\boldsymbol{w}^2$ if $O(\boldsymbol{w}^1) \le O(\boldsymbol{w}^2)$, $D(\boldsymbol{w}^1) \le D(\boldsymbol{w}^2)$, and at least one of these inequalities is strict. A feasible vector $\boldsymbol{w}^1$ is non-dominated if there is no $\boldsymbol{w}^2 \in W$ dominating $\boldsymbol{w}^1$. The problem is to find a *Pareto set*, i.e. a set of all non-dominated feasible solutions within the model (1)–(5), given the functions $O(\cdot)$, $D(\cdot)$, $L(\cdot)$, and $NO(\cdot)$. Since the Pareto set could be computationally hard to find, it is reasonable in practice to aim at finding a set of feasible solutions that is "close" to the Pareto set. A set of feasible solutions $S \subseteq W$ is called a *Pareto set approximation* if there is no $\boldsymbol{w}^1$, $\boldsymbol{w}^2 \in S$ such that $\boldsymbol{w}^1$ dominates $\boldsymbol{w}^2$. Further, in Sect. 4.2, we discuss performance indicators that characterize the quality of Pareto set approximation from different points of view.

3   ILP-Based Heuristic

The heuristic presented in this section is an ILP-based approach. It uses values of the traffic flow changes depending on the modifications of link weights. Though this heuristic can be applied on its own, of great interest is the hybridization of this scheme with the well-known multi-objective evolutionary algorithms (MOEAs). Both of these approaches have their advantages and disadvantages. MOEAs have proven themselves capable of finding solutions that are very close to the Pareto set. However, they may struggle to deal with a constrained problem, whereas the model-based approach explicitly considers these constraints. The latter heuristic allows obtaining a very diverse Pareto set approximation, but it may be inaccurate. Thus, a two-stage scheme, where MOEAs are used for post-optimization of the Pareto set approximation found by the model-based approach, should provide good results.

3.1   Overload Minimizing ILP Model

Given the current vector of link weights $\boldsymbol{w}$, we could compute how a weight modification of link $e \in E, E \subseteq A$ affects loads on the links. We consider only a subset of all links because the simulation is computationally expensive and we want to minimize the number of simulator calls. The structure of the set E will be presented below in Sect. 3.2. Let h denote the step size. And let $\boldsymbol{w}^{e+}$ denote the weight vector obtained from $\boldsymbol{w}$ by increasing its e-th component as follows: $w_e^{e+} = \min(w_e^{\max}, w_e + h)$. Similarly, the e-th component of decreased weight vector $\boldsymbol{w}^{e-}$ equals to $w_e^{e-} = \max(w_e^{\min}, w_e - h)$. The load change for each arc $a \in A$ can be computed within a single simulation run as

$$l_a^{e+} = l_a(\boldsymbol{w}^{e+}) - l_a(\boldsymbol{w}), \quad l_a^{e-} = l_a(\boldsymbol{w}^{e-}) - l_a(\boldsymbol{w}).$$

To formulate the overload minimization ILP model, we introduce Boolean variables (x_a), $a \in A$, indicating if the corresponding arc is overloaded or not, and Boolean variables (λ_e^+), (λ_e^-), indicating whether or not we increase or decrease the weight of the corresponding link by h, respectively.

The overload minimization (OM) model is written as follows:

$$\min_{(x_a),(\lambda_e^+),(\lambda_e^-)} \sum_{a \in A} x_a, \tag{6}$$

$$\sum_{e \in E} (\lambda_e^+ + \lambda_e^-) \le k, \tag{7}$$

$$\lambda_e^+ + \lambda_e^- \le 1, \quad e \in E, \tag{8}$$

$$l_a(\boldsymbol{w}) + \sum_{e \in E} l_a^{e+}\lambda_e^+ + \sum_{e \in E} l_a^{e-}\lambda_e^- \le c_a + o_a(\boldsymbol{w}^0)Mx_a, \quad a \in A, \tag{9}$$

$$\sum_{a \in A} \left(l_a(\boldsymbol{w}) + \sum_{e \in E} l_a^{e+}\lambda_e^+ + \sum_{e \in E} l_a^{e-}\lambda_e^- \right) \le L(\boldsymbol{w}^0), \tag{10}$$

$$\lambda_e^+, \lambda_e^-, x_a \in \{0, 1\}, \quad a \in A, e \in E. \tag{11}$$

The objective function (6) represents the number of overloaded links, which is to be minimized. Conditions (7) limit the number of modifications of the current weight vector $\boldsymbol{w}$. It is necessary because the effects of changing the weight vector components affect each other, and the simultaneous changing of multiple components causes unpredictable load changes. Since the load changes caused by increasing and decreasing the link weight are not opposite to each other, the model could try to apply both modifications of a single weight component for optimization purposes. But the constraints (8) forbid this behaviour. For $a \in A : o_a(\boldsymbol{w}^0) = 0$, the constraints (9) state that links, which are not overloaded initially, cannot be overloaded after the weight modification. For $a \in A : o_a(\boldsymbol{w}^0) = 1$, these constraints, involving a sufficiently large constant M, ensure that variable x_a is set to one if the capacity of the corresponding link a is exceeded. Finally, the inequality (10) guarantees that the total load after modification of the weights is not higher than the initial total load value $L(\boldsymbol{w}^0)$.

3.2   Variable Neighbourhood Model-Based Search

We propose an algorithm that is similar to the variable neighbourhood descent scheme with pipe neighbourhood change step [19]. A neighbourhood $N(\boldsymbol{w})$ of solution $\boldsymbol{w}$ includes solutions that can be achieved by changing no more than k elements of $\boldsymbol{w}$ by the value of h as described in the previous subsection. We can find the best solution in the neighbourhood by solving model (6)–(11) if we do not consider the influence of the modifications on each other. For different k and h, there are different neighbourhood structures. We introduce sets K and H for the values of k and h. The set K depends on two parameters: $k_{\max} \in \mathbb{Z}_{>0}$ and categorical parameter RT $\in \{\texttt{fixed}, \texttt{decremental}, \texttt{exp}\}$ standing for range type:

RT	K
fixed	$\{k_{\max}\}$
decremental	$\{k_{\max}, k_{\max} - 1, \ldots, 1\}$
exp	$\{k_{\max}, \lfloor k_{\max}/2^1 \rfloor, \lfloor k_{\max}/2^2 \rfloor, \ldots, 1\}$

Set H depends on parameters $h_{\min}$ and $h_{\max}$ as follows:

$$H(h_{\min}, h_{\max}) = \{2^i h_{\min} | i \in 0, 1, \ldots, \lfloor \log_2 \frac{h_{\max}}{h_{\min}} \rfloor \}.$$

The algorithm divides all the links into groups based on the initial weights: $E_g = \{i \in A \,|\, 10^g \leq w_i^0 < 10^{g+1}\}$, $G = \{E_g | g = 0, \ldots, \lfloor \log_{10} \max_{i \in A} w_i^0 \rfloor\}$. Here, G is the set of groups, and E_g is the set of link indices for the group. For each group, we use its own initial step $h_{\min,g} = \lceil \min \{w_i^0 | i \in E_g\}/4 \rceil$ and $h_{\max} = 64 h_{\min}$. In our experiments, we also consider the case when there is no division into groups and $|G| = 1$. In this case, we set $h_{\min,1} = 1$ and $h_{\max} = 0.5 \max_{i \in A} w_i^0$. If the solution returned after solving the OM model appears to be infeasible, we try to make it feasible greedily by increasing the weights of newly overloaded links. We perform at most 200 increasing iterations. In order to speed up the search and add diversification to the algorithm, we evaluate only $p \in (0, 1]$ percent of the neighbours. In other words, we consider a random subset of E_g that is denoted as $E_{g,p}$ hereinafter. The scheme of the variable neighbourhood model-based search (VNMS) is presented in Algorithm 1.

Algorithm 1. Variable Neighborhood Model-based Search

```
 1: function VNMS
 2:     for g ∈ 1 ... |G| do
 3:         H ← H(h_min,g, h_max)
 4:         for h ∈ H do
 5:             i ← 0
 6:             while i ≤ |K| do
 7:                 w' ← solveModel(w, h, E_g,p, K[i])
 8:                 if w' is infeasible then
 9:                     greedyFix(w')
10:                 if w' is not better than w then
11:                     i ← i + 1
12:                 else
13:                     w ← w'
14:     return w
```

The local improvement procedure solveModel(w, h, E, k) consists in solving OM model (6)–(11) with the corresponding values of w, h, E, and k. The algorithm starts with $k = k_{max}, h = h_{min}, g = 1$. If the new solution obtained after solving the model is not better than the old one, we update the parameters in the following way. First, we decrease k. If it becomes less than k_{min}, we reset it to k_{max} and double the value of h. If h gets larger than h_{max}, it is set to h_{min}, and we move to the next group. When the algorithm is done with the last group, it stops. However, in the experiments, we restart the algorithm from the beginning while the evaluation budget is not exhausted.

4   Computational Experiments

All experiments in this section were performed on a computer equipped with Intel Core i7-8700 CPU 3.20 GHz and 32 GB of RAM, running Microsoft Windows 10 Pro operating system. We used the PuLP optimization library for Python and the CBC solver to solve the model described in the previous section. The solver was run on a single thread, whereas all other operations were performed in parallel using all available cores.

4.1   Test Instances

We have generated several test instances from the real-world one that was provided to us in order to analyze the efficiency of the algorithm. The original instance has $|A| = 628$ links and 1324 origin-destination (OD) pairs. The generated instances have the same graph structure and OD matrix as the original instance. The difference lies in the vector of initial weights w^0. The initial weights can be generated inversely proportional to the link capacities, i.e. $w_a \propto \frac{1}{c_a}, a \in A$, in accordance with Cisco recommendation [9]. Four instances were generated this way by setting different weight coefficients for the proportion. Namely, these are instances cisco_1, cisco_2, cisco_3, and cisco_4, corresponding to the maximal weight in w^0 being equal to 10, 100, 1000, and 10 000, respectively. The other way to change the original vector w^0 is by dividing it by some number. This way we generated instances divided_1, divided_2, and divided_3, corresponding to the maximal weight in w^0 being around 10, 100, and 1000, respectively (in the original instance it is 15 000). Then, we generated instances with uniformly distributed weights uniform_1, uniform_2, uniform_3, uniform_4, where the indices mean the same as before. The last generated instance was shuffle. For this instance, we just randomly shuffled the original weights. Some properties of the instances are presented in Table 1.

4.2   Performance Indicators

In addition to a visual Pareto front comparison, it is useful to apply performance indicators designed specifically for multi-objective optimization problems [4] when comparing the Pareto set approximations. To describe the indicators, we

Table 1. Values of $O(\boldsymbol{w}^0)$ and $L(\boldsymbol{w}^0)$ for the instances.

Instance	$O(\boldsymbol{w}^0)$	$L(\boldsymbol{w}^0)$
original	208	7.18×10^6
cisco_1	211	7.05×10^6
cisco_2	204	7.38×10^6
cisco_3	204	7.39×10^6
cisco_4	204	7.39×10^6
divided_1	234	6.96×10^6
divided_2	213	7.10×10^6
divided_3	208	7.18×10^6
uniform_1	228	7.07×10^6
uniform_2	227	6.81×10^6
uniform_3	226	7.45×10^6
uniform_4	204	7.16×10^6
shuffle	242	7.98×10^6

will use the notation S or S_k, $k \in \mathbb{N}$, for Pareto set approximations. These approximations are assumed to be obtained by the algorithms under study. Some of the indicator definitions use a special approximation R called *reference set*, which is supposed to have a distinguished quality. It can be computed, for example, by some slow and accurate algorithm within a large computational budget. The vector of optimization criteria would be denoted as $F = (f_i)$, $i \in I$ for the sake of convenience. In our case, $I = \{1, 2\}$, $f_1(\cdot) = O(\cdot)$ and $f_2(\cdot) = D(\cdot)$. We examine the following indicators:

- *Hypervolume* indicator is described as the volume of the space in the objective space dominated by the Pareto front approximation S and delimited from above by a reference point. As the reference point, we use $(O(\boldsymbol{w}^0), \max_{w \in R} D(\boldsymbol{w}))$.
- *Contribution* is the fraction of the points from reference set R that are in S.
- *Generational distance* (GD) is the distance from S to R:

$$GD(S, R) = \frac{1}{|S|} \left(\sum_{\boldsymbol{w}^s \in S} \min_{\boldsymbol{w}^r \in R} ||F(\boldsymbol{w}^s) - F(\boldsymbol{w}^r)||^p \right)^{\frac{1}{p}}.$$

- *Inverted generational distance* (IGD) is the distance from R to S:

$$IGD(S, R) = GD(R, S).$$

We use $p = 2$ for GD and IGD.

- ϵ-*indicator* is the value necessary for S to additively ϵ-dominate R. We say that link weight vector $\boldsymbol{w}^1$ additively ϵ-dominates $\boldsymbol{w}^2$ if

$$f_i(\boldsymbol{w}^1) \leq \epsilon + f_i(\boldsymbol{w}^2) \quad \forall i \in I.$$

- *Maximum Pareto front error* (MPFE) is the maximal distance for a point from S to R:

$$MPFE(S, R) = \max_{\boldsymbol{w}^s \in S} \min_{\boldsymbol{w}^r \in R} ||F(\boldsymbol{w}^s) - F(\boldsymbol{w}^r)||.$$

- R_1 *and* R_2 *indicators.* Their description is given below.
  Let S_1 and S_2 be two Pareto set approximations, U a set of utility functions $u : \mathbb{R}^m \to \mathbb{R}$. For each $u \in U$ and $s = 1, 2$, let associate $u^\star(S_s) = \min_{\boldsymbol{w} \in S_s} u(F(\boldsymbol{w}))$. The two indicators measure to which extent S_1 is better than S_2 over the set of utility functions U.

$$C(S_1, S_2, u) = \begin{cases} 1 & \text{if } u^\star(S_1) < u^\star(S_2), \\ 1/2 & \text{if } u^\star(S_1) = u^\star(S_2), \\ 0 & \text{if } u^\star(S_1) > u^\star(S_2). \end{cases}$$

$$R_1(S_1, S_2, U) = \frac{1}{|U|} \sum_{u \in U} C(S_1, S_2, u).$$

$$R_2(S_1, S_2, U) = \frac{1}{|U|} \sum_{u \in U} \left(u^\star(S_1) - u^\star(S_2)\right).$$

If $R_1(S_1, S_2, U) > 0.5$, then S_1 is considered to be better than S_2. Likewise, $R_2(S_1, S_2, U) < 0$ corresponds to S_1 showing better results than S_2.
- *Spacing* indicator is computed with

$$SP(S) = \sqrt{\frac{1}{|S| - 1} \sum_{i=1}^{|S|} \left(\bar{d} - d_i\right)^2},$$

where $d_i = \min_{\boldsymbol{w}^j \in S \setminus \boldsymbol{w}^i} ||F(\boldsymbol{w}^i) - F(\boldsymbol{w}^j)||_{\ell_1}$ is the distance between point $\boldsymbol{w}^i \in S$ and the closest point of the Pareto set approximation produced by the same algorithm, and $\bar{d}$ is the mean of the d_i.
- $O_{\min}$ is computed as $O_{\min}(S) = \min_{\boldsymbol{w} \in S} O(\boldsymbol{w})$.
- *Cardinality* is equal to $|S|$.

4.3   Multi-objective Evolutionary Algorithms

Population-based algorithms are among the most popular choices for multi-objective optimization [21]. Thus, it is reasonable to study the performance of the most widely-used multi-objective evolutionary algorithms (MOEAs) on the problem formulated in (1)–(5). We consider the following multi-objective algorithms from the Java library MOEA framework [18]:

- NSGA-II, nondominated sorting genetic algorithm II [11];
- SPEA2, strength Pareto evolutionary algorithm 2 [28];
- PESA-II, Pareto envelope region-based selection algorithm [10];
- PAES, pareto archived evolutionary strategy (1+1 ES) [20].

We have also tested other MOEAs but they have shown worse results than the algorithms listed above.

It is worth noting that population-based algorithms perform much better if they are provided with two or more initial solutions $\{w^0, w^1, \ldots, w^m\}$ instead of only w^0. Figure 1 illustrates the behaviour of the algorithms with a population size greater than 1 for the different w^1 provided. All the results correspond to the original instance.

As can be seen from the figure, the quality of the results achieved by the algorithms is highly dependent on the initial solutions. It is desirable to have in the initial population a solution that has as small as possible value of $O(w)$ and is not very distant. The latter is necessary to be able to find intermediate feasible solutions.

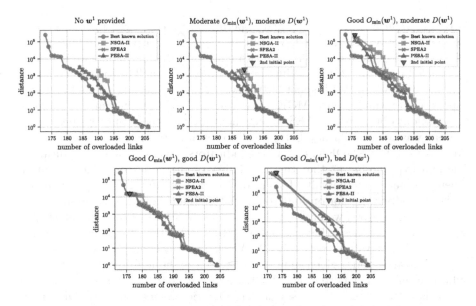

Fig. 1. Impact of the initial solutions provided on the algorithms' performances.

The MOEAs struggle to find such an initial solution using a small computational budget due to the constraints in the problem (Fig. 1). We can explicitly address the constraints and initialize the population of MOEAs with good-quality solutions by employing a global optimization method, such as the one presented in Sect. 3.

Comparative Analysis. To study the efficiency of the MOEAs, we run them 10 times on each of the generated instances (Sect. 4.1) and compare them in terms of the performance indicators (Sect. 4.2). Two options for the algorithms are considered:

– run it independently on the whole computational budget;
– use a different scheme for generating initial solutions, that are then optimized by the algorithm on the remaining computational budget.

Since PAES is better suited for finding good results in terms of the number of overloaded links O_{min} in a short time, we use it in a two-stage approach to generate initial solutions for NSGA-II, SPEA2, and PESA-II. The computational experiments show that PAES requires on average 50 000 evaluations to obtain a 85% quality solution in terms of O_{min} for 85% of the instances. Thus, this budget is further used for generating initial solutions. Starting from the initial solutions provided by PAES, the MOEAs require around 70 000 evaluations to obtain a 80% quality solution in terms of IGD indicator for 80% of the instances. That is, the overall computational budget is equal to 120 000 evaluations.

Impact of Hybridization. In the first experiment, we compared the two options mentioned above. To this end, we ran 10 times each of the three algorithms (NSGA-II, SPEA2, and PESA-II) in two modes. In the first mode, MOEA used the whole budget of 120 000 evaluations for optimization. The second mode was a hybrid one, where PAES spent 50 000 evaluations on getting initial Pareto set approximation, and then a MOEA was run for 70 000 evaluations to improve the solutions obtained by PAES. Using the results for these runs, we performed two one-way Mann-Whitney U tests [23] for each instance, each algorithm pair (MOEA vs. PAES+MOEA) and all the indicators, except R_1 and R_2 indicators. The null hypothesis was that the probability of the indicator value for option 1 being better than the value for option 2 is not greater than the probability of the indicator value for option 2 being better than the value for option 1. For R_1 and R_2 indicators, we performed one-sample Wilcoxon signed-rank tests [27] with the null hypothesis that one option is not better than the other in terms of R_1, R_2 indicator values (see Sect. 4.2). We say that a hybrid algorithm wins at a given instance and a given indicator if we reject the null hypothesis that the hybrid scheme is not better than MOEA and accept the alternative hypothesis that it is better. Similarly, we say that the hybrid scheme loses to MOEA if we accept the alternative hypothesis that MOEA is better at a given instance and indicator. In case we cannot accept either hypothesis, we say that it is a tie. Table 2 shows the number of wins and losses of a hybrid scheme versus a single MOEA. The first number of a pair delimited by a colon mark is the number of wins for the hybrid scheme, and the second number is the number of wins for the MOEA.

We can see that the hybrid scheme performs better in almost all instances. This demonstrates the possible benefits of the two-stage approach for the problem under consideration.

Table 2. The number of wins for PAES+MOEA vs. MOEA pairwise comparison

	NSGA-II	PESA-II	SPEA2
Hypervolume	10:1	7:2	8:1
Contribution	1:2	0:6	0:2
GD	5:3	3:6	6:6
IGD	13:0	9:0	11:0
ϵ-indicator	13:0	13:0	13:0
MPFE	3:1	1:6	2:5
R_1	13:0	12:0	12:0
R_2	13:0	13:0	13:0
Spacing	8:2	6:3	8:1
O_{min}	13:0	13:0	13:0
Cardinality	11:0	12:1	12:1

MOEA Comparison. In the second experiment, we compared the hybrid schemes with different MOEAs to each other. We performed similar statistical tests for each algorithm pair (PESA-II vs. NSGA-II; SPEA2 vs. NSGA-II, and SPEA2 vs. PESA-II). The results of these comparisons are presented in Table 3. The meaning behind the number pairs delimited by a colon mark is the same as for the previous table. The first number corresponds to the wins of the first algorithm in the pair, and the second number corresponds to the wins of the second algorithm in the pair.

Table 3. Numbers of wins for MOEA pairwise comparison

	PESA-II vs. NSGA-II	SPEA2 vs. NSGA-II	SPEA2 vs. PESA-II
Hypervolume	8:0	6:0	0:6
Contribution	5:0	6:0	1:2
GD	9:0	2:0	0:6
IGD	6:0	2:0	0:5
ϵ-indicator	3:0	1:0	0:1
MPFE	5:1	1:0	1:4
R_1	5:1	0:2	0:4
R_2	2:1	0:3	0:5
Spacing	1:3	0:0	4:0
O_{min}	0:0	0:0	0:0
Cardinality	2:0	0:0	0:0

We can see that both PESA-II and SPEA2 outperform NSGA-II. Also, PESA-II shows better results than the SPEA2 algorithm. So, from this comparison, we can conclude that PESA-II got the best results out of the three compared algorithms.

4.4 Tuning of VNMS Parameters

To determine good values for VNMS parameters, we decided to use the hyper-parameter optimization tool SMAC3 [22].

For the two-stage approach, the primary goal of the scheme was to obtain a good Pareto set approximation containing points with a small number of over-loaded links in a short time. The computational budget for the scheme was set to 5 min. This budget roughly corresponds to 50 000 evaluations used by PAES in the previous subsection. The best number of overloaded links O_{min} was chosen as the quality indicator. We noticed that some parameters affect the final solution more than others. We discovered that set K significantly influences the quality of the results, so at the first stage we optimized RT and k_{max}. The SMAC3 tool found RT = exp and $k_{max} = 8$ to be the most promising parameters. So, at the second stage, we fixed them to these values and wanted to decide whether to divide links into groups. It was discovered that division into groups improves the final results. The third parameter to decide was randomization probability. The best value of this parameter appeared to be $p = 0.33$. These settings showed good average results for all instances. The VNMS using these parameters is further referred as VNMS$_{fast}$.

The second way of using the VNMS scheme is to construct the whole Pareto set using solely it. For this goal, we selected inverted generational distance as a quality metric. The computational budget was 12 min, which corresponds to 120 000 evaluations used in total by the hybrid schemes in the previous subsection. SMAC3 found the best parameters to be RT = decremental, $k_{max} = 4$, $p = 0.5$, and non-use of groups. The VNMS using these parameters is further referred as VNMS$_{long}$.

4.5 Comparison of the Schemes

In this section, we compare the schemes presented in the last two subsections. Namely, that is PAES+PESA-II, VNMS$_{fast}$+PESA-II, and VNMS$_{long}$. The first two schemes are two-stage, where 5/12 of the budget is devoted to generating the initial solutions, and the remaining budget is spent by PESA-II using these solutions at the start. For brevity, we refer to these two schemes as PPESA and VPESA, respectively. We selected PESA-II for the two-stage schemes since it provides better results in comparison to other MOEAs for the test instances under consideration. The total computational budget for all three schemes is approximately the same and equals 12 min or 120 000 evaluations. The results of the comparisons are presented in Table 4.

As can be seen from the table, the two-stage scheme using VNMS$_{fast}$ for generating initial solutions performs the best. VNMS$_{long}$ also obtains solid results.

Table 4. Numbers of wins for pairwise comparison of the selected schemes

	PPESA vs. VPESA	VPESA vs. VNMS$_{long}$	VNMS$_{long}$ vs. PPESA
Hypervolume	1:11	11:0	6:6
Contribution	1:6	10:0	1:8
GD	4:6	12:0	1:11
IGD	1:12	11:0	9:3
ϵ-indicator	0:11	1:3	0:12
MPFE	5:3	8:0	1:7
R_1	0:10	1:2	10:0
R_2	0:10	1:2	11:0
Spacing	1:6	13:0	0:13
O_{min}	0:12	0:3	13:0
Cardinality	0:10	13:0	2:10

Although the hybrid scheme using MOEAs approximates the Pareto front near the initial solution much better than VNMS$_{long}$, the latter significantly outperforms MOEA methods (including PAES) in finding solutions with small values of $O(\boldsymbol{w})$. That results in IGD, ϵ-indicator, and O_{min} being better, and comparable results in terms of hypervolume indicator. VPESA scheme takes advantage of both approaches, which leads to performance improvement.

5   Conclusion

We consider the new bi-objective black-box optimization problem for managing a traffic flow in a network. The problem deals with the reconfiguration of network link weights, used by a routing protocol so that the number of overloaded links is minimized and the difference between the old weights and the new ones is as small as possible. The new weights must not increase the total flow and must not introduce new overloaded links. We develop a specialized scheme for solving this problem and perform comparative experiments for this scheme and general-purpose heuristic algorithms. The following results were obtained.

The experiments have shown that SPEA2 and PESA2 algorithms from the MOEA Framework perform the best among a variety of out-of-box population-based heuristics, used in the literature. The tests revealed that, although the mentioned algorithms can build a good Pareto front, they struggle to find solutions with a small number of overloaded links. So, they greatly benefit from decent initial points with this feature. These algorithms can efficiently fill the gaps between given points.

We designed an integer programming model (6)–(11) for finding the best solution in a large neighbourhood. For this model, we compute variations of link load after modifications of a single weight. The model searches for the combination of several modifications that minimizes the number of overloaded links.

Then, we propose an iterative algorithm, VNMS, for the initial problem based on this model.

The VNMS algorithm can be used for two purposes: to find a solution with a small number of overloaded links in a short time, or to find the whole Pareto set. The second option demonstrates results that are close to the ones obtained by a hybrid of population-based methods with the well-known PAES algorithm, demonstrating good results in solving a single-objective problem to minimize the number of overloaded links. A combination of the VNMS with population-based algorithms shows the best results for the initial problem since it takes advantage of both approaches.

Acknowledgements. The study was carried out within the framework of the state contract of the Sobolev Institute of Mathematics (project FWNF-2022-0019).

References

1. Abbasi, M., Guleria, A., Devi, M.: Traffic engineering in software defined networks: a survey. J. Telecommun. Inf. Technol. **4**, 3–14 (2016)
2. Altın, A., Fortz, B., Thorup, M., Ümit, H.: Intra-domain traffic engineering with shortest path routing protocols. Ann. Oper. Res. **204**(1), 65–95 (2013). https://doi.org/10.1007/s10479-012-1270-7
3. Athanasiou, G., Tsagkaris, K., Vlacheas, P., Karvounas, D., Demestichas, P.: Multi-objective traffic engineering for future networks. IEEE Commun. Lett. **16**(1), 101–103 (2012). https://doi.org/10.1109/LCOMM.2011.110711.112071
4. Audet, C., Bigeon, J., Cartier, D., Le Digabel, S., Salomon, L.: Performance indicators in multiobjective optimization. Eur. J. Oper. Res. **292**(2), 397–422 (2021). https://doi.org/10.1016/j.ejor.2020.11.016
5. Audet, C., Hare, W.: Derivative-Free and Blackbox Optimization. SSORFE, Springer, Cham (2017). https://doi.org/10.1007/978-3-319-68913-5
6. Balon, S., Skivée, F., Leduc, G.: How well do traffic engineering objective functions meet TE requirements? In: Boavida, F., Plagemann, T., Stiller, B., Westphal, C., Monteiro, E. (eds.) NETWORKING 2006. LNCS, vol. 3976, pp. 75–86. Springer, Heidelberg (2006). https://doi.org/10.1007/11753810_7
7. Bhatia, R., Hao, F., Kodialam, M., Lakshman, T.: Optimized network traffic engineering using segment routing. In: Proceedings IEEE INFOCOM 2015, pp. 657–665 (2015). https://doi.org/10.1109/INFOCOM.2015.7218434
8. Blanchy, F., Melon, L., Leduc, G.: Routing in a MPLS network featuring preemption mechanisms. In: ICT 2003: 10th International Conference on Telecommunications, vol. 1, pp. 253–260 (2003). https://doi.org/10.1109/ICTEL.2003.1191228
9. Cisco Systems, I.: Internetworking Technologies Handbook, 3rd edn. Cisco Press (2000)
10. Corne, D.W., Jerram, N.R., Knowles, J.D., Oates, M.J.: PESA-II: region-based selection in evolutionary multiobjective optimization. In: Proceedings of the 3rd Annual Conference on Genetic and Evolutionary Computation (GECCO 2001), pp. 283–290 (2001)
11. Deb, K., Pratap, A., Agarwal, S., Meyarivan, T.: A fast and elitist multiobjective genetic algorithm: NSGA-II. IEEE Trans. Evol. Comput. **6**(2), 182–197 (2002). https://doi.org/10.1109/4235.996017

12. El-Alfy, E.S.: Flow-based path selection for internet traffic engineering with NSGA-II. In: ICT 2010: 17th International Conference on Telecommunications (2010). https://doi.org/10.1109/ICTEL.2010.5478839
13. Elwalid, A., Jin, C., Low, S., Widjaja, I.: MATE: MPLS adaptive traffic engineering. In: Proceedings IEEE INFOCOM 2001, vol. 3, pp. 1300–1309 (2001). https://doi.org/10.1109/INFCOM.2001.916625
14. Ericsson, M., Resende, M., Pardalos, P.: A genetic algorithm for the weight setting problem in OSPF routing. J. Comb. Optim. **6**, 299–333 (2002). https://doi.org/10.1023/A:1014852026591
15. Fortz, B., Rexford, J., Thorup, M.: Traffic engineering with traditional IP routing protocols. IEEE Commun. Mag. **40**, 118–124 (2002). https://doi.org/10.1109/MCOM.2002.1039866
16. Fortz, B., Thorup, M.: Internet traffic engineering by optimizing OSPF weights. In: Proceedings IEEE INFOCOM 2000, vol. 2, pp. 519–528 (2000). https://doi.org/10.1109/INFCOM.2000.832225
17. Fortz, B., Thorup, M.: Optimizing OSPF/IS-IS weights in a changing world. IEEE J. Sel. Areas Commun. **20**, 756–767 (2002). https://doi.org/10.1109/JSAC.2002.1003042
18. Hadka, D.: MOEA framework (2023). https://moeaframework.org
19. Hansen, P., Mladenović, N., Todosijević, R., Hanafi, S.: Variable neighborhood search: basics and variants. EURO J. Comput. Optim. **5**(3), 423–454 (2016). https://doi.org/10.1007/s13675-016-0075-x
20. Knowles, J.D., Corne, D.W.: Approximating the nondominated front using the pareto archived evolution strategy. Evol. Comput. **8**(2), 149–172 (2000). https://doi.org/10.1162/106365600568167
21. Konak, A., Coit, D.W., Smith, A.E.: Multi-objective optimization using genetic algorithms: a tutorial. Reliab. Eng. Syst. Saf. **91**(9), 992–1007 (2006). https://doi.org/10.1016/j.ress.2005.11.018
22. Lindauer, M., et al.: SMAC3: a versatile Bayesian optimization package for hyper-parameter optimization. J. Mach. Learn. Res. **23**(54), 1–9 (2022)
23. Mann, H.B., Whitney, D.R.: On a test of whether one of two random variables is stochastically larger than the other. Ann. Math. Stat. **18**(1), 50–60 (1947). https://doi.org/10.1214/aoms/1177730491
24. Pereira, V., Sousa, P., Rocha, M.: A comparison of multi-objective optimization algorithms for weight setting problems in traffic engineering. Nat. Comput. **21**, 507–522 (2022). https://doi.org/10.1007/s11047-020-09807-1
25. Piòro, M., Szentesi, Á., Harmatos, J., Jüttner, A., Gajowniczek, P., Kozdrowski, S.: On open shortest path first related network optimisation problems. Perform. Eval. **48**(1), 201–223 (2002). https://doi.org/10.1016/S0166-5316(02)00036-6
26. Sousa, P., Cortez, P., Rio, M., Rocha, M.: Traffic engineering approaches using multicriteria optimization techniques. In: Masip-Bruin, X., Verchere, D., Tsaousidis, V., Yannuzzi, M. (eds.) WWIC 2011. LNCS, vol. 6649, pp. 104–115. Springer, Heidelberg (2011). https://doi.org/10.1007/978-3-642-21560-5_9
27. Wilcoxon, F.: Individual comparisons by ranking methods. Biometrics Bull. **1**(6), 80–83 (1945)
28. Zitzler, E., Laumanns, M., Thiele, L.: SPEA2: improving the strength pareto evolutionary algorithm. Tech. rep., Inst. f. Technische Informatik und Komm./Computer Eng. and Networks Lab. (2001)

Global Optimization

Optimizing a Feedback in the Form of Nested Saturators to Stabilize the Chain of Three Integrators

Alexander Pesterev$^{(\boxtimes)}$ and Yury Morozov

Institute of Control Sciences, Moscow 117997, Russia
alexanderpesterev.ap@gmail.com

Abstract. The problem of stabilizing the chain of three integrators by a piecewise continuous constrained control is studied. A feedback law in the form of three nested saturators specified by six—three model and three design—parameters is proposed. Global stability of the closed-loop system is studied, and an optimization problem of determining the feedback coefficients ensuring the greatest convergence rate near the equilibrium while preserving global asymptotic stability is stated. It is shown that the loss of global stability results from arising hidden attractors, which come to existence when the convergence rate becomes greater than or equal to a critical value depending on the control resource. A numerical procedure for constructing hidden attractors is developed. The bifurcation value of the convergence rate, which is an exact upper bound of the parameter values ensuring global asymptotic stability of the closed-loop system, is determined numerically by solving an algebraic system of four equations.

Keywords: Chain of integrators · Nested saturators · Hidden oscillations and Attractors · Global stability

1 Introduction

The problem of stabilizing chains of integrators, as well as that of tracking a desired target trajectory, was widely discussed in the literature during last several decades (see, e.g., [1–5] and references therein). The interest to the problem is motivated by the fact that control scheme for chain of integrators can be easily extended to larger classes of systems (see, for example, [4]). Moreover, in many applications the nominal models have the form of chain of integrators, for instance, mechanical planar systems.

This study is a sequel of paper [1], where the problem of stabilizing the third-order integrator by means of an unbounded control under the additional condition of asymptotic tracking a desired trajectory when approaching the equilibrium state was discussed. A piecewise continuous control law based on that

© The Author(s), under exclusive license to Springer Nature Switzerland AG 2023
N. Olenev et al. (Eds.): OPTIMA 2023, LNCS 14395, pp. 129–142, 2023.
https://doi.org/10.1007/978-3-031-47859-8_10

in the form of two nested saturators for the second-order integrator was proposed, and it was proved that the system closed by this feedback is globally asymptotically stable for any positive feedback coefficients.

In this paper, a more complicated task of stabilizing the chain of three integrators by means of a constrained control is discussed, and an advanced control law in the form of three nested saturators is proposed. Unlike in the case of an unbounded control, positiveness of the feedback coefficients guarantees only local asymptotic stability. This brings us at the problem of finding sets of the feedback coefficients that ensure global asymptotic stability. Moreover, there arises an *optimization problem of determining coefficients that ensure the fastest convergence rate while preserving the property of being globally asymptotically stable.*

One of the difficulties associated with this setting is the large number of parameters describing the controller: three fixed model parameters and three design parameters to be selected. By turning to a dimensionless model and confining ourselves to selecting the three design coefficients from a one-parameter family, we managed to reduce the number of the parameters to two dimensionless ones: the control resource and the desired rate of exponential convergence near the equilibrium. The optimization problems thus reduces to finding the greatest exponential rate of the deviation decrease near the equilibrium for which the given closed-loop system is globally asymptotically stable or, vice versa, given a desired convergence rate, determining the minimum control resource that ensures global asymptotic stability.

It has been established that the loss of global stability when increasing the desired rate of exponential convergence results from arising the so-called *hidden oscillations* [6–8], which appear when the rate exceeds certain value. It is this value that is the upper bound of the range of the parameter ensuring global stability of the system. Then, it follows that solution of the optimization problem stated above reduces to determining the bifurcation value of the design parameter (convergence rate), which is found by constructing the corresponding hidden attractor. As is known [9], numerical localization, computation, and analytical study of hidden attractors are much more difficult problems compared to those in the case of limit cycles and self-excited attractors, since hidden attractors are not connected with equilibrium states and their basins of attractions are "hidden" in the phase space of the system, so that no information about the equilibria can be directly used with standard computational procedures. Due to great complexity of these tasks, they cannot be solved analytically and the analysis should heavily rely on numerical simulations. Thus, the goal of this study is to develop a numerical procedure for determining the bifurcation value of the design parameter and constructing the corresponding hidden attractor.

2   Problem Statement

We consider the problem of stabilizing a chain of three integrators,

$$\dot{x}_1 = x_2, \ \dot{x}_2 = x_3, \ \dot{x}_3 = U(x), \tag{1}$$

where $x \equiv [x_1, x_2, x_3]^T$, by means of a piecewise continuous control $U(x)$ subject to the control constraint

$$|U(x)| \leq U_{max}. \tag{2}$$

Such a statement is quite natural from the standpoint of practice. Indeed, suppose that (1) governs a mechanical system, with x_1, x_2, and x_3 being the position, velocity, and acceleration of the system. The system is controlled by varying the traction applied to the system (e.g., through a step motor), and the rate of traction variation is limited.

To stabilize the system, we apply an advanced control law in the form of three nested saturators. Feedbacks in the form of nested saturators were studied and used for stabilizing integrators in many publications (see, e.g., [1–3,10–12] and references therein). The general case of the n nested saturators for stabilizing the nth-order integrator was discussed, for example, in [2,3]. Global stability was proved for the special case of saturation functions with the limit values satisfying certain inequalities [2] (Theorem 2.1), which are not fulfilled for the feedback proposed in this paper. The authors are not aware of works the results of which could be used for establishing stability of system (1) with constrained control resource.

The particular form of the feedback proposed in this study, as well as the optimization problem associated with this setting, will be specified later in this section after introducing some necessary background results.

2.1 Background

Second-Order Integrator. In [10,13], the problem of stabilizing the second-order integrator

$$\dot{x}_1 = x_2, \quad \dot{x}_2 = U_1(x_1, x_2) \tag{3}$$

subject to a phase constraint on the second variable by means of a continuous constrained control was considered. It has been proven [14] that the feedback in the form of nested saturators

$$U_1(x_1, x_2) = -\mathrm{sat}_{k_4}(k_3(x_2 + k_2 \mathrm{sat}_1(k_1 x_1))), \tag{4}$$

where $\mathrm{sat}_d(w)$ is the nonsmooth saturation function, $\mathrm{sat}_d(w) = w$ for $|w| \leq d$ and $\mathrm{sat}_d(w) = d\,\mathrm{sign}(w)$ for $|w| > d$, globally stabilizes the integrator for any positive k_i, $i = 1, \ldots, 4$, and that $|x_2(t)| \leq k_2$ $\forall t > 0$ as long as $|x_2(0)| \leq k_2$. In (4), the parameters k_4 and k_2 are assumed given. The former denotes the control resource and the latter constrains variation of variable x_2 (in mechanical applications, maximum traction developed by the control system and the maximum allowed velocity, respectively) within prescribed limits: $|x_2(t)| \leq k_2$.

Unlike k_2 and k_4, k_1 and k_3 are not a priori given and can arbitrarily be selected by the designer of the control system. They affect performance of the control system, in particular, the rate of convergence near the equilibrium point and its type, and can be used to optimize certain characteristics of the closed-loop system (see, e.g., [10,13]).

Stabilizing the Chain of Three Integrators in the Case of Uncon-strained Control Resource. A more complicated problem of stabilizing the chain of three integrators (1) by a piecewise continuous unconstrained control $U(x)$ with the additional condition of asymptotic tracking the trajectory of the second-order integrator stabilized by control (4) was studied in [1]. It was proved that control of the form

$$U(x) = \dot{U}_1(x_1, x_2) + \gamma(U_1(x_1, x_2) - x_3), \ \gamma > 0, \tag{5}$$

where $\dot{U}_1$ is the time derivative of function U_1 by virtue of system (3), globally asymptotically stabilizes the system.

2.2 Proposed Feedback

In view of the above, it seems natural to seek a feedback based on the uncon-strained one, namely, to apply control of form (5) in the regions where the control resource is sufficient and its saturated value elsewhere. In this study, we propose the feedback in the form of three nested saturators based on a modified version of (5):

$$U(x) = -\mathrm{sat}_{k_6}\left(k_5(x_3 - U_1(x_1, x_2)) - \frac{dU_1}{dt}\right). \tag{6}$$

Here, k_6 is the control resource, $U_1(x_1, x_2)$ is given by (4), and dU_1/dt is the time derivative of function U_1 by virtue of system (1) (recall that, in (5), $\dot{U}_1$ is the time derivative of U_1 by virtue of the second-order system (3)).

In this setting, coefficients k_2, k_4, and k_6 are characteristics of the system (in mechanical applications, they can stand, e.g., for maximum allowed veloc-ity, maximum traction, and maximum rate of traction variation) and are fixed, whereas k_1, k_3, and k_5 can vary to ensure certain desirable properties of the transient process and are selected by the designer of the control system. To dis-tinguish between them, we refer to these groups of coefficients as *model* and *design parameters,* respectively.

Unlike in the two-dimensional case, where global asymptotic stability takes place for any positive coefficients, for the third-order integrator, positiveness of these coefficients guarantees only local stability. Thus, the first task is to determine (or estimate) the range of parameters that ensure global stability of the closed-loop system (1), (6). This problem is addressed in Sect. 5.

Remark. To avoid confusion, it should be noted that the controls in the two- and three-dimensional problems are different, so that one should distinguish between control resources in these two settings. In the former, the control resource is k_4 (traction). In the three-dimensional setting, the control resource is k_6, whereas k_4 is the desired maximum value of the third variable.

2.3 Optimization Problem Statement

The most important requirement on the control system is that it must ensure, if possible, global asymptotic stability of the closed-loop system. This brings us at the problem of determining domains in the space of coefficients that ensure global stability. On the other hand, of great importance is how fast the system comes to the equilibrium. Numerical experiments show that these two features contradict one another: an attempt to increase the convergence rate without increasing the control resource may result in loss of stability. When the convergence rate exceeds certain critical value depending on the control resource, the system cannot be stabilized from an arbitrary initial position and possesses only local stability. As the convergence rate increases, the domain of attraction of the equilibrium state decreases. Thus, we arrive at the problem of finding coefficients k_1, k_3, and k_5 that, for given parameters k_2, k_4, and k_6, ensure the greatest convergence rate while preserving the property of global asymptotic stability.

Given the great number of the parameters, both above-mentioned tasks are too complicated. The problem is greatly simplified if we turn to a dimensionless system to reduce the number of the parameters affecting the solution by two, which is addressed in the next section. The number of the parameters can further be reduced if we confine our consideration to a one-parameter family of coefficients k_1, k_3, and k_5, as discussed in Sect. 4.

3   Dimensionless Model

First, to simplify the notation, we will rewrite all formulas in terms of the saturation functions with unitary saturation limit $\mathrm{sat}_1(\cdot)$ by taking advantage of the well-known formula

$$\mathrm{sat}_d(w) \equiv d\,\mathrm{sat}_1\left(\frac{w}{d}\right)$$

and will omit the subscript 1. Then, feedbacks (4) and (6) are written as

$$U_1(x_1, x_2) = -k_4\,\mathrm{sat}\left(\frac{k_3}{k_4}(x_2 + k_2\mathrm{sat}(k_1x_1))\right) \tag{7}$$

and

$$U(x) = -k_6\,\mathrm{sat}\left(\frac{k_5}{k_6}(x_3 - U_1(x_1, x_2)) - \frac{1}{k_6}\frac{dU_1}{dt}\right), \tag{8}$$

Further, introducing the dimensionless variables $\tilde{x}_1 = k_4x_1/k_2^2$, $\tilde{x}_2 = x_2/k_2$, $\tilde{x}_3 = x_3/k_4$ and time $\tilde{t} = k_4t/k_2$ and turning to the new variables in (1), (7), (8), we get dimensionless coefficients $\tilde{k}_i$:

$$\tilde{k}_1 = k_1k_2^2/k_4, \quad \tilde{k}_2 = 1, \quad \tilde{k}_3 = k_2k_3/k_4, \quad \tilde{k}_4 = 1, \quad \tilde{k}_5 = k_2k_5/k_4, \quad \tilde{k}_6 = k_2k_6/k_4^2.$$

Thus, the dimensionless system is described by only one model parameter, namely, the dimensionless control resource $\tilde{k}_6$.

In what follows, only dimensionless systems are dealt with, therefore we will use the same symbols as the dimensional ones to denote dimensionless variables and parameters omitting the tilde to avoid messy notation. The dot notation is used to denote the derivatives with respect to the dimensionless time. Thus, by turning to the dimensionless notation, we have got rid of two parameters and arrived at studying stability of system (1), (8) with function $U_1(x_1, x_2)$ given by

$$U_1(x_1, x_2) = -\mathrm{sat}(k_3(x_2 + \mathrm{sat}(k_1x_1))). \tag{9}$$

4   Local Stability

From (9), it is seen that function U_1 is piecewise linear:

$$U_1 = \begin{cases} -k_1 k_3 x_1 - k_3 x_2, & (x_1, x_2) \in D_1, \\ -k_3(x_2 + \text{sign}(x_1)), & (x_1, x_2) \in D_2, \\ -\text{sign}(x_2 + \text{sat}(k_1 x_1)), & (x_1, x_2) \in D_3, \end{cases} \qquad (10)$$

where D_1, D_2, and D_3 are sets shown in Fig. 1. The set D_1 includes all points where both saturators are not saturated; D_2 consists of the points where the internal saturator reaches saturation, whereas the external one does not; and D_3 is the set where the external saturation function is saturated (see [1,10,13,14] for more detail). The dashed lines are the boundaries of sets D_1 and D_2. The solid broken line is the set of points where the argument of the external saturation function vanishes, and the red solid line is the asymptote of the linear system $\dot{x}_1 = x_2$, $\dot{x}_2 = -k_3 x_2 - k_3 k_1 x_1$.

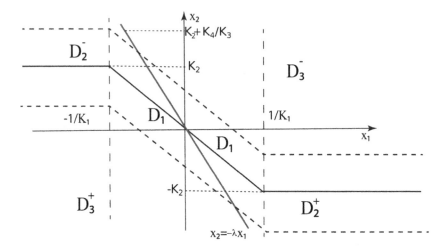

Fig. 1. Partitioning of the plane (x_1, x_2) into sets D_1, D_2, and D_3. (Color figure online)

Taking the derivative of U_1 by virtue of system (1),

$$\frac{dU_1}{dt} = \frac{\partial U_1}{\partial x_1} x_2 + \frac{\partial U_1}{\partial x_2} x_3,$$

in the sets D_i, $i = 1, 2, 3$, we get

$$\frac{dU_1}{dt} = \begin{cases} -k_1 k_3 x_2 - k_3 x_3,, & (x_1, x_2) \in D_1, \\ -k_3 x_3, & (x_1, x_2) \in D_2, \\ 0, & (x_1, x_2) \in D_3. \end{cases} \qquad (11)$$

Clearly, the third-order system under study is a linear switched affine system. Switching from one linear system to another occurs when one of the three saturators reaches saturation or, vice versa, becomes unsaturated. In the neighborhood of the origin, where all saturators are not saturated (such a neighborhood obviously exists), system (1), (8), (9) takes the form

$$\dot{x}_1 = x_2, \ \dot{x}_2 = x_3, \ \dot{x}_3 = -(k_3 + k_5)x_3 - k_3(k_1 + k_5)x_2 - k_1k_3k_5x_1. \quad (12)$$

From the right-hand side of this system, it is seen that the zero solution of (12) is stable for any positive coefficients.

If the system under study is globally asymptotically stable, the deviation near the equilibrium tends to zero with the exponential rate determined by the lowest (in magnitude) root of the characteristic equation of system (12). Then, without loss of generality, we can select the design parameters k_1, k_3, and k_5 from the one-parameter family parametrized by the exponential rate of the deviation decrease λ. This can be done if, for any $\lambda > 0$, these coefficients satisfy the following system of the algebraic equations

$$k_3 + k_5 = 3\lambda, \ k_3(k_1 + k_5) = 3\lambda^2, \ k_1k_3k_5 = \lambda^3. \quad (13)$$

Indeed, in this case, system (12) takes the form

$$\dot{x}_1 = x_2, \ \dot{x}_2 = x_3, \ \dot{x}_3 = -3\lambda x_3 - 3\lambda^2 x_2 - \lambda^3 x_1. \quad (14)$$

The characteristic equation of system (14) has one repeated root $-\lambda$, and the deviation x_1 decreases exponentially with the rate λ.

It can be checked by the direct substitution that Eqs. (13) are satisfied, if we take

$$k_1 = \lambda/2, \ k_3 = 2\lambda, \ k_5 = \lambda. \quad (15)$$

Thus, the original problem of finding three design coefficients of the feedback stabilizing the system described by three model parameters have been reduced to that of determining one parameter λ of the feedback to stabilize the system described by one parameter. The optimization problem formulated in Sect. 2 can now be formulated as follows:

Optimization Problem. *For a given control resource k_6, find the maximum value of λ for which system (1), (8), (9), with the coefficients specified by (15), is globally asymptotically stable.*

Note that, even after such simplification, the problem is still too complicated to solve it analytically. Our analysis will rely on results of numerical simulations, which will allow us to make reasonable assumptions on the way the system loses global stability, and refer to conclusions on solvability of systems of algebraic equations obtained numerically.

5   Study of Global Stability

First, we prove that, for any control resource, the system can always be made globally stable by taking a sufficiently small λ.

Proposition. For any control resource k_6, there exists $\tilde{\lambda}$ such that system (1), (8), (9) is globally stable for any $\lambda \leq \tilde{\lambda}$.

The proof is omitted to save room. Note only that it is based on proving boundedness of a solution of the system and takes advantage of the fact that the argument of the external saturation function decreases as λ diminishes. The latter implies that one can always take λ so small that the saturation is not reached, thus reducing the problem to studying the system with unbounded control, which is known to be globally stable [1].

Note that this assertion holds independent of whether the two internal saturation functions are saturated or not. This result is important because it substantiates correctness of the optimization problem stated above. Given that, for any control resource, there is a range of λ ensuring global stability of the closed-loop system, it makes sense to seek for an upper bound of this interval.

Clearly, the estimate $\tilde{\lambda}$ obtained based on the requirement of an unsaturated control is too conservative, since the system can generally be stable when the control does reach saturation. Our goal is to find an exact upper bound of $\lambda(k_6)$ for which global stability takes place.

Our numerical experiments for different values of the control resource k_6 showed that the loss of global stability when increasing λ is accompanied with arising periodic motions called by G.A. Leonov *hidden oscillations* or *hidden attractors* [7]. The attractor is called hidden if its basin of attraction does not intersect with arbitrarily small neighborhoods of unstable equilibria states [8,9]. Numerical localization, computation, and analytical study of hidden attractors are much more difficult problems compared to those in the case of limit cycles and self-excited attractors, since, in this case, no information about the equilibria can be directly used with standard computational procedures. Whereas the latter can easily be detected and visualized by trajectories in numerical experiments with initial data belonging to neighborhoods of unstable equilibrium points, hidden attractors are not connected with such states and their basins of attractions are "hidden" in the phase space of the system [9].

In our numerical experiments, hidden oscillations in system (1), (8), (9) came to existence when λ exceeded certain critical value $\lambda_{cr}(k_6)$. When $\lambda < \lambda_{cr}(k_6)$, solutions of system (1), (8), (9) for any initial conditions go to the origin; i.e., the system is globally stable. When $\lambda \geq \lambda_{cr}(k_6)$, global stability is lost: depending on the initial condition, system (1), (8), (9) is either stabilized at the origin or performs a periodic motion. The number of different periodic motions is uncountable. The discussion of the structure of the set of hidden attractors is beyond the scope of this paper, since our goal is to determine the minimum value of λ when the global stability is lost. Note only that all periodic trajectories lie on the surface of a cylinder located along the x_1-axis densely covering it. When $\lambda = \lambda_{cr}(k_6)$, the cylinder degenerates into a single hidden attractor.

The subsequent analysis of global stability is based on the assumptions (i) that it is the origination of hidden oscillations that destroys global stability of the system, so that the system is globally stable in the lack of such oscillations, and (ii) that the projection of a cycle on the plane (x_1, x_2) is a simple closed curve without self-intersections. These assumptions are substantiated by results of numerous simulations, but we cannot formally prove them.

Based on the above, our goal, thus, is to prove that system (1), (8), (9) admits periodic solutions and, for any control resource, to determine the range of λ for which such solutions exist. The minimal value of this range is the upper bound of the range of λ ensuring global asymptotic stability. The existence of periodic motions will be proved by direct construction of cycles corresponding to such solutions.

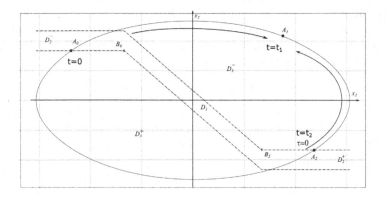

Fig. 2. Construction of the periodic solution of system (1), (8), (9). Projection onto the plane (x_1, x_2). (Color figure online)

Suppose that there exists a cycle and consider its projection on the plane (x_1, x_2) (Fig. 2). Let us denote by A_0 and A_2 the points on the plane (x_1, x_2) where the trajectory intersects the lower and upper boundaries of the set $D_1 \cup D_2$, respectively (dashed lines in Fig. 2). Let us set the initial time ($t = 0$) at the moment when the projection of the trajectory passes through the point A_0, and let t_2 denote the time when the system comes to the point A_2. From the symmetry considerations, the second part of the cycle projection (from A_2 to A_0) is symmetric with respect to the line $A_0 A_2$. Let us assume now that $U(x) \equiv \pm k_6$ at any point of the cycle segment the projection of which onto the plane (x_1, x_2) lies in the set D_1 or D_2. This assumption will be a posteriori validated when the cycles are constructed. Then, it follows that $U(x) \equiv -k_6$ on the initial part of the cycle and that the control becomes unsaturated at a point lying in the set D_3^-. Let us denote the switching point and the time moment when this happens as A_1 and t_1, respectively (Fig. 2).

On the time interval $[0, t_1]$, system (1), (8), (9) takes the form

$$\dot{x}_1 = x_2, \quad \dot{x}_2 = x_3, \quad \dot{x}_3 = -k_6. \tag{16}$$

Integrating system (16), we obtain

$$x_3(t) = -k_6 t + x_3(0), \tag{17}$$

$$x_2(t) = -\frac{k_6 t^2}{2} + x_3(0)t + x_2(0), \tag{18}$$

$$x_1(t) = -\frac{k_6 t^3}{6} + x_3(0)\frac{t^2}{2} + x_2(0)t + x_1(0). \tag{19}$$

To find solution in the interval $[t_1, t_2]$, we will integrate the system in the inverse time $\tau = t_2 - t$. To distinguish the solution in the inverse time from that in the direct time, we denote the phase variables x_1, x_2, x_3 in the new system as y_1, y_2, y_3. Replacing differentiation with respect to time in (1) by that with respect to τ and taking into account that $U_1(x_1, x_2) = -1$, we obtain the following system of equations:

$$y_1' = -y_2, \ y_2' = -y_3, \ y_3' = \lambda(y_3 + 1), \tag{20}$$

where the prime denotes differentiation with respect to τ. From the symmetry considerations, it follows that the initial conditions (specified at point A_2) are given by $y_i(0) = -x_i(0)$, $i = 1, 2, 3$. Integrating system (20) in the interval $[0, t_2 - t_1]$, we obtain

$$y_3(\tau) = (1 - x_3(0))e^{\lambda\tau} - 1, \tag{21}$$

$$y_2(\tau) = -\frac{1 - x_3(0)}{\lambda}(e^{\lambda\tau} - 1) + \tau - x_2(0), \tag{22}$$

$$y_1(\tau) = \frac{1 - x_3(0)}{\lambda^2}(e^{\lambda\tau} - 1) - \frac{\tau^2}{2} + \left(x_2(0) - \frac{1 - x_3(0)}{\lambda}\right)\tau - x_1(0). \tag{23}$$

The condition that the two solutions must coincide at the point A_1 (sewing condition) yields three equations:

$$x_i(t_1) = y_i(t_2 - t_1), \ i = 1, 2, 3, \tag{24}$$

which are supplemented by the saturation condition $\lambda(y_3(t_2 - t_1) + 1) = k_6$ at the point A_1. Substituting the right-hand side of (21) for y_3 in the last condition and denoting $\theta = t_2 - t_1$, $\varepsilon = 1 - x_3(0)$, and $\mu = 1/\lambda$ to shorten the notation, we get

$$\varepsilon e^{\lambda\theta} = k_6\mu. \tag{25}$$

Conditions (24), (25), which any cycle must satisfy, constitute the system of four transcendental equations. To simplify it, we substitute the right-hand side of (25) for $(1 - x_3(0))e^{\lambda\theta}$ in the right-hand sides of Eqs. (24) given by (21)–(23), which will allow us to get rid of the exponential terms in (24). Now, (24) is the system of the following three polynomial equations of the first, second, and third orders in the unknowns $\mu, x_1(0), x_2(0), \varepsilon, t_1$, and θ:

$$k_6\mu^3 + \frac{k_6 t_1^3}{6} - \varepsilon\mu^2 - \varepsilon\mu\theta - (1 - \varepsilon)\frac{t_1^2}{2} - \frac{\theta^2}{2} + (\theta - t_1)x_2(0) - 2x_1(0) = 0,$$

$$k_6\mu^2 - \varepsilon\mu - \frac{k_6 t_1^2}{2} + (1 - \varepsilon)t_1 - \theta + 2x_2(0) = 0, \tag{26}$$

$$k_6\mu + \varepsilon + k_6 t_1 - 2 = 0.$$

Altogether, we have the system of three algebraic equations (26) and one transcendental equation (25) in six unknowns. One unknown–$x_2(0)$–can be determined from the equation of the boundary of the set $D_1 \cup D_2$. Indeed, if the
starting point lies on the boundary between D_3^+ and D_2 (like in Fig. 2), then

$$x_2(0) = 1 - \frac{1}{2\lambda}. \tag{27}$$

If the point A_1 lies on the boundary of the set D_1, then

$$x_2(0) = -\frac{\lambda}{2}x_1(0) - \frac{1}{2\lambda}. \tag{28}$$

Nevertheless, the number of the unknowns is still greater than the number
of the equations by one. This obstacle can be overcome as follows.

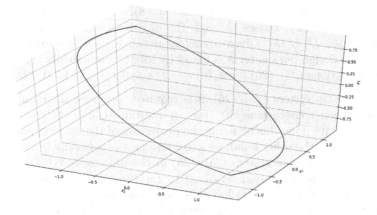

Fig. 3. The minimal cycle. (Color figure online)

Let us fix the position of the starting point on the plane (x_1, x_2) to be the
angular point B_0 of the boundary (see Fig. 2) and find out whether there exists
a cycle whose projection onto the plane (x_1, x_2) passes through B_0 and B_2.
Substituting $x_1(0) = -2/\lambda$ and $x_2(0) = 1-1/(2\lambda)$ into Eqs. (26), we arrive at the
system of four nonlinear equations in the four unknowns. Solving it numerically,
we find the value of the design parameter λ corresponding to the cycle passing
through the angular points B_0 and B_2 of the boundary if such a cycle exists. The
cycle itself is constructed by formulas (17)–(19) and (21)–(23). Our numerical
experiments showed that the system obtained has solutions for any tested values
of k_6, at least, in the range $(0, 8)$. Figures 3 and 4 show the cycle obtained and
its projection on the plane (x_1, x_2) for $k_6 = 1$. The segments of the cycle with
the saturated control are marked by the green color, those with unsaturated
control are depicted in blue. For this model, $\lambda \approx 1.852$, $\varepsilon = 0.010$, $t_1 = 1.449$,
$t_2 = 2.130$. The period of the oscillations is thus $T = 2t_2 = 4.260$ units of time.

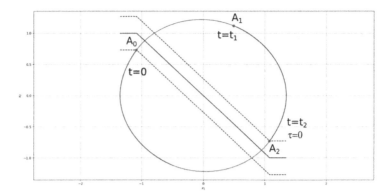

Fig. 4. Projection of the minimal cycle onto the plane (x_1, x_2). (Color figure online)

Proposition. The minimal cycle of system (1), (8), (9) is that whose projection onto the plane (x_1, x_2) passes through the angular points B_0 and B_2 of the boundary of set $D_1 \cup D_2$. Given a control resource k_6, the bifurcation value $\lambda_{cr}(k_6)$ of the design parameter λ is found by solving numerically the system of nonlinear equations (25)–(26) for $x_1(0) = -2/\lambda$ and $x_2(0) = 1 - 1/(2\lambda)$.

Proof. To resolve system (25), (26), it is required to reduce the number of the unknowns. When constructing the cycle in question to find the corresponding value of λ, we have done this by fixing the projection of the starting point on the plane (x_1, x_2). On the other hand, we can consider λ as a parameter, set it equal to some value, and solve system (25), (26) in the unknowns $x_1(0), \varepsilon, t_1$, and θ (recall that $x_2(0)$ is uniquely determined by Eq. (28) or (27)). The proposition is true if system (25), (26), (28) has no solutions for any $\lambda < \lambda_{cr}(k_6)$ satisfying the condition $-2/\lambda < x_1(0) < 0$. It seems unlikely that the insolvability of this system can be proved analytically. We verified this proposition numerically: in our experiments, no solutions to system (25), (26), (28) have been found for any tested values of $\lambda < \lambda_{cr}(k_6)$.

Of interest is also dependence of the maximum convergence rate λ_{cr} on the control resource k_6. Results of our numerical experiments demonstrate that, at least in the tested range $k_6 \in (0.5, 7.5)$, it is almost (but not strongly) linear as shown in Fig. 5.

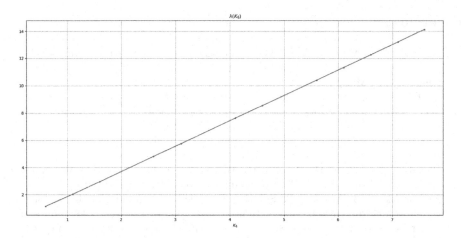

Fig. 5. Dependence of the bifurcation value λ_{cr} on the control resource k_6.

6   Conclusions

The problem of stabilizing the chain of three integrators by means of piece-wise continuous constrained control has been considered. A feedback law in the form of three nested saturators specified by six—three model and three design—parameters has been proposed. By turning to a dimensionless notation and selecting the design coefficients from the one-parameter family, the number of the parameters affecting the solution has been reduced to two: the dimensionless control resource and exponential convergence rate near the equilibrium. The optimization problem of finding the greatest convergence rate that, for a given control resource, ensures global stability of the closed-loop system has been stated.

It has been established that the loss of global stability when increasing the desired rate of exponential convergence is associated with arising hidden oscillations, which reduced solution of the optimization problem to determining the bifurcation value of the convergence rate and constructing the corresponding hidden attractor. It has been shown that the bifurcation value of the convergence rate and, thus, solution of the optimization problem can be found by solving a system of nonlinear equations consisting of three polynomial and one transcendental equations. A numerical procedure for constructing hidden attractors, which underlies determination of the critical value of the convergence rate, has been developed. By means of this procedure, the bifurcation value of the convergence rate has been determined numerically, and the hidden attractor corresponding to this value has been constructed.

References

1. Pesterev, A.V., Morozov, Y.V.: The best ellipsoidal estimates of invariant sets for a third-order switched affine system. In: Olenev, N., Evtushenko, Y., Jaćimović, M., Khachay, M., Malkova, V., Pospelov, I. (eds.) OPTIMA 2022. LNCS, vol. 13781, pp. 66–78. Springer, Cham (2022). https://doi.org/10.1007/978-3-031-22543-7_5

2. Teel, A.R.: Global stabilization and restricted tracking for multiple integrators with bounded controls. Syst. Control Lett. **18**, 165–171 (1992)

3. Teel, A.R.: A nonlinear small gain theorem for the analysis of control systems with saturation. IEEE Trans. Autom. Control **41**, 1256–1270 (1996)

4. Polyakov, A., Efimov, D., Perruquetti, W.: Robust stabilization of MIMO systems in finite/fixed time. Int. J. Robust Nonlinear Control **26**, 69–90 (2016)

5. Kurzhanski, A.B., Varaiya, P.: Solution examples on ellipsoidal methods: computation in high dimensions. In: Kurzhanski, A.B., Varaiya, P. (eds.) Dynamics and Control of Trajectory Tubes. SCFA, vol. 85, pp. 147–196. Springer, Cham (2014). https://doi.org/10.1007/978-3-319-10277-1_4

6. Moreno, I., Suárez, R.: Existence of periodic orbits of stable saturated systems. Syst. Control Lett. **51**, 293–309 (2004)

7. Leonov, G.A., Kuznetsov, N.V.: Hidden oscillations in dynamical systems. 16 Hilbert's problem, Aizerman's and Kalman's conjectures, hidden attractors in Chua's circuits. J. Math. Sci. **201**, 645–662 (2014)

8. Andrievsky, B.R., et al.: Hidden oscillations in stabilization system of flexible launcher with saturating actuators. IFAC Proc. **46**, 37–41 (2013)

9. Kuznetsov, N.V.: Theory of hidden oscillations and stability of control systems. J. Comput. Syst. Sci. Int. **59**, 647–668 (2020)

10. Pesterev, A., Morozov, Y.: Optimizing coefficients of a controller in the point stabilization problem for a robot-wheel. In: Olenev, N.N., Evtushenko, Y.G., Jaćimović, M., Khachay, M., Malkova, V. (eds.) OPTIMA 2021. LNCS, vol. 13078, pp. 191–202. Springer, Cham (2021). https://doi.org/10.1007/978-3-030-91059-4_14

11. Hua, M.-D., Samson, C.: Time sub-optimal nonlinear PI and PID controllers applied to longitudinal headway car control. Int. J. Control **84**, 1717–1728 (2011)

12. Marconi, L., Isidori, A.: Robust global stabilization of a class of uncertain feedforward nonlinear systems. Syst. Control Lett. **41**, 281–290 (2000)

13. Pesterev, A., Morozov, Y., Matrosov, I.: On optimal selection of coefficients of a controller in the point stabilization problem for a robot-wheel. In: Olenev, N., Evtushenko, Y., Khachay, M., Malkova, V. (eds.) OPTIMA 2020. CCIS, vol. 1340, pp. 236–249. Springer, Cham (2020). https://doi.org/10.1007/978-3-030-65739-0_18

14. Morozov, Y.V., Pesterev, A.V.: Global stability of a fourth-order hybrid affine system. J. Comput. Syst. Sci. Int. **62**, 595–606 (2023)

Discrete and Combinatorial Optimization

One Segregation Problem for the Sum of Two Quasiperiodic Sequences

Liudmila Mikhailova$^{(\boxtimes)}$ [iD]

Sobolev Institute of Mathematics, 4 Koptyug Avenue, 630090 Novosibirsk, Russia
mikh@math.nsc.ru

Abstract. The subject of the study is a noise-proof segregation problem for the sequence being the sum of two independent quasiperiodic sequences. The problem is stated for the case when every quasiperiodic sequence is formed from the known number of identical given subsequences-fragments. A posteriori approach to this problem leads to solving an unexplored discrete optimization problem. A polynomial-time algorithm that guarantees the optimal solution to this optimization problem is proposed. Additionally, there are some examples of numerical simulation for illustration.

Keywords: Discrete optimisation problem · Quasiperiodic sequence · Detection · Segregation · Polynomial-time solvability · One-microphone signal separation

1 Introduction

By a numerical quasiperiodic sequence, we mean any sequence that includes subsequences-fragments (subsequences formed by consecutive sequence elements) that have some predetermined characteristic properties with the following restrictions on their relative positions: 1) the fragments do not intersect each other; 2) they are included in the sequence as a whole; 3) the interval between the initial positions of two consecutive fragments is limited from below and from above by the given constants. We can interpret numerical quasiperiodic sequences as time series obtained as a result of quasiperiodic pulse trains uniform sampling. In these pulse trains, fluctuations in the interval between pulses, as well as variations in the shape of the pulse relative to a given pattern, are allowed.

For example, such pulse trains arise when monitoring natural objects (underwater, aerospace, underground, bio-medical, etc.) in the case of quasiperiodic repeatability of their states. This type of state repeatability implies that 1) the object can't be in two different states simultaneously; 2) the distance between

The study presented was supported by the Russian Academy of Science (the Program of basic research), project FWNF-2022-0015.

two successive state repetitions belongs to the given interval; 3) a typical state allows some fluctuations from one repetition to another. The problem that motivated this study is that it is difficult, and sometimes impossible, to isolate the signal from a single observed object. Therefore, the result of monitoring is the sum of signals received from several objects—the sum of quasiperiodic pulse trains. Note that even if the fluctuations of state are prohibited and only two pulse trains are summed up, such a sum is no longer quasiperiodic.

There are many algorithms for quasiperiodic sequence processing (detection and identification of fragments, recognition of the sequence, and so on) with various assumptions about the structure of the sequence and the property of an individual fragment (see, for example, [1–3]). What all these algorithms have common is that each of them deals with a single noisy quasiperiodic sequence, which reflects monitoring of a single isolated object. In the current paper, we consider a more complicated structure of the sequence to be processed—it is the noisy sum of two quasiperiodic sequences. This structure of the observable sequence reflects the monitoring of a mixture of signals from two close objects. Such structure of the sequence has not been studied previously. The problem is stated for the assumption that every summed sequence contains the given number of identical subsequences-fragments. The segregation problem under consideration is to restore the quasiperiodic sequences from their noisy sum. In other words, we have to determine two collections of initial indices corresponding to the beginnings of the fragments in every unobservable sequence. Since the subsequences-fragments are identical in every sequence, this information is sufficient to restore the desired unobservable sequences.

The considered segregation problem allows the interpretation as the blind signal separation problem (BSS) (see, for example, [4–6]). BSS problem is to restore initial unobserved signals using some observations of their mix with a lack of information about the source signals or the mixing process. Here we have only one observation for the mixture of two signals taken with equal weights. A big part of the algorithms for BSS, such as independent component analysis (ICA) and principal component analysis (PCA), is not applicable due to only one observable sequence. The variant of the BSS problem, when only one observation of the mixture of signals is available, is known as a one-microphone signal or source separation problem.

Traditional approaches to this problem, as well as to the BBS problem, are based on statistical principles, spectral analysis, time-frequency mask-estimation, and so on (see, for example, [7–9]) as a rule accompanied by filtering out noise with subsequent analysis of the received unmixed signals (detecting separate fragments of every unmixed signal, analysing and identifying these fragments). This traditional sequential (multistage) approach has well-known advantages—the reuse of available mathematical tools and the low complexity as a rule of the resulting algorithms. Also, it has a known disadvantage—the lack of the optimality guarantee of the final solution, even if the solution at each stage is optimal.

In the current paper, an alternative a posteriori approach is implemented, so there is no separation of the decision process into stages. This approach is less common than multistage because solving each new problem includes solving a unique, usually unexplored discrete optimization problem. Examples of applying a posteriori approach to problems of quasiperiodic sequence processing can be found in [1–3], and the works cited there.

The main goal of the research is to construct an efficient algorithm with theoretical guarantees of quality (accuracy and time complexity), solving the segregation problem and the corresponding discrete optimization problem.

2   Data Generation Model

There is a general description of a quasiperiodic sequence in the previous section. To give a formal definition of a quasiperiodic sequence containing the given number of identical subsequences-fragments, we need some designations. Let $X = (x_0, \ldots, x_{N-1})$ be a N-length sequence of real numbers. Assume that it contains M subsequences-fragments of the same length q and every fragment coincide with the subsequence-fragment $U = (u_0, \ldots, u_{q-1})$ of the given sequence u_i, $i = 0, \pm 1, \pm 2, \ldots$, of real numbers, where

$$u_j = 0, \ n \notin \{0, \ldots, q-1\},$$
$$\sum_{j=0}^{q-1} u_j^2 > 0. \tag{1}$$

We will call U a *reference sequence*, subsequences-fragments coinciding with it—*reference fragments*. Let $\mathcal{M} = \{n_1, \ldots, n_M\} \subset \{0, \ldots, N-1\} = \mathcal{N}$ be the collection of numbers of elements of X corresponding to the beginnings of the reference sequences. Assume that the length of an interval between the beginnings of two successive fragments in the sequence is bounded above and below by the constants $T_{\max}$ and $T_{\min}$ and the last fragment is not cut by the bounds of the sequence:

$$q < T_{\min} \leq n_m - n_{m-1} \leq T_{\max}, \quad m = 2, \ldots, M,$$
$$n_M \leq N - q. \tag{2}$$

It is easy to see that (2) is compatible, if inequalities

$$q \leq T_{\min} \leq T_{\max} < N,$$
$$1 \leq M \leq \lfloor (N - q)/T_{\min} \rfloor, \tag{3}$$

are met for the components of a collection $\mathcal{P} = (N, q, M, T_{\min}, T_{\max})$ of positive integers. Call this collection a *collection of quasiperiodicity parameters*.

For a fixed collection $\mathcal{P}$ of quasiperiodicity parameters, define a non-empty set

$$\Omega = \Omega(\mathcal{P}) = \{(n_1, \ldots, n_M) \ : \ n_1 \geq 0,$$
$$T_{\min} \leq n_m - n_{m-1} \leq T_{\max}, \ m = 2, \ldots, M, \ n_M \leq N - q\}.$$

This set gathers all collections $\mathcal{M}$ satisfying (2)—possible collections of the initial indices for the fragments in a quasiperiodic sequence.

From the above notations and definitions, it follows that if $\mathcal{P}$ and U are given, the collection $\mathcal{M} \in \Omega(\mathcal{P})$ uniquely determines the sequence X, i.e., $X = X(\mathcal{M}) = X(\mathcal{M} \,|\, U, \mathcal{P})$, while the n-th element of the sequence can be written in the form

$$x_n = x_n(\mathcal{M}) = x_n(\mathcal{M} \,|\, U, \mathcal{P}) = \sum_{m=1}^{M} u_{n-n_m}, \quad n = 0, \ldots, N - 1. \qquad (4)$$

This sequence is said to be a *quasiperiodic sequence generated by* U. It contains M identical subsequences-fragments $(x_{n_m}, \ldots, x_{n_m+q-1})$, $m = 1, \ldots, M$, coinciding with U. These fragments are included in a sequence as a whole, don't intersect, and satisfy the quasiperiodicity restrictions on the interval between them. Unite all such sentences into a set

$$\mathcal{X} = \mathcal{X}(U, \mathcal{P}) = \big\{ X \in \mathbb{R}^N \,:\, X = X(\mathcal{M} \,|\, U, \mathcal{P}), \, \mathcal{M} \in \Omega(\mathcal{P}) \big\}$$

of all admissible quasiperiodic sequences engendered by U and $\mathcal{P}$.

If $M < \lfloor N/(q+1) \rfloor$, which is common for applications,

$$\| \mathcal{X}(U, \mathcal{P}) \| = \| \Omega(\mathcal{P}) \| > 2^M.$$

It means that if $M = \mathcal{O}(N)$, then we have an exponentially sized set $\mathcal{X}(U, \mathcal{P})$ of quasiperiodic sequences engendered by U and $\mathcal{P}$. In fact, the assumption that the number of fragments increases with increasing sequence length is quite natural.

There are two examples of quasiperiodic sequences in Fig. 1: examples of reference sequences in the left part of the figure and examples of quasiperiodic sequences generated by the corresponding reference sequence in the right part. The first sequence contains ten fragments, the second one—nine fragments.

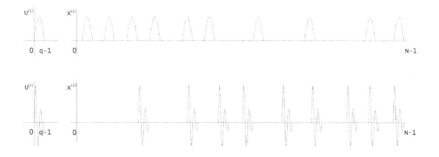

Fig. 1. Examples of quasiperiodic sequences.

Let $\mathcal{P}^{(1)} = (N, q, M_1, T_{\min}^{(1)}, T_{\max}^{(1)})$ and $\mathcal{P}^{(2)} = (N, q, M_2, T_{\min}^{(2)}, T_{\max}^{(2)})$ be the collections of quasiperiodicity parameters, $U^{(1)} = (u_0^{(1)}, \ldots, U_{q-1}^{(1)})$ and

$U^{(2)} = (u_0^{(2)}, \ldots, U_{q-1}^{(2)})$ be not coinciding reference sequences of the same length q. Assume that we observe a sequence $Y = (y_0, \ldots, y_{N-1})$, which is the element-wise sum of two unobservable quasiperiodic sequences $X^{(1)} = (x_0^{(1)}, \ldots, x_{N-1}^{(1)}) = X(\mathcal{M}^{(1)} \,|\, U^{(1)}, \mathcal{P}^{(1)}) \in \mathcal{X}(U^{(1)}, \mathcal{P}^{(1)})$, where $\mathcal{M}^{(1)} \in \Omega(\mathcal{P}^{(1)})$, and $X^{(2)} = (x_0^{(2)}, \ldots, x_{N-1}^{(2)}) = X(\mathcal{M}^{(2)} \,|\, U^{(2)}, \mathcal{P}^{(1)}) \in \mathcal{X}(U^{(2)}, \mathcal{P}^{(2)})$, where $\mathcal{M}^{(2)} \in \Omega(\mathcal{P}^{(2)})$, and some sequence $(e_0, \ldots, e_{N-1}) \in \mathbb{R}^N$, reflecting possible noise distortion:

$$y_n = x_n^{(1)} + x_n^{(2)} + e_n. \tag{5}$$

An example of the unobservable sum $X^{(1)} + X^{(2)}$ is presented in Fig. 2, here the sequences presented in Fig. 1 are taken as the summands. An example of this sequence distorted by some additive noise—an example of the observable sequence Y, is depicted in Fig. 3.

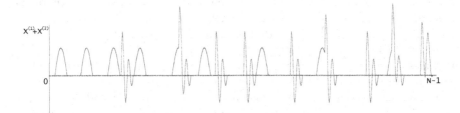

Fig. 2. An example of the sum of quasiperiodic sequences.

Fig. 3. An example of the observable sequence.

The *segregation problem for the sum of two quasiperiodic sequences* is to restore the unobservable sequences $X^{(1)}$ and $X^{(2)}$ when processing the observable sequence Y. It means we have to determine two collections $\mathcal{M}^{(1)}$ and $\mathcal{M}^{(2)}$ on the base of Y.

In the current paper, the segregation problem is stated with additional restrictions

$$T_{\min}^{(1)} \geq 2q, \quad T_{\max}^{(2)} \geq 2q, \tag{6}$$

on the components of $\mathcal{P}^{(1)}$ and $\mathcal{P}^{(2)}$. These restrictions guarantee that the fragments in the sequence are not too close to each other. So, there is not greater than one fragment from one sequence for every fragment of another sequence such that their elements are summed when calculating (5).

3   Approximation Problem and Optimization Problem

Since the sequences $X^{(1)}$, $X^{(2)}$, and Y can be considered as vectors in N-dimensional space, we can formulate the segregation problem (the problem of restoring the unobservable sequences based on their sum) in the form of the following approximation problem:

$$\|Y - (X^{(1)} + X^{(2)})\|^2 \longrightarrow \min_{X^{(1)} \in \mathcal{X}(U^{(1)}, \mathcal{P}^{(1)}), X^{(2)} \in \mathcal{X}(U^{(2)}, \mathcal{P}^{(2)})} . \qquad (7)$$

As it is mentioned above, with sufficiently natural constraints on the components of the collections of quasiperiodicity parameters, we have exponentially sized sets $\mathcal{X}(U^{(1)}, \mathcal{P}^{(1)})$ and $\mathcal{X}(U^{(2)}, \mathcal{P}^{(2)})$. Hence, taking into account that $X^{(1)}$ and $X^{(2)}$ are independent, we have exponentially sized set $\mathcal{X}(U^{(1)}, \mathcal{P}^{(1)}) \times \mathcal{X}(U^{(2)}, \mathcal{P}^{(2)})$ of admissible solutions to approximation problem (7). It is clear that brute-force searching through this set is hardly possible in a reasonable time.

Taking into account that $X^{(1)}$ and $X^{(2)}$ depends on $\mathcal{M}^{(1)}$ and $\mathcal{M}^{(2)}$ respectively, with fixed reference sequences $U^{(1)}$ and $U^{(2)}$ as well as the quasiperiodicity parameters $\mathcal{P}^{(1)}$ and $\mathcal{P}^{(2)}$, the approximation problem (7) can be rewritten in the following equivalent form:

$$\left\| Y - \left(X(\mathcal{M}^{(1)} \,|\, U^{(1)}, \mathcal{P}^{(1)}) + X(\mathcal{M}^{(2)} \,|\, U^{(2)}, \mathcal{P}^{(2)}) \right) \right\|^2 \longrightarrow \min_{\mathcal{M}^{(1)}, \mathcal{M}^{(2)}} . \qquad (8)$$

Expanding the squares of the norm in (8) and taking into account (1), (2), and (4), we obtain by simple calculations that

$$\|Y - (X^{(1)} + X^{(2)})\|^2$$

$$= \sum_{n=0}^{N-1} y_n^2 + M_1 \sum_{i=0}^{q-1} (u_i^{(1)})^2 + M_2 \sum_{j=0}^{q-1} (u_j^{(2)})^2 - 2 \sum_{m=1}^{M_1} \sum_{i=0}^{q-1} u_i^{(1)} y_{n_m^{(1)}+i}$$

$$- 2 \sum_{m=1}^{M_2} \sum_{j=0}^{q-1} u_j^{(2)} y_{n_m^{(2)}+j} + 2 \sum_{m=1}^{M_1} \sum_{s=1}^{M_2} \sum_{i=0}^{q-1} u_i^{(1)} u_{i+n_m^{(1)}-n_s^{(2)}}^{(2)} .$$

Since the numbers M_1 and M_2 of fragments in every sequence are assumed to be known, the three first terms on the right-hand side are constants. So we have the following discrete optimization problem equivalent to approximation problem (7).

Problem 1. *Given:* the numerical sequence $Y = (y_1, \ldots, y_N)$, the collections $\mathcal{P}^{(1)}$ and $\mathcal{P}^{(2)}$ of quasiperiodicity parameters satisfying (6), and two numerical sequences $U^{(1)} \in \mathbb{R}^q$ and $U^{(2)} \in \mathbb{R}^q$. *Find:* the collections $\mathcal{M}^{(1)} =$

$(n_1^{(1)}, \ldots, n_{M_1}^{(1)})$ and $\mathcal{M}^{(2)} = (n_1^{(2)}, \ldots, n_{M_2}^{(2)})$ of indices of the sequence Y that minimize the objective function

$$G(\mathcal{M}^{(1)}, \mathcal{M}^{(2)}) = -$$

$$\sum_{m=1}^{M_1} \sum_{i=0}^{q-1} y_{n_m^{(1)}+i} u_i^{(1)} - \sum_{s=1}^{M_2} \sum_{j=0}^{q-1} y_{n_s^{(2)}+j} u_j^{(2)} + \sum_{m=1}^{M_1} \sum_{s=1}^{M_2} \sum_{i=0}^{q-1} u_i^{(1)} u_{i+n_m^{(1)}-n_s^{(2)}}^{(2)},$$

under the constraints (2) on the elements of the sought collections $\mathcal{M}^{(1)}$, and $\mathcal{M}^{(2)}$.

4   Auxiliary Problem and Algorithm

In order to construct algorithm solving Problems 1, we need to consider the following auxiliary problem.

Problem 2. *Given:* the collections $\mathcal{P}^{(1)} = (N, q, M_1, T_{\min}^{(1)}, T_{\max}^{(1)})$ and $\mathcal{P}^{(2)} = (N, q, M_2, T_{\min}^{(2)}, T_{\max}^{(2)})$ of positive integers such that

$$2q \leq T_{\min}^{(i)} \leq T_{\max}^{(i)} < N, \quad i \in \{1, 2\},$$
$$1 \leq M_i \leq \lfloor (N - q)/T_{\min} \rfloor,$$

the numerical sequences $f_1(n)$, $f_2(n)$, $n = 0, \ldots, N - 1$, the numerical sequence $h(n)$, $n = 0, \pm 1, \pm 2, \ldots$, such that $h(n) = 0$, if $|n| \geq q$.

Find: the collections $\mathcal{M}^{(1)} = \{n_1^{(1)}, \ldots, n_{M_1}^{(1)}\} \subset \mathcal{N}$ and $\mathcal{M}^{(2)} = \{n_1^{(2)}, \ldots, n_{M_2}^{(2)}\} \subset \mathcal{N}$ that minimize the objective function

$$F(\mathcal{M}^{(1)}, \mathcal{M}^{(2)}) = \sum_{m=1}^{M_1} f_1(n_m^{(1)}) + \sum_{k=1}^{M_2} f_2(n_k^{(2)}) + \sum_{m=1}^{M_1} \sum_{s=1}^{M_2} h(n_m^{(1)} - n_s^{(2)}) \quad (9)$$

under the constraints (2) on the elements of the collections $\mathcal{M}^{(1)}$ and $\mathcal{M}^{(2)}$.

To write down the recurrent formulas solving Problem 2, we need the following trivial lemma, which defines the sets of admissible values for the elements of the sought collections.

Lemma 1. *Let* $\mathcal{M} = (n_1, \ldots, n_M) \in \Omega(\mathcal{P})$, *where the elements of the collection* $\mathcal{P} = (N, q, M, T_{\min}, T_{\max})$ *satisfy (3). Then*
(1) the set of possible values for n_m *is*

$$\omega_m(\mathcal{P}) = \{(m-1)T_{\min}, \ldots, N - (M-m)T_{\min}\}, \quad m = 1, \ldots, M;$$

(2) if $n_m = n$, $n \in \omega_m(\mathcal{P})$, *then the set of possible values for* n_{m-1} *is*

$$\gamma_{m-1}(n, \mathcal{P}) = \{\max\{(m-2)T_{\min}, n - T_{\max}\}, \ldots, n - T_{\min}\}, \quad m = 2, \ldots, M.$$

The following lemma provides recurrent formulas to solve Problem 2.

Lemma 2. *Let the conditions of Problem 2 hold. Then the optimal value F^* of the objective function of the problem is given by the formula*

$$F^* = \min_{n \in \omega_{M_1}^{(1)}} \min_{s \in \omega_{M_2}^{(2)}} F_{M_1, M_2}(n, s), \tag{10}$$

and the values of $F_{i,j}(n, s)$ are calculated with the recurrent formula

$$F_{i,j}(n, s) = \begin{cases} f_1(n) + f_2(s) + h(n - s), \quad i = 1, j = 1, n \in \omega_1^{(1)}, s \in \omega_1^{(2)}, \\ f_1(n) + h(n - s) + \min_{k \in \gamma_{i-1}^{(1)}(n)} F_{i-1,1}(k, s), \\ \qquad i = 2, \ldots, M_1, \ j = 1, \ n \in \omega_i^{(1)}, \ s \in \omega_1^{(2)}, \\ f_2(s) + h(n - s) + \min_{t \in \gamma_{j-1}^{(2)}(s)} F_{1,j-1}(n, t), \\ \qquad i = 1, \ j = 2, \ldots, M_2, \ n \in \omega_1^{(1)}, \ s \in \omega_j^{(2)}, \\ f_1(n) + h(n - s) + \min_{k \in \gamma_{i-1}^{(1)}(n)} F_{i-1,j}(k, s), \\ \qquad i = 2, \ldots, M_1, j = 2, \ldots, M_2, n \in \omega_i^{(1)}, s \in \omega_j^{(2)}, n \geq s, \\ f_2(s) + h(n - s) + \min_{t \in \gamma_{j-1}^{(2)}(s)} F_{i,j-1}(n, t), \\ \qquad i = 2, \ldots, M_1, j = 2, \ldots, M_2, n \in \omega_i^{(1)}, s \in \omega_j^{(2)}, n < s, \end{cases} \tag{11}$$

where $\omega_m^{(i)} = \omega_m(\mathcal{P}^{(i)})$, $m = 1, \ldots, M_i$, $i \in \{1, 2\}$, and $\gamma_{m-1}^{(i)}(n) = \gamma_{m-1}(n, \mathcal{P}^{(i)})$, $n \in \omega_m^{(i)}$, $m = 2, \ldots, M_i$, $i \in \{1, 2\}$.

In order to find the optimal collections $\mathcal{M}^{(1)}$ and $\mathcal{M}^{(2)}$ define

$$I_{i,j}^{(1)}(n, s) = \arg \min_{k \in \gamma_{i-1}^{(1)}(n)} F_{i-1,j}(k, s), \ i = 2, \ldots, M_1, \ j = 1, \ldots, M_2,$$
$$I_{i,j}^{(2)}(n, s) = \arg \min_{t \in \gamma_{j-1}^{(2)}(s)} G_{i,j-1}(n, t), \ i = 1, \ldots, M_1, \ j = 2, \ldots, M_2, \tag{12}$$

$$P_{i,j}(n, s) = \begin{cases} 1, i = 2, \ldots, M_1, j = 1, \\ 2, i = 1, j = 2, \ldots, M_2, \\ 1, i = 2, \ldots, M_1, j = 2, \ldots, M_2, n \geq s, \\ 2, i = 2, \ldots, M_1, j = 2, \ldots, M_2, n < s, \end{cases} \tag{13}$$

for all $n \in \omega_i^{(1)}$, $s \in \omega_j^{(2)}$. Also calculate the components of four auxiliary collections $(\nu_1^{(1)}, \ldots, \nu_{M_1+M_2-1}^{(1)})$, $(\nu_1^{(2)}, \ldots, \nu_{M_1+M_2-1}^{(2)})$, $(k_1^{(1)}, \ldots, k_{M_1+M_2-1}^{(1)})$, and $(k_1^{(2)}, \ldots, k_{M_1+M_2-1}^{(2)})$, of positive integers. The last components of these collections are

$$\nu_{M_1+M_2-1}^{(1)} = \arg \min_{n \in \omega_{M_1}^{(1)}} \left\{ \min_{s \in \omega_{M_2}^{(2)}} F_{M_1, M_2}(n, s) \right\},$$
$$\nu_{M_1+M_2-1}^{(2)} = \arg \min_{s \in \omega_{M_2}^{(2)}} F_{M_1, M_2}(\nu_{M_1+M_2-1}^{(1)}, s), \tag{14}$$
$$k_{M_1+M_2-1}^{(1)} = M_1, \quad k_{M_1+M_2-1}^{(2)} = M_2,$$

while the rest of components are obtained by the following recurrent formulas for $i = M_1 + M_2 - 1, \ldots, 2$:

$$
\nu_{i-1}^{(1)} = \begin{cases} I_{k_i^{(1)}, k_i^{(2)}}^{(1)}(\nu_i^{(1)}, \nu_i^{(2)}), & \text{if } P_{k_i^{(1)}, k_i^{(2)}}(\nu_i^{(1)}, \nu_i^{(2)}) = 1, \\ \nu_i^{(1)}, & \text{if } P_{k_i^{(1)}, k_i^{(m)}}(\nu_i^{(1)}, \nu_i^{(2)}) = 2, \end{cases}
$$

$$
\nu_{i-1}^{(2)} = \begin{cases} \nu_i^{(2)}, & \text{if } P_{k_i^{(1)}, k_i^{(2)}}(\nu_i^{(1)}, \nu_i^{(2)}) = 1, \\ I_{k_i^{(1)}, k_i^{(2)}}^{(2)}(\nu_i^{(1)}, \nu_i^{(2)}), & \text{if } P_{k_i^{(1)}, k_i^{(2)}}(\nu_i^{(1)}, \nu_i^{(2)}) = 2, \end{cases} \tag{15}
$$

$$
k_{i-1}^{(1)} = \begin{cases} k_i^{(1)} - 1, & \text{if } P_{k_i^{(1)}, k_i^{(2)}}(\nu_i^{(1)}, \nu_i^{(2)}) = 1, \\ k_i^{(1)}, & \text{if } P_{k_i^{(1)}, k_i^{(2)}}(\nu_i^{(1)}, \nu_i^{(2)}) = 2, \end{cases}
$$

$$
k_{i+1}^{(2)} = \begin{cases} k_i^{(2)}, & \text{if } P_{k_{i+1}^{(1)}, k_{i+1}^{(2)}}(\nu_i^{(1)}, \nu_i^{(2)}) = 1, \\ k_i^{(2)} - 1, & \text{if } P_{k_{i+1}^{(1)}, k_{i+1}^{(2)}}(\nu_i^{(1)}, \nu_i^{(2)}) = 2. \end{cases} \tag{16}
$$

Corollary 1. *The optimal solution to Problem 2 are the collections* $\mathcal{M}^{(1)} = (n_1^{(1)}, \ldots, n_{M_1}^{(1)})$ *and* $\mathcal{M}^{(2)} = (n_1^{(2)}, \ldots, n_{M_2}^{(2)})$ *obtained as*

$$
\begin{aligned}
n_m^{(1)} &= \nu_{\varkappa^{(1)}}^{(1)}, m = 1, \ldots, M_1, \\
n_m^{(2)} &= \nu_{\varkappa^{(2)}}^{(2)}, m = 1, \ldots, M_2,
\end{aligned} \tag{17}
$$

where

$$
\varkappa_m^{(i)} = \min\{j \in \{1, \ldots, M_i\} : k_j^{(i)} = m\}, \; m = 1, \ldots, M_i, \; i \in \{1, 2\}. \tag{18}
$$

The proofs of Lemma 2 and Corollary 1 are based on the direct derivation of the recurrent formulas. The finiteness of the sequence h(n) together with restrictions (6) plays a key role when constructing formulas. Thanks to this, it is possible to avoid double summation in the right part of (9).

Based on the considered auxiliary problem and formulas for its solution, we write down the following algorithm.

Algorithm $\mathcal{A}$.

INPUT: collections $\mathcal{P}^{(1)}$ and $\mathcal{P}^{(2)}$ of quasiperiodicity parameters, numerical sequences $U^{(1)}$, $U^{(2)}$, Y.

Forward pass.

STEP 1. For every $n = 0, \ldots, N - q$, put

$$
f_1(n) = -\sum_{s=0}^{q-1} u_s^{(1)} y_{n+s}, \quad f_2(n) = -\sum_{s=0}^{q-1} u_s^{(2)} y_{n+s}.
$$

For every $n = 0, \pm 1, \pm 2, \ldots$, put

$$
h(n) = \begin{cases} \sum_{s=0}^{q-1} u_s^{(1)} u_{s+n}^{(2)}, & |n| < q, \\ 0, & |n| \geq q. \end{cases}
$$

STEP 2. Calculate $F_{i,j}(n, s)$, $i = 1, \ldots, M_1$, $j = 1, \ldots, M_2$, $n \in \omega_i^{(1)}$, $s \in \omega_j^{(2)}$ using (11) and F^* using (10). Calculate $I_{i,j}^{(1)}(n, s)$, $i = 2, \ldots, M_1$, $j = 1, \ldots, M_2$, $I_{i,j}^{(2)}(n, s)$, $i = 1, \ldots, M_1$, $j = 2, \ldots, M_2$, and $P_{i,j}(n, s)$, $i = 1, \ldots, M_1$, $j = 1, \ldots, M_2$ (except $i = j = 1$), for all $n \in \omega_i^{(1)}$ and $s \in \omega_j^{(2)}$ by (12) and (13). Put $G_A = F^*$.

Backward pass.

STEP 3. Find auxiliary collections $(\nu_1^{(1)}, \ldots, \nu_{M_1+M_2-1}^{(1)})$, $(\nu_1^{(2)}, \ldots, \nu_{M_1+M_2-1}^{(2)})$, $(k_1^{(1)}, \ldots, k_{M_1+M_2-1}^{(1)})$, and $(k_1^{(2)}, \ldots, k_{M_1+M_2-1}^{(2)})$ by (14), (15), and (16). Calculate $\mathcal{M}^{(1)}$ and $\mathcal{M}^{(2)}$ by (17), (18). Put $\mathcal{M}_A^{(1)} = \mathcal{M}^{(1)}$, $\mathcal{M}_A^{(2)} = \mathcal{M}^{(2)}$.

OUTPUT: The collections $\mathcal{M}_A^{(1)}$, $\mathcal{M}_A^{(1)}$, and the value G_A.

The main result of the research is the following theorem.

Theorem 1. *Algorithm $\mathcal{A}$ finds an exact solution to Problem 1 in time $\mathcal{O}(M_1 M_2 T_{\max} N^2)$.*

The optimality of the obtained solution follows from Lemma 1. The description presented allows to estimate time-complexity of Algorithm $\mathcal{A}$.

5   Numerical Simulation

Theorem 1 proves the optimality of the solution to Problem 1 obtained by Algorithm $\mathcal{A}$, so the numerical simulation results are presented for illustration only. There are two examples of processing modeled noisy sum of two quasiperiodic sequences. In each of the examples, for every quasiperiodic sequence signals having abstract geometrical form (such as decaying sinusoids, the half-period of sinusoids, steps, and so on) are chosen as reference sequences in order to illustrate the potential applicability of the algorithm as one of the steps to constructing noise-resistant algorithms for processing signals of various nature without focusing on specific possible applications.

Figure 4 shows the input data of Algorithm $\mathcal{A}$: two reference sequences $U^{(1)}$ and $U^{(2)}$ at the top, and the sequence Y to be processed—at the bottom.

Figure 5 presents two unobservable quasiperiodic sequences generated by reference sequences $U^{(1)}$ and $U^{(2)}$ (the first and the second rows) and their unobservable sum (the third row). Recall that the observable sequence Y is the element-wise sum of these two sequences and a sequence of independent identically distributed Gaussian random variables (white noise).

At last, Fig. 6 presents the results of Algorithm $\mathcal{A}$ operation. There are two quasiperiodic sequences recovered by the rule (4) using collections $\mathcal{M}_A^{(1)}$ and $\mathcal{M}_A^{(2)}$ obtained as Algorithm $\mathcal{A}$ output. For clarity, the frames on the graphs mark the locations of fragments in an unobservable sequence, which allows you to visually assess the quality of the algorithm.

This example is computed for $N = 600$, $q = 20$, $M_1 = 8$, $M_2 = 9$, $T_{\min}^{(1)} = T_{\min}^{(2)} = 40$, $T_{\max}^{(1)} = T_{\max}^{(2)} = 135$, maximum amplitude value is 137, the noise level $\sigma = 90$.

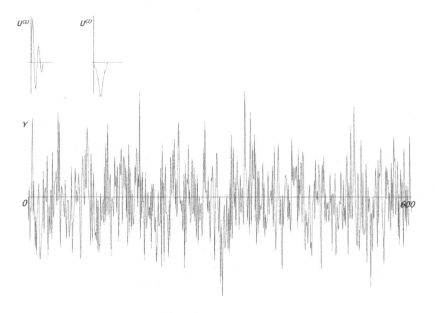

Fig. 4. Example 1. Input data.

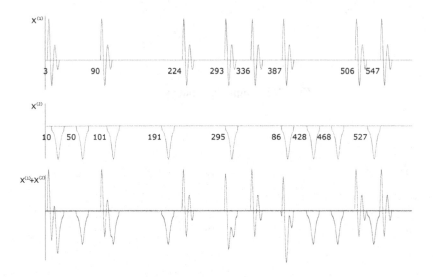

Fig. 5. Example 1. Unobservable data.

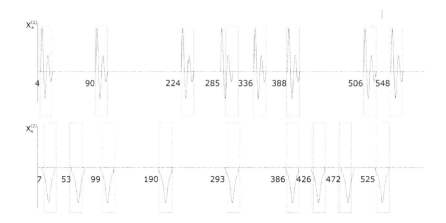

Fig. 6. Example 1. Results of processing.

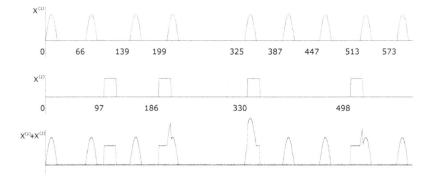

Fig. 7. Example 2. Input data.

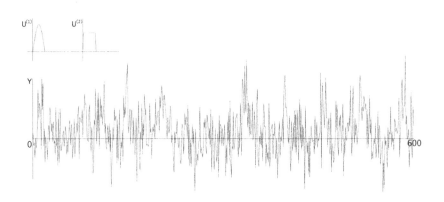

Fig. 8. Example 2. Unobservable data.

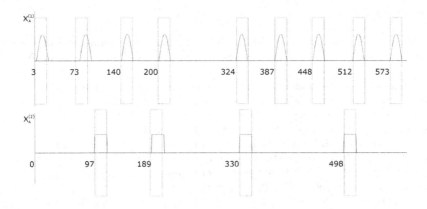

Fig. 9. Example 2. Results of processing.

Similarly, Fig. 7, 8, and 9 give another example of the algorithm operation. This example is computed for $N = 600$, $q = 20$, $M_1 = 9$, $M_2 = 4$, $T_{\min}^{(1)} = 60$, $T_{\max}^{(1)} = 300$, $T_{\min}^{(2)} = 40$, $T_{\max}^{(2)} = 200$, maximum amplitude value is 86, the noise level $\sigma = 60$.

The numerical simulation results clearly confirm the optimality of the solution found by the algorithm. In such a way, the algorithm suggested is a suitable tool for noise-resistant processing of data in the form of the sum of two quasiperiodic sequences. The visual comparison of the two pairs of graphs (of the unobservable sequences $X^{(1)}$ and $X^{(2)}$ with the recovered sequences $X_{\mathcal{A}}^{(1)}$ and $X_{\mathcal{A}}^{(2)}$) demonstrates only insignificant differences.

6 Conclusion

In this paper, the algorithm that finds the optimal solution to one previously unexplored discrete optimization problem has been constructed and justified. This optimization problem arises (in the framework of the a posteriori approach) when solving the applied problem of segregation of two quasiperiodic sequences on the base of their noisy sum. The algorithm obtained is proved to be polynomial-time. There are some examples of numerical simulation in the paper that illustrates the algorithm processing.

A modification of Problem 1, in which the number of subsequences-fragments in one or both sequences is unknown, remains to be studied. The problems of joint separation and recognition are also of interest. The nearest plans are connected with exploring these problems.

References

1. Kel'manov, A., Khamidullin, S.: Posterior detection of a given number of identical subsequences in a quasi-periodic sequence. Comput. Math. Math. Phys. **41**, 762–774 (2001)
2. Kel'manov, A.V., Khamidullin, S.A.: Simultaneous a posteriori detection and identification of a predetermined number of quasi-periodic fragments in a sequence based on their segments. Pattern Recogn. Image Anal. **16**(3), 344–357 (2006)
3. Kel'manov, A., Khamidullin, S., Mikhailova, L., Ruzankin, P.: Polynomial-time solvability of one optimization problem induced by processing and analyzing quasiperiodic ECG and PPG signals. In: Jaćimović, M., Khachay, M., Malkova, V., Posypkin, M. (eds.) OPTIMA 2019. CCIS, vol. 1145, pp. 88–101. Springer, Cham (2020). https://doi.org/10.1007/978-3-030-38603-0_7
4. Comon, P., Jutten, C. (eds.): Handbook of Blind Source Separation. Independent Component Analysis and Applications. Academic Press, Cambridge (2010)
5. Chien, J.-T.: Source Separation and Machine Learning. Academic Press, Cambridge (2019)
6. Schobben, D.W.E.: Blind signal separation, an overview. In: Schobben, D.W.E. (ed.) Real-time Adaptive Concepts in Acoustics, pp. 75–86. Springer, Dordrecht (2001). https://doi.org/10.1007/978-94-010-0812-9_6
7. Cardoso, J.-F.: Blind signal separation: statistical principles. In: Proceedings of the IEEE, vol. 86, no. 10, pp. 2009–2025 (1998). https://doi.org/10.1109/5.720250
8. Bach, F.R., Jordan, M.I.: Blind one-microphone speech separation: a spectral learning approach. In: Saul, L.K., Weiss, Y., Bottou, L. (eds.) Advances in Neural Information Processing Systems, vol. 17, pp. 65–72. MIT Press, Cambridge (2005)
9. Roweis, S.: One microphone source separation. In: Proceedings of the 13th International Conference on Neural Information Processing System, pp. 763–769. MIT Press, Cambridge (2001)

Parameter Estimation via Time Modeling for MLIR Implementation of GEMM

Alexey Romanov🅸🅳, Andrei Turkin🅸🅳, Oleg Myakinin🅸🅳, Fiodar Tsupko🅸🅳, and Jiexing Gao$^{(\boxtimes)}$🅸🅳

Huawei Technologies, Central Research Institute, 2012 Labs, Shenzhen, China
{romanov.alexey2,andrei.turkin,miakinin.oleg1,
tsupko.fiodar,gaojiexing}@huawei.com

Abstract. We consider the problem of identifying optimal parameters for two implementations of the general matrix multiplication (GEMM). Optimal parameters are chosen based on time modeling for two GEMM implementations, which is done by analyzing the structure of each implementation and the characteristics of the hardware. Each implementation has specific packing strategies that influence data movement and time of data access. The data movement, as well as constraints for the registers and each level of the two-level cache, is considered to ensure proper data usage. Based on the proposed models, an exhaustive search procedure for microkernel and tiling parameters was used to obtain the best parameters for each of the considered implementations of GEMM for multi-level intermediate representation (MLIR). The results show that the performance of MLIR-based code generation for these GEMM implementations, when different matrix sizes are used, is comparable with the performance that can be obtained for Basic Linear Algebra Subprograms (BLAS).

Keywords: Computational optimization · Code generation · GEMM · MLIR

Introduction

High-performance target-specific code generation is the essential part of the intermediate representation, for instance, when multi-level intermediate representations (MLIRs) [4] are used to progressively decrease the computation of data flow graphs [1]. Matrix-matrix multiplication is a fundamental operator in deep learning that has a significant performance impact on training and executing deep learning models [8]; thus, its optimization is a high-priority task. As the authors of [1] note, an out-of-the box MLIR shows 25% of the peak machine performance, while oneMKL[1] and OpenBLAS[2] show nearly 92% of the peak

---

[1] https://software.intel.com/content/www/us/en/develop/tools/oneapi/components/onemkl.html.

[2] https://www.openblas.net/.

N. Olenev et al. (Eds.): OPTIMA 2023, LNCS 14395, pp. 159–173, 2023.
https://doi.org/10.1007/978-3-031-47859-8_12

performance, and BLIS[3] shows almost 85% of the performance, which makes it challenging to obtain high-performance matrix multiplication for MLIRs. The authors of [1] show how to achieve a near-MKL performance, although how to obtain the best performance for any matrix size is an important question, which the paper is focused on, while using the model-based approach as it is in [6].

This paper shows how to model the execution time to identify the optimal parameters for matrix multiplication in MLIR based on hardware-specific constraints. It is organized as follows. In Sect. 1, we introduce a model that describes computations as they are presented in GotoBLAS [2,3] and are further refactored in BLIS [3,7]. In Sect. 4, we discuss time modeling for two implementations matrix multiplication and define the constraints for each cache level to state the optimization problem formally. Section 5 speculates on the results of modeling when an x86 architecture is considered and its parameters are used to explicitly define the aforementioned constraints. Section 6 concludes the paper.

1   Modeling Calculation Time for GEMM

This section introduces two models for matrix multiplication when three cache levels are used. Each of these models depends on 1) the structure of the implementation $\mathcal{I}$, 2) the architecture parameters $\mathcal{A}$, and 3) the algorithm parameters $\mathcal{P}$.

As in [3], we considered the GotoBLAS [2] implementation structure with two different approaches to packing: Implementation $\mathcal{I}_1$ of the form that was described in [7] with contiguous packing, which the authors of [5] used to calculate its parameters (see Listing 1, $\mathcal{I} = \mathcal{I}_1$), and Implementation $\mathcal{I}_2$, which has a modified form to represent computations with one additional packing and no contiguous placement (see Listing 1, $\mathcal{I} = \mathcal{I}_2$):

$$\mathcal{T}_{\mathcal{I}_1}(\mathcal{A}, \mathcal{P}) = \sum_{I_{B_C}(\mathcal{P})} \left(T_{\mathcal{A}}(B_C) + \sum_{I_{A_C}(\mathcal{P})} \left(T_{\mathcal{A}}^{\mathcal{I}_1}(A_C) + \sum_{I_{B_p}(\mathcal{P})} S_{\mathcal{A}}^{\mathcal{I}_1}(c_r) \right) \right), \quad (1)$$

$$\mathcal{T}_{\mathcal{I}_2}(\mathcal{A}, \mathcal{P}) = \sum_{I_{B_C}(\mathcal{P})} \left(T_{\mathcal{A}}(B_C) + \sum_{I_{A_C}(\mathcal{P})} \left(T_{\mathcal{A}}^{\mathcal{I}_2}(A_C) + \sum_{I_{B_p}(\mathcal{P})} S_{\mathcal{A}}^{\mathcal{I}_2}(B_p, c_r) \right) \right), \quad (2)$$

where $S_{\mathcal{A}}^{\mathcal{I}_1}(c_r) = \sum_{l \in I_{A_p}(\mathcal{P})} T_{\mathcal{A}}^{\mathcal{I}_1}(c_r)$, $S_{\mathcal{A}}^{\mathcal{I}_2}(B_p, c_r) = T_{\mathcal{A}}^{\mathcal{I}_2}(B_p) + \sum_{l \in I_{A_p}(\mathcal{P})} T_{\mathcal{A}}^{\mathcal{I}_2}(c_r)$, $I_{B_C}(\mathcal{P})$, $I_{A_C}(\mathcal{P})$, and $I_{B_p}(\mathcal{P})$ are index sets, $T(\cdot)$ is the time estimate for a specific operation with corresponding data.

The function $T_{\mathcal{A}}^{\bullet}(A_C)$ estimates the time spent packing a block of matrix A, which is denoted as A_C. The size of the block is $m_c \times k_c$ and it is packed as a set of horizontal micropanels, each of which is A_r that has the size of $m_r \times k_c$. Data within each micropanel is arranged in column-major order, in which each column is a_r.

---

[3] https://github.com/flame/blis.

```
Input  : A, B, n_c, m_c, k_c, n_r, m_r, I
Output: C
1  for j_c ← 1 to n − 1 in steps n_c do
2  |   for p_c ← 1 to k − 1 in steps k_c do
3  |   |   Packing: B(p_c : p_c + k_c − 1, j_c : j_c + n_c − 1) → B_C
4  |   |   for i_c ← 1 to m − 1 in steps m_c do
5  |   |   |   Packing: A(i_c : i_c + m_c − 1, p_c : p_c + k_c − 1) → A_C
6  |   |   |   for j_r ← 1 to n_c − 1 in steps n_r do
7  |   |   |   |   if I = I_2 then
8  |   |   |   |   |   Packing: B(j_r : j_r + p_c − 1, p_c : p_c + k_c − 1) → B_r
9  |   |   |   |   end
10 |   |   |   |   for i_r ← 1 to m_c − 1 in steps m_r do
11 |   |   |   |   |   for p_r ← 1 to k_c − 1 in steps 1 do
12 |   |   |   |   |   |   C (i_r : i_r + m_r − 1, j_r : j_r + n_r − 1) + =
13 |   |   |   |   |   |   A (i_r : i_r + m_r − 1, p_r) · B(p_r, j_r : j_r + n_r − 1)
14 |   |   |   |   |   end
15 |   |   |   |   end
16 |   |   |   end
17 |   |   end
18 |   end
19 end
```

Algorithm 1: GEMM Implementations $\mathcal{I}_1, \mathcal{I}_2$.

The function $T_{\mathcal{A}}(B_C)$ estimates the time spent packing a block of matrix B, which is denoted as B_C and has the following size: $k_c \times n_c$. This block is represented as a set of vertical micropanels B_r, each of which has the size of $k_c \times n_r$. The data within each micropanel is arranged in row-major order, in which each row is b_r.

As an additional component of the model for $\mathcal{I}_2$, $T_{\mathcal{A}}^{\mathcal{I}_2}(B_r)$ estimates the time spent copying the corresponding micropanel of B_C. Each a_r and b_r are used to calculate c_r, which takes $T_{\mathcal{A}}(c_r)$. The time is spent on (1) accessing a_r, b_r, c_r, (2) calculating the result, and (3) storing c_r.

Because the number of A_C and B_C blocks, as well as the number of low-level operations, which involves the aforementioned use of a_r, b_r c_r, depend on the algorithm parameters, one can write the formula as:

$$\mathcal{T}_{\mathcal{I}_1}(\mathcal{A}, \mathcal{P}) = k_{B_C}^{\mathcal{P}} \left(T_{\mathcal{A}}(B_C) + k_{A_C}^{\mathcal{P}} \left(T_{\mathcal{A}}^{\mathcal{I}_1}(A_C) + k_{B_p}^{\mathcal{P}} k_{A_p}^{\mathcal{P}} T_{\mathcal{A}}^{\mathcal{I}_1}(c_r) \right) \right), \qquad (3)$$

and

$$\mathcal{T}_{\mathcal{I}_2}(\mathcal{A}, \mathcal{P}) = k_{B_C}^{\mathcal{P}} \left(T_{\mathcal{A}}(B_C) + k_{A_C}^{\mathcal{P}} \left(T_{\mathcal{A}}^{\mathcal{I}_2}(A_C) + k_{B_p}^{\mathcal{P}} \left(T_{\mathcal{A}}^{\mathcal{I}_2}(B_p) + k_{A_p}^{\mathcal{P}} T_{\mathcal{A}}^{\mathcal{I}_2}(c_r) \right) \right) \right), \quad (4)$$

where coefficients $k_{B_C}^{\mathcal{P}} = |I_{B_C}(\mathcal{P})|$, $k_{A_C}^{\mathcal{P}} = |I_{A_C}(\mathcal{P})|$, $k_{A_p}^{\mathcal{P}} = |I_{A_p}(\mathcal{P})|$, and $k_{B_p}^{\mathcal{P}} = |I_{B_p}(\mathcal{P})|$.

Thus, in order to estimate calculation time, these coefficients must be specified as well as time estimates for block packing, i.e. $T_{\mathcal{A}}^{\bullet}(A_C)$, $T_{\mathcal{A}}(B_C)$, $T_{\mathcal{A}}^{\bullet}(B_p)$ as well as for calculating c_r, which is $T_{\mathcal{A}}^{\bullet}(c_r)$.

Based on the structure of the aforementioned implementations $\mathcal{I}_1$ and $\mathcal{I}_2$, the aforementioned coefficients of a time model are specified as follows:

$$k_{B_C}^{\mathcal{P}} = \left\lceil \frac{k}{k_c} \right\rceil ; \ k_{A_C}^{\mathcal{P}} = \left\lceil \frac{m}{m_c} \right\rceil ; \ k_{B_p}^{\mathcal{P}} = \left\lceil \frac{n_c}{n_r} \right\rceil ; \ k_{A_p}^{\mathcal{P}} = \left\lceil \frac{m_c}{m_r} \right\rceil .$$

In the above notation, the optimization problem for finding the optimal algorithm parameters $\mathcal{P}$ for the implementations $\mathcal{I}_1, \mathcal{I}_2$ can be formulated as follows:

$$T_{\mathcal{I}}(\mathcal{A}, \mathcal{P}_{opt}) = \min_{\mathcal{P}} T_{\mathcal{I}}(\mathcal{A}, \mathcal{P}), \ \mathcal{I} \in \{\mathcal{I}_1, \mathcal{I}_2\}.$$

Before estimating the packing and calculation times, let us assume that the number of words in a vectorized part of data, $\mathfrak{f}_o$, is equal to or less than the number of words in one cache line, which is denoted as $\mathfrak{f}_c$, i.e., $\mathfrak{f}_o \le \mathfrak{f}_c$. The number of vectors in one cache line is denoted as $\mathfrak{c}$, so the previously mentioned assumption can be written as follows: any cache line contains one or more vectors, i.e., $\mathfrak{c} \ge 1$.

2   Time Modeling for Implementation $\mathcal{I}_1$

An expression that assesses the time spent packing A_C can be written based on two terms: one for the data movement from memory to the registers and one for storing the packed A_C,

$$T_{\mathcal{A}}^{\mathcal{I}_1}(A_C) = T_{\mathcal{A},l}^{\mathcal{I}_1}(A_C) + T_{\mathcal{A},s}^{\mathcal{I}_1}(A_C), \tag{5}$$

where $T_{\mathcal{A},l}^{\mathcal{I}_1}(A_C)$ denotes the time estimate for loading, and $T_{\mathcal{A},s}^{\mathcal{I}_1}(A_C)$ denotes the time estimate for storing the packed A_C.

Because it is necessary to pack A_C so its elements are reordered, there are two versions of this function. This section discusses the version related to implementation $\mathcal{I}_1$.

Let us extract a contiguous part of A_p that has the size of $m_r \times \mathfrak{f}_c$, where $\mathfrak{f}_c$ is the size of the cache line in words, and is denoted as A_{box}. Each access to elements $A_{box}^1 = (A_{box}^{(1,1)}, \ldots, A_{box}^{(m_r,1)})$ of the first column of A_{box} produces cache misses in all the cache levels because the data is at the memory level, so $m_r t_{M0}$ is a time estimate for the column, i.e.: $T_a(A_{box}^1) = m_r t_{M0}$, where t_{M0} denotes the time that is spent on moving a data element from memory to the registers.

To calculate the time spent accessing the reminder of A_{box}, the following should be noted: for each access to cache data element $A_{box}^{(i,1)}$, $i \in \overline{1, m_r}$ assumes that all the neighbors in L_1 that are in the same cache line, i.e., $A_{box}^{(i,2)}, \ldots, A_{box}^{(i,\mathfrak{f}_c)}$, are accessed.

Considering the assumption that there is enough space for A_{box} in L_1, it is possible to calculate the access time for the reminder of A_{box}: $A_{box}^2, \ldots, A_{box}^{\mathfrak{f}_c}$ and add it to the previously calculated time for the first column of A_{box} to obtain the following estimate: $T_a(A_{box}) = T_a(A_{box}^1) + \sum_{j \in \overline{2,f_c}} T_a(A_{box}^j) = m_r t_{M0} + m_r t_{10} (f_c - 1)$.

The equation for $T_a(A_{box})$ enables calculation of the time that is necessary to spend on accessing the whole A_p: $T_a(A_p) = \left\lceil \dfrac{k_c}{f_c} \right\rceil T_a(A_{box})$, which leads to a time estimation for A_C: $T_a(A_C) = k_{A_p}^P T_a(A_p) = \left\lceil \dfrac{m_c}{m_r} \right\rceil T_a(A_p)$.

The yielding estimate can be written as:

$$T_a(A_C) = m_r \left\lceil \frac{m_c}{m_r} \right\rceil \left\lceil \frac{k_c}{f_c} \right\rceil (t_{M0} + t_{10} (f_c - 1)).$$

As in the previous section, where access time is modeled, let us assess the time spent storing A_{box} is written as follows: $T_s(A_{box}) = m_r \eth_v t_s$, where $\eth_v = \left\lceil \dfrac{c}{v_s} \right\rceil$, v_s denotes the number of vectors that can be copied back from registers and t_s is the time that must be spent on that cycle.

The previously found estimation for A_{box} enables us to assess the time for A_p, which can be written as follows: $T_s(A_r) = \left\lceil \dfrac{k_c}{f_c} \right\rceil T_s(A_{box})$. This estimation enables us to assess time for the whole A_C when it is copied back: $T_s(A_C) = k_{A_p}(P)T_s(A_r)$, which gives the following result

$$T_s(A_C) = m_r \left\lceil \frac{m_c}{m_r} \right\rceil \left\lceil \frac{k_c}{f_c} \right\rceil \left\lceil \frac{c}{v_s} \right\rceil t_s.$$

Packing for B_C is done by loading its elements from memory to the registers and copying them to compose B_C, so an expression for the time estimate can be written based on these terms as follows:

$$T_A(B_C) = T_{A,l}(B_C) + T_{A,s}(B_C),$$

where $T_{A,l}(B_C)$ denotes the time estimate for loading, and $T_{A,s}(B_C)$ denotes the time estimate to store the packed B_C.

Because there is no discrepancy between packing procedures for B_C when Implementation $\mathcal{I}_1$ and $\mathcal{I}_2$ are considered the estimate is valid for both. Indeed, packing B_C is done in the order of the initial placement, i.e., in row-major order, which is preserved during the loading and storing cycles when either $\mathcal{I}_1$ or $\mathcal{I}_2$ is considered.

Before modeling loading time for B_C, let us assume that the first access to any row of Br is limited by the maximum number of vectors, r_o, which can be delivered to registers during the first cycle of accessing data from the cache lines. If something is left in a cache line, it is accessed in later access-store cycles; therefore, the estimate for the time spent on this process is written as follows: $T_a(b_r^1) = t_{M0} + (\eth_r - 1) t_{10}$, where b_r^1 is the first part of b_r, which has the length of the cache line, $\eth_r = \left\lceil \dfrac{c}{r_o} \right\rceil$ is a limitation of data movement from L_1 to registers, which represents the number of access-store cycles to deliver cache line to the registers, $\mathfrak{c}$ denotes the number of vectors that can be placed in one cache line and $\mathfrak{r}_o$ is the number of vectors that can be accessed and stored in registers at once, i.e., in one cycle.

This estimate enables us to assess the access time for b_r as follows: $T_a(b_r) = \left\lceil \frac{n_r}{\mathfrak{f}_\mathfrak{c}} \right\rceil T_a(b_r^1)$, which is used to estimate the access time for a micro-panel B_r: $T_a(B_r) = k_c T_a(b_r)$.

The later estimate enables calculation of the access time for the whole B_C as follows: $T_a(B_C) = k_{B_r}(\mathcal{P})T_a(B_r) = \left\lceil \frac{n_c}{n_r} \right\rceil T_a(B_r)$, which gives the following result:

$$T_a(B_C) = k_c \left\lceil \frac{n_c}{n_r} \right\rceil \left\lceil \frac{n_r}{\mathfrak{f}_\mathfrak{c}} \right\rceil \left(t_{MO} + \left(\left\lceil \frac{\mathfrak{c}}{\mathfrak{r}_o} \right\rceil - 1 \right) t_{10} \right). \tag{6}$$

Modeling time for storing B_C can be estimated as follows. If $\mathfrak{f}_o$ is the number of words in a vector to be copied back, then time estimate for b_r is written as follows: $T_s(b_r) = \left\lceil \frac{n_r}{\mathfrak{f}_o \mathfrak{v}_s} \right\rceil t_s$, which enables us to write a time estimate for a micro-panel B_r: $T_s(B_r) = k_c T_s(b_r)$.

Thus, the time estimated for the whole block B_C is written as follows: $T_s(B_C) = \left\lceil \frac{n_c}{n_r} \right\rceil T_s(B_r)$, which gives the following result:

$$T_s(B_C) = k_c \left\lceil \frac{n_c}{n_r} \right\rceil \left\lceil \frac{n_r}{\mathfrak{f}_o \mathfrak{v}_s} \right\rceil t_s. \tag{7}$$

2.1   Modeling Low-Level Calculations

The time spent on microkernel calculations of GEMM includes the following components:

- T_m^A – time that is spent accessing A_r every time it is needed for calculations
- T_m^B – time that is spent accessing B_r every time it is needed for calculations
- $T_m^{C_L}$ – time that is spent accessing c_r every time it is needed for calculations
- $T_m^{C_S}$ – time that is spent on saving c_r every time the result is obtained
- T_m^P – time spent on main calculations,

therefore, micro-kernel calculations take

$$T_0(a_r, b_r, c_r; \mathcal{A}) = \sum_{u \in \{A,B,C_L,C_S,P\}} T_m^u.$$

Let us calculate the high-level common components $T_m^{C_L}$, $T_m^{C_S}$, T_m^P.

Loading Time for c_r. A part of the yielding matrix c_r has the size $m_r \times n_r$, in words, which can be written in vectors as follows $m_r \times \overline{n}_r$, where $\overline{n}_r$ satisfies $n_r = \mathfrak{f}_0 \overline{n}_r$. Each row of c_r can be represented as a number of lines C_l, each of which has the size of a cache line in words, i.e. $\mathfrak{f}_\mathfrak{c} = \mathfrak{c}\mathfrak{f}_o$. Thus, the following characteristics of vectorized c_r can be calculated:

- The number of cache lines in one vectorized row, which is calculated as $\Lambda(\overline{n_r}) = \left\lfloor \frac{\overline{n_r}}{c} \right\rfloor$,
- The indicator of an incomplete line, which can written as $I_l(\overline{n_r}) = \left\lceil \frac{\overline{n_r}}{c} \right\rceil - \Lambda(\overline{n_r})$,
- The number of vectors in the last incomplete line is $\overline{n_r} - \Lambda(\overline{n_r})c$.

Let us estimate the time required to load the elements of a line C_l. If c_r is in memory, access to the first element, C_l^0 of line C_l must produce cache misses at all cache levels, so this access loads the whole line C_l to L_1 and $\mathfrak{r}_0$ vectors to the registers, i.e., $T_m(C_l^0) = t_{M0}$.

The reminder of line C_l, i.e., $c - \mathfrak{r}_0$ vectors, is already in L_1 and its loading time is estimated as $\left(\left\lceil \frac{c}{\mathfrak{r}_0} \right\rceil - 1 \right) t_{10}$, which enables the calculation of the time estimate for the whole line as $T_m(C_l) = t_{M0} + \left(\left\lceil \frac{c}{\mathfrak{r}_0} \right\rceil - 1 \right) t_{10}$.

The summation of all complete lines gives the following: $T_m(C_l^c) = \Lambda(\overline{n_r})T_m(C_l)$. The incomplete line time estimate is written as $T_m(C_l^i) = t_{M0} + \left(\left\lceil \frac{\overline{n_r} - \Lambda(\overline{n_r})c}{\mathfrak{r}_0} \right\rceil - 1 \right) t_{10}$. Thus, the time estimate for one row of c_r is estimated as follows:

$$T_m(c_r^i) = T_m(C_l^c) + I_l(\overline{n_r})T_m(C_l^i),$$

which gives the total time required for loading the whole c_r:

$$T_m^{C_L} = m_r T_m(c_r^i). \tag{8}$$

Storing Time for c_r. When storing c_r we consider the time spent to load data from registers to L_1 only, i.e. the time spent storing these data to other cache levels or memory is ignored. Thus, the storing time is estimated as follows:

$$T_m^{C_S} = \left\lceil \frac{m_r \overline{n_r}}{\mathfrak{v}_s} \right\rceil t_s, \tag{9}$$

where $\mathfrak{v}_s$ is the limit of the vectorized data movement from the registers to L_1.

Processing Time for c_r. Processing time is limited by another architectural parameter $\mathfrak{v}_o$, which is number of vectors that can be processed simultaneously; therefore, the processing time for c_r is written as follows:

$$T_m^P = k_c \left\lceil \frac{m_r \overline{n_r}}{\mathfrak{v}_0} \right\rceil t_0. \tag{10}$$

Loading Time for a_r. Let us extract a number of parts, A_l^u, $u \in \overline{1, \Lambda_{A_p}}$, from A_p, each of which has a size of the cache line. Their number is calculated as $\Lambda_{A_p} = \left\lceil \frac{m_r k_c}{f_c} \right\rceil$. It should be noted that when the first element of A_l^u, which is a contiguous part of A_p the length of which is the cache line size, is accessed, the reminder of it is delivered to L_1 immediately. Moreover, when processing one micro-panel C_r, the number of loads cycles corresponding to L_2 is Λ_{A_r}, and the total number of accesses to A_p is $k_c m_r$, so the number of accesses to L_1 is $(k_c m_r - \Lambda_{A_p})$.

Time estimate for this process is:

$$T_m^A = T_m(A_p) = \Lambda_{A_p} t_{20} + (k_c m_r - \Lambda_{A_p}) t_{10}. \tag{11}$$

Loading Time for b_r. As there is no special packing for B_p, it is necessary to estimate the time T_m^B of access to B_p, which is done in two steps, as follows: (1) assess the time for the first access to B_p, which is assumed to be in L_3 as well as the whole B_C, from the microkernel and (2) assess the time for later accesses, the number of which is calculated as $(k_{A_p}^{\mathcal{P}} - 1)$ by assuming that B_p is in L_1. Let us extract a number of parts, B_l^u, $u \in \overline{1, \Lambda_{B_p}}$ from B_p, each of which has the same cache line size. Their number is calculated as $\Lambda_{B_p} = \left\lceil \frac{\overline{n_r} k_c}{c} \right\rceil$.

Let us assume that B is already vectorized, so the first access to B_p delivers exactly c vectors from L_3 to L_1, and the exact vector of B_l is loaded to the registers.

Thus, it is necessary to spend t_{30} to get the first vector of B_l^u to the registers. As the number of parts is Λ_{B_p}, the time estimate for delivering all the first vectors of B_l^u, $u \in \overline{1, \Lambda_{B_p} r}$ is $\Lambda_{B_p} t_{30}$. The reminder of B_l^u is loaded vector-by-vector to make it possible to copy them from L_1 to the registers, which takes t_{10} for each. The total number of vectors in B_p is $\overline{n_r} k_c$; after the first parts are delivered to the registers, the remainder is $(\overline{n_r} k_c - \Lambda_{B_p})$. Thus, the time estimate for B_p when one panel of c_r is being processed is:

$$T_m^{B_1} = T_m^1(B_p) = \Lambda_{B_p} t_{30} + (\overline{n_r} k_c - \Lambda_{B_p}) t_{10}. \tag{12}$$

If it is assumed that all data of the B_p are in cache L_1, it is possible to write the following estimate for the remaining calculations:

$$T_m^{B_2} = T_m^2(B_p) = (\overline{n_r} k_c) t_{10}. \tag{13}$$

Time Estimate for Micro-kernel. As there are two components of T_m^B, it is necessary to split $T_{\mathcal{A}}^{\mathcal{I}_1}(c_r)$ as follows:

$$T_{\mathcal{A}}^{\mathcal{I}_1,v}(c_r) = \sum_{u \in \{A, B_v, C_L, C_S, \mathcal{P}\}} T_m^u, v \in \overline{1, 2}.$$

Thus, the time estimate for the microkernel is written as follows:

$$\hat{T}_{\mathcal{A}}^{\mathcal{I}_1}(c_r) = T_{\mathcal{A}}^{\mathcal{I}_1,1}(c_r) + (k_{A_p}(\mathcal{P}) - 1) T_{\mathcal{A}}^{\mathcal{I}_1,2}(c_r),$$

and Eq. (3) for the time model is rewritten as follows:

$$T_{\mathcal{I}}(\mathcal{A}, \mathcal{P}) = k_{B_C}^{\mathcal{P}} \left(T_{\mathcal{A}}(B_C) + k_{A_C}^{\mathcal{P}} \left(T_{\mathcal{A}}^{\mathcal{I}_1}(A_C) + k_{B_p}^{\mathcal{P}} \hat{T}_{\mathcal{A}}^{\mathcal{I}_1}(c_r) \right) \right). \tag{14}$$

3 Time Modeling for Implementation $\mathcal{I}_2$

It worth noting that packing time for B_C does not depend on implementation, so the corresponding estimates are equal: $T_{\mathcal{I}_2,\mathcal{A}}(B_C) = T_{\mathcal{I}_1,\mathcal{A}}(B_C)$. Indeed, the packing of B_C is done in the order of its initial placement, i.e., in row-major order, which enables it to move the rows of B_C vector-by-vector, each of which is placed in a cache line. The packing time for A_C differs from what was written previously for $\mathcal{I}_1$ because $\mathcal{I}_2$ needs no reordering. Thus, all of A_C can be moved to the registers vector-by-vector instead of word-by-word copying, as in $\mathcal{I}_1$.

3.1 Modeling Packing Time for A_C

As for the case of $\mathcal{I}_1$, let us extract a contiguous part of A_r that has the size of $m_r \times f_c$, where f_c is the size of the cache line in words and is denoted as A_{box}. The time spent moving the elements from $A_{box}^1 = (A_{box}^{(1,1)}, \ldots, A_{box}^{(m_r,1)})$ to the registers is equal to $m_r t_{M0}$. The reminder of A_{box} is delivered to the registers vector-by-vector; thus, the time estimate for A_{box} is written as follows: $T_a(A_{box}) = T_a(A_{box}^1) + \sum_{j \in \overline{2,f_c}} T_a(A_{box}^j) = m_r t_{M0} + m_r t_{10} (\mathfrak{d}_{\mathfrak{r}} - 1)$, where $\mathfrak{d}_{\mathfrak{r}} = \left\lceil \frac{c}{\mathfrak{r}_o} \right\rceil$ is a limitation of data movement from L_1 to registers.

This result enables us to estimate the time spent on accessing all A_C as follows:

$$T_a(A_C) = m_r \left\lceil \frac{m_c}{m_r} \right\rceil \left\lceil \frac{k_c}{f_c} \right\rceil \left(t_{M0} + t_{10} \left(\left\lceil \frac{c}{\mathfrak{r}_o} \right\rceil - 1 \right) \right). \tag{15}$$

The time spent storing $T(A_C)$ is the same as for the case of $\mathcal{A}_1$, which enables us to write an expression for $T(A_C)$:

$$T_{\mathcal{A}}(A_C) = m_r \left\lceil \frac{m_c}{m_r} \right\rceil \left\lceil \frac{k_c}{f_c} \right\rceil \left(t_{M0} + t_{10} \left(\left\lceil \frac{c}{\mathfrak{r}_o} \right\rceil - 1 \right) + \left\lceil \frac{c}{\mathfrak{v}_s} \right\rceil t_s \right). \tag{16}$$

3.2 Modeling Packing Time for B_p

Let us assume that B_C is in L_3, so the time for moving B_p to the registers is assessed as follows. First, let us assess the access time for any row of B_r. The maximum number of vectors that can be delivered to the registers from a cache line is $\mathfrak{r}_o$; therefore, the estimate for the time spent moving the first row of B_p is written as $T_a(b_r^1) = t_{30} + t_{10} (\mathfrak{d}_{\mathfrak{r}} - 1)$, where b_r^1 is the first row of b_r, which has the length of a cache line, $\mathfrak{d}_{\mathfrak{r}} = \left\lceil \frac{c}{\mathfrak{r}_o} \right\rceil$ is a limitation of data movement from L_1 to registers, which represents the number of access-store cycles to deliver cache-line to the registers, c denotes the number of vectors that can be placed in one cache line and $\mathfrak{r}_o$ is the number of vectors that can be accessed and stored in the registers at once, i.e. in one cycle.

This estimate enables us to assess the access time for B_r as follows:

$$T_a(B_r) = k_c \left\lceil \frac{n_r}{f_c} \right\rceil \left(t_{30} + t_{10} \left(\left\lceil \frac{c}{\mathfrak{r}_o} \right\rceil - 1 \right) \right).$$

Thus, the packing time can be assessed as follows:

$$T_{\mathcal{A}}(B_p) = k_c \left(\left\lceil \frac{n_r}{f_c} \right\rceil \left(t_{30} + t_{10} \left(\left\lceil \frac{c}{\tau_o} \right\rceil - 1 \right) \right) + \left\lceil \frac{n_r}{f_o v_s} \right\rceil t_s \right). \quad (17)$$

3.3   Modeling Low-Level Calculations

Considering the implementation differences, the only discrepancy between the time estimates for the low-level calculations is in T_m^B. Indeed, in the case of $\mathcal{I}_2$, B_p is in L_1 cache as it is explicitly packed. In the case of $\mathcal{I}_1$, its elements must be delivered to the registers from L_3, where the whole B_C is, before making any further calculations. Thus, the time estimate for B_p, when one panel of C_r is being processed, is written as follows:

$$T_m^B = T_m(B_p) = \overline{n_r} k_c t_{10}.$$

As for the case of $\mathcal{I}_1$, Eq. (4) for the time model is rewritten as follows to obtain the yielding form of $T_{\mathcal{I}_2}(\mathcal{A}, \mathcal{P})$:

$$T_{\mathcal{I}_2}(\mathcal{A}, \mathcal{P}) = k_{B_C}^{\mathcal{P}} \left(T_{\mathcal{A}}(B_C) + k_{A_C}^{\mathcal{P}} \left(T_{\mathcal{A}}^{\mathcal{I}_2}(A_C) + k_{B_p}^{\mathcal{P}} \left(T_{\mathcal{A}}^{\mathcal{I}_2}(B_p) + T_{\mathcal{A}}^{\mathcal{I}_2}(c_r) \right) \right) \right), \quad (18)$$

where $T_{\mathcal{A}}^{\mathcal{I}_2}$ is calculated as: $T_{\mathcal{A}}^{\mathcal{I}_2}(c_r) = \sum_{u \in \{A, B, C_L, C_S, P\}} T_m^u$.

4   Cache Constraints Implementation

This section combines the aforementioned assumptions for tile placements to present constraints for an exhaustive search procedure, which is used to calculate the parameters for the previously discussed implementations.

To calculate c_r on a microkernel level so that corresponding calculations fit to registers, the following constraint must be applied:

$$m_r \overline{n_r} + m_r + 1 \leq l_o,$$

where l_o is the number of registers that are used for the calculations.

The first term of this constraint is the number of registers used to preserve c_r during the computations; the second term is the number of vector registers used to preserve a_r; the third one denotes the number of vectors from b_r that are used in one cycle of c_r calculations (see, for instance [1]).

Based on the aforementioned procedure, the constraint for the first cache level is written as follows:

$$\left\lceil \frac{m_r k_c}{f_c} \right\rceil + \left\lceil \frac{k_c \overline{n_r}}{c} \right\rceil + 2 m_r \left\lceil \frac{\overline{n_r}}{c} \right\rceil \leq l_1,$$

where the first term of this constraint is the number of cache lines that are reserved for A_p; the second term is the number of cache lines that are reserved for B_p; the third term is the doubled number of cache lines that are reserved for c_r: one for calculations on the microkernel level and one for the result obtained when the aforementioned calculations are done.

At the second cache level, it is necessary to preserve the following:

$$\left\lceil \frac{m_c k_c}{\mathfrak{f}_c} \right\rceil + \left\lceil \frac{k_c \overline{n_r}}{\mathfrak{c}} \right\rceil + 2 m_r \left\lceil \frac{\overline{n_r}}{\mathfrak{c}} \right\rceil \leq \mathfrak{l}_2,$$

where the first term is the number of cache lines that are reserved for A_C and the second term is the number of cache lines that are reserved for B_C; the third term is the doubled number of cache lines that are reserved for c_r: one for calculations and another for the results of these calculations.

5 Experimental Results

The proposed time model was applied to calculate parameters for Intel Coffee Lake[4]/AVX-2 architecture when 32-bit words are used. The parameters for the model were initialized as follows:

$$t_{M0} = 140,\ t_{30} = 32,\ t_{20} = 12,\ t_{10} = 4,\ t_{01} = 1$$
$$\mathfrak{l}_0 = 16,\ \mathfrak{l}_1 = 512,\ \mathfrak{l}_2 = 4096,\ \mathfrak{l}_3 = 32768$$
$$\mathfrak{v}_0 = 2,\ \mathfrak{t}_0 = 1,\ \mathfrak{t}_s = 1,\ \mathfrak{v}_s = 1$$
$$\mathfrak{c} = 2,\ \mathfrak{f}_0 = 4,\ \mathfrak{f}_c = 8$$
$$\mathfrak{r}_0 = 2$$

Thus, the constraints are written as follows:

$$L_0:\ m_r \overline{n_r} + m_r + 1 \leq 16$$
$$L_1:\ \lceil m_r k_c / 8 \rceil + \lceil k_c \overline{n_r} / 2 \rceil + 2 m_r \lceil \overline{n_r} / 2 \rceil \leq 512 \qquad (19)$$
$$L_2:\ \lceil m_c k_c / 8 \rceil + \lceil k_c \overline{n_r} / 2 \rceil + 2 m_r \lceil \overline{n_r} / 2 \rceil \leq 4096$$

Based on L_0 constraints and the previously mentioned hardware parameters, microkernels $M_{m_r, \overline{n_r}}$ that use the maximum amount of space in registers were found to be used in an exhaustive search procedure: $M_{1,14}, M_{2,6}, M_{3,4}, M_{4,2}, M_{5,2}$.

Implementation $\mathcal{I}_1$ and $\mathcal{I}_2$ were done in MLIR [4], and an *mlir-cpu-runner* was used to run the program. Each GEMM implementation is called 11 times to obtain the runtime value for each pair of matrices with specific sizes: m, n, k. The parameters $(m_c, n_c, k_c, m_r, n_r)$ for each call are calculated via an exhaustive search procedure to ensure that, for example, all possible values are used to obtain the optimal ones. Every runtime value t is used to calculate performance p as follows: $p_i = \frac{2mnk}{t} \cdot 10^{-9}$, where $i \in \overline{1,11}$.

---

[4] See, for instance, https://en.wikichip.org/wiki/intel/microarchitectures/coffee_lake.

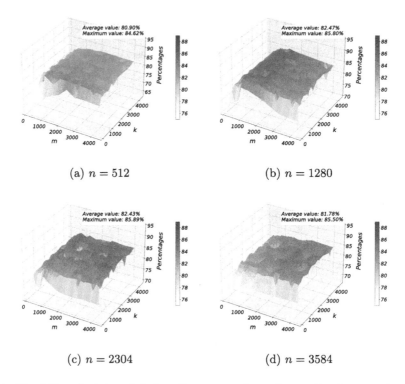

(a) $n = 512$ (b) $n = 1280$

(c) $n = 2304$ (d) $n = 3584$

Fig. 1. The bilinearly interpolated performance values when optimal parameters are calculated based on the matrix sizes for $\mathcal{I}_1$. For small values of k, there is a decrease in the performance for all values of n, which is the result of the rounding that occurs when k_{A_B} is calculated.

The median of the values $p_1, .., p_{11}$ is used as a performance estimate for the series of calls. The number of elements in this series is calculated based on the mean value to ensure that the deviation is less than 2.5%.

The program is run 15 times for different matrix sizes, each of which is from the grid: $m, n, k \in \{\delta, 2\delta, ..., 32\delta\}$, where $\delta = 128$. As the implementation does not process reminders when the number of tiles is not an integer, the initial matrix is padded to fulfill the following conditions: $0 \equiv m \pmod{m_c}, 0 \equiv m_c \pmod{m_r}, 0 \equiv n_c \pmod{n_r}, 0 \equiv k \pmod{k_c}, 0 \equiv k_c \pmod{k_u}, n = n_c$, where k_u is an unrolling factor [1].

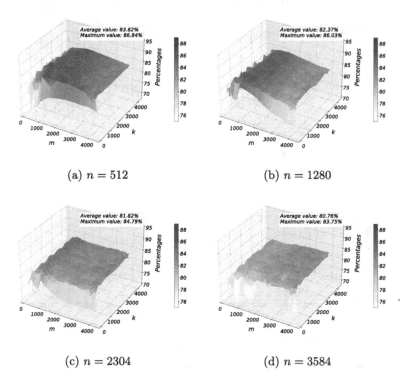

Fig. 2. The bilinearly interpolated performance values when optimal parameters are calculated based on the matrix sizes for $\mathcal{I}_2$.

The equality $n = n_c$ takes place due to the fact that L_3 cache size is not involved in any constraint used for the optimization problem.

All of the experiments were run on Ubuntu 20.04 LTS with LLVM 15.0.0, and a CPU core of i7-8700 was isolated to obtain valid experimental results.

Figure 1 and Fig. 2 show the change in the performance value for different sizes of the input matrix that is achieved for implementation $\mathcal{I}_1$ and $\mathcal{I}_2$ when optimal values of the parameters are calculated based on their corresponding models.

Both figures indicate that there is a decrease in performance for all values of n when k is relatively small, which is the result of the rounding that occurs when k_{A_B} is calculated. This calculation involves rounding, which influences the choice of optimal values. As k_c must be chosen based on the aforementioned constraints, the lower the k, the smaller the number of choices. Processing the reminder of a matrix would be a remedy to the problem, although this is not the case for the analyzed models and for $\mathcal{I}_1$ and $\mathcal{I}_2$.

The performance difference for $\mathcal{I}_1$ and $\mathcal{I}_2$ can be assessed by using Fig. 3, which shows performance values for fixed k values. When value of n is small, $n = 896$ and m is large, the implementation $\mathcal{I}_1$ outperforms $\mathcal{I}_2$. For small values

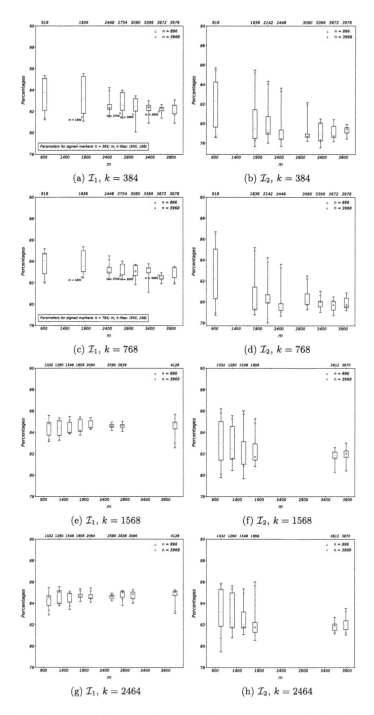

Fig. 3. The performance difference when optimal parameters are calculated for different matrix sizes based on $\mathcal{I}_1$ and $\mathcal{I}_2$. Each plot shows the performance values for values of $k \in \{384, 768, 1568, 2464\}$.

of m for the previously fixed n, the implementations show vice versa behavior: performance of $\mathcal{I}_2$ is better than the performance of $\mathcal{I}_1$. On the other hand, when n is large, the performance of $\mathcal{I}_1$ is better no matter which m is used.

It should be noted that the implementation $\mathcal{I}_2$ demonstrates large variance of n for small values of m compared to model $\mathcal{I}_1$ (see Fig. 3 (b, d, f, and h)). With the increase of m, the variance of the values n decreases, which could be a result of packing procedure and should be analyzed further.

6   Conclusion

The study discussed in this paper proposed two models to find the best parameters for matrix multiplication. These models are based on two implementations of matrix multiplication procedures that represent different ways of packing matrix parts. The first implementation $\mathcal{I}_1$ packs two blocks for the left-hand and right-hand sides of the matrix multiplication. The second implementation $\mathcal{I}_2$ has additional packing, which is used to pack a smaller part of the right-hand side. These two implementations are used to obtain the performance values when the optimal parameters are calculated with the corresponding models that are used. The results show that this modeling can be used to obtain a better performance for MLIR implementation and is comparable to BLIS, although additional effort should be made to ensure that the performance for each implementation is comparable to MKL or OpenBLAS.

References

1. Bondhugula, U.: High performance code generation in MLIR: an early case study with GEMM (2020). https://doi.org/10.48550/arXiv.2003.00532
2. Goto, K., Geijn, R.V.D.: Anatomy of high-performance matrix multiplication. ACM Trans. Math. Softw. **34**, 12:1–12:25 (2008)
3. Huang, J., Geijn, R.: BLISlab: a sandbox for optimizing GEMM. arXiv, p. 1609.00076 (2016). https://doi.org/10.48550/arXiv.1609.00076
4. Lattner, C., et al.: MLIR: scaling compiler infrastructure for domain specific computation. In: 2021 IEEE/ACM International Symposium on Code Generation and Optimization (CGO), Seoul, Korea, pp. 2–14 (2021)
5. Low, T.M., Igual, F., Smith, T., Quintana-Ortí, E.S.: Analytical modeling is enough for high-performance BLIS. ACM Trans. Math. Softw. (TOMS) **43**, 1–18 (2016)
6. Lu, L., et al.: TENET: a framework for modeling tensor dataflow based on relation-centric notation (2021). https://doi.org/10.48550/arXiv.2105.01892
7. Zee, F.G.V., van de Geijn, R.A.: BLIS: a framework for rapidly instantiating BLAS functionality. ACM Trans. Math. Softw. **41**, 14:1–14:33 (2015)
8. Zhang, H., Cheng, X., Zang, H., Park, D.H.: Compiler-level matrix multiplication optimization for deep learning. arXiv, p. 1909.10616 (2019). https://doi.org/10.48550/arXiv.1909.10616

Reliable Production Process Design Problem: Compact MILP Model and ALNS-Based Primal Heuristic

Roman Rudakov[1]($^{(\boxtimes)}$), Daniil Khachai[2], Yuri Ogorodnikov[1], and Michael Khachay[1]

[1] Krasovsky Institute of Mathematics and Mechanics, Ekaterinburg, Russia
r.a.rudakov@gmail.com, {yogorodnikov,mkhachay}@imm.uran.ru
[2] Kedge Business School, Bordeaux, France
daniil.khachai@kedgebs.com

Abstract. Supply chain resilience is one of the most relevant topics of operations research and production management, which is aimed to risk mitigation in the global manufacturing, logistics, and trade. Conventional approach for resilient supply chain design involves the stochastic modeling and scenery-based description of anticipated failures in transportation networks. However, the stochastic approach can be insufficiently adequate in a situation of an unexpected failure or interruption. In this paper, we introduce the Reliable Production Process Design Problem (RPPDP), where the goal is to guarantee a suitable behaviour of the given highly distributed manufacturing system with respect to an (almost) arbitrary potential failure. This problem appears to be strongly NP-hard, similarly to the famous Constrained Shortest Path Tour and Shortest Simple Path with t Must Pass Nodes combinatorial problems. In order to find (close to) optimal solutions of the problem in question efficiently, we propose a compact Mixed Integer Linear Program (MILP) model and an Adaptive Large Neighborhood Search (ALNS) based primal heuristic. Results of their extensive numerical evaluation on top of the Gurobi branching framework, against the instances derived from the PCGTSPLIB library show high performance of the proposed methods.

Keywords: Reliable Production Process Design Problem · MILP models · Adaptive Large Neighborhood Search

1 Introduction

Recently, the hyper-competitive marketplace environment calls for an extensive research and development of sustainable and resilient supply chains. Manufacturing process supervisors face numerous challenges including insufficient resilience of global supply chains, unpredictable delivery times, lack vital resources, and total disruption of the entire network [30]. An important approach to address these challenges [6,8] exploits stochastic models, where possible disruptions of a transportation network are described in terms of given scenarios. However, such an approach deteriorates if the interruption in question was not anticipated. Therefore, in some cases, a more suitable solution would be the one that minimizes the risks and remain preferably deterministic. To the best of our knowledge, those assured risk mitigation techniques for the deterministic production processes remain rather weakly studied. In this paper, we propose a novel deterministic modeling framework aimed to defend the manufacturing system from various non-anticipated production or logistic failures.

Consider a production process consisting of m *operations*, each of them should be carried out by a dedicated *Manufacturing Unit (MU)*, in some given (partial or linear) order. Each MU can perform a single operation, which induces their natural grouping into m *clusters*. All MUs belonging to the same cluster are assumed to be mutually interchangeable. Production of each MU is used as an input of other MUs performing the descendant operations and can be transferred through some *Transportation Hubs (TH)* corresponding to the topology of a given transportation network. Each TH has a given *opening cost* specifying a charge for using this hub in a production plan, and a *capacity* constraining the number of paths that can pass through it. Transportation between all the MUs and THs are fared with given *transportation costs*.

Production Plan (PP) is a subgraph that consists of virtual *source* and *destination* points s and t, respectively, single representative (MU) per each operation cluster, and directed paths connecting them in consistence within the given order and the corresponding capacity constraints. The goal, for a given number k, is to construct a family of k node-disjunctive production plans (each pair of PPs has no common nodes except s and t) that minimizes a maximum cost of these plans. In the sequel, we refer to this problem as Reliable Production Process Design Problem (RPPDP).

2   Related Work

To the best of our knowledge, the RPPDP problem appears to be similar to the known Constrained Shortest Path Tour Problem (CSPTP), introduced by Ferone et al. in [10]. An instance of the CSPTP is given by an edge-weighted digraph $G = (V \cup \{s,t\}, A)$, an ordered set of disjoint vertex subsets $\mathfrak{T} = \{T_1, \ldots, T_k\}$, and two weighting functions $c, q \colon A \to \mathbb{R}^+$, where c specifies transportation costs of the arcs and q their capacity bounds. The goal is to construct a shortest capacitated path connecting s and t nodes, visiting every subset of $\mathfrak{T}$, such that this path contains a hitting set of $T_1, \ldots, T_k$, which is traversed in a given linear order.

Ferone et al. proved that CSPTP is NP-hard [10], and proposed branch-and-bound and GRASP algorithms. Further, in paper [9], these authors improved their branching approach. Finally, Martin et al. [22] proposed the new branch-and-price algorithm relied on an improved MILP model allowing simple decomposition of the considered problem.

The Shortest Simple Path Problem with k Must Pass Nodes (SSPP-k-MPN) [27] also appears to be close to the RPPDP introduced in this paper. An instance of the SSPP-k-MPN is given by an edge-weighted digraph $G = (V, E, c)$ with a non-ordered subset $T \subset V$ of size k, and dedicated origin and destination nodes s and t. The goal is to find the shortest elementary path from s to t visiting all the nodes from T exactly once. Like the major part of combinatorial optimization problems, the SSPP-k-MPN is strongly NP-hard, even for an arbitrary fixed $k \geq 1$ [21]. There are known a number of MILP models both for the SSPP-k-MPN [1,2] and its several modifications [11,12]. To the best of our known, the branch-and-bound algorithm proposed in [21] appears to be state-of-the-art for this problem and its setting, where the desired solution should be protected by some backup path [24].

Finally, we should remind the Generalized Traveling Salesman Problem with Precedence Constraints (PCGTSP) [14], which has many common features with the RPPDP.

In the PCGTSP, we are given by a directed graph G, whose nodeset is partitioned into pairwise disjoint clusters C_i, $i = \overline{1, m}$. In addition, the instance of the PCGTSP is extended with a directed graph Π representing a partial order of the clusters visiting. The goal is to construct a shortest elementary tour that departs from and arrives to the first cluster and visits all the clusters in one node exactly with respect to the given order specified by Π. The PCGTSP has many important industrial applications, e.g. in optimizing the processes of multi-hole drilling [7], efficient programming of Coordinate Measuring Machines (CMM) applied for industrial inspection processes [16,28], and metal sheet cutting [7].

Being an extension of the well-known Generalized Traveling Salesman Problem (GTSP), PCGTSP is strongly NP-hard [25]. As many other combinatorial optimization problems, PCGTSP admits a number of approximation algorithms and branching schemes. Thus, there are known efficient algorithms for several specific precedence constraints [4,5,18,20], the PCGLNS meta-heuristic [19], branch-and-bound and DP-and-bound algorithms based on Balas instance pre-processing [3], Held and Karp branching framework [23], and the combinatorial lower bounds [29]. The most recent result for the PCGTSP is the state-of-the-art branch-and-cut algorithm, which exploits an extensive polyhedral study of the problem in question [17]. As a result, the number of instances from the well-known PCGTSPLIB benchmark library solved to optimality has increased twice (to 24 out of 40), while approximation gaps for the remaining instances were significantly improved as well.

As it follows from the literature, there are several attempts to model the production processes in terms of the aforementioned combinatorial problems. Unfortunately, none of them fits well to the highly resilient production plan generation. In particular, each of these problems is aimed to construct a unique (sub)optimal subgraph (path, route, tour, etc.), which makes the obtained solutions fragile to possible faults in a transportation network. Furthermore, these subgraphs have insufficiently simple structure.

Contribution of our paper is two-fold:

(i) we introduce the RPPDP and prove its NP-hardness;

(ii) we propose a compact MILP model and primal heuristic, which is based on the well-known Adaptive Large Neighborhood Search approach; we prove its numerical performance against generated test instances of RPPDP

The rest of the paper is structured as follows. In Sect. 3, we introduce a mathematical statement of the RPPDP and establish its complexity status. Further, in Sect. 4, we discuss our compact MILP model. In turn, Sect. 5, we describe the proposed primal heuristic. Results of the numerical evaluation that proves high performance of the proposed techniques are reported in Sect. 6. Finally, in Sect. 7, we summarize the obtained results and discuss some open questions.

3   Problem Statement

An instance of the RPPDP is given by a triple (G, Π, k), where

(i) $G = (V, E, c)$ is an edge-weighted digraph, such that
   - $V = \{s, t\} \cup \mathfrak{M} \cup \mathfrak{H}$,
   - s and t are the source and destination points of an arbitrary production plan,
   - $\mathfrak{M} = \mathfrak{M}_0 \cup \ldots \cup \mathfrak{M}_{m+1}$, where each *cluster* $\mathfrak{M}_j$, $j = \overline{1, m}$ consists of the nodes associated with operation j, $\mathfrak{M}_0 = \{s\}$, and $\mathfrak{M}_{m+1} = \{t\}$,
   - $\mathfrak{H}$ is a set of transportation hubs, each hub $h \in \mathfrak{H}$ is supplemented with capacity bound q_h and opening cost C_h,
   - the weighting function $c \colon E \to \mathbb{R}_+$ assigns transportation cost c_e to each arc $e \in E$;
(ii) $\Pi = (\{0, 1, \ldots, m + 1\}, A)$ is an auxiliary directed acyclic graph (DAG) establishing an order on the set of clusters $\mathfrak{M}$, where 0 and $m+1$ are assumed to be the minimum and maximum elements and correspond to the 'virtual' clusters $\mathfrak{M}_0$ and $\mathfrak{M}_{m+1}$, respectively;
(iii) an integer $k \geq 1$ is the required number of production plans.

The goal is to construct a family of production plans $\mathcal{P} = \{P_1, P_2, \ldots, P_k\}$, such that

   - each plan is a DAG obtained from the digraph Π by replacement each its node i with some node $v_i \in \mathfrak{M}_i$ and each arc (i, j) with an elementary route $v_i, h_1, \ldots, h_p, v_j$, all whose interim nodes are transportation hubs;
   - $\mathcal{P} = \arg\min\{max\{cost(P_i) \colon i = \overline{1, k}\}\}$, where

$$cost(P_i) = \sum_{e \in P_i} c_e + \sum_{h \in P_i} C_h.$$

A simple example of the introduced problem is presented in Fig. 1. Complexity status of the RPPDP is defined by the following theorem.

Theorem 1. *The RPPDP problem is strongly NP-hard.*

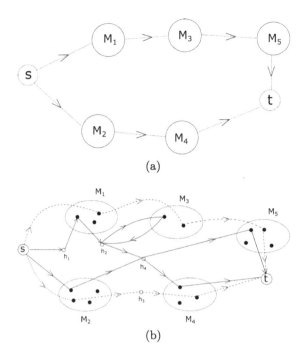

Fig. 1. Sample RPPDP instance for $m = 5$ and $k = 2$. (1a) order graph Π (1b) corresponding graph G. Solid and dashed lines specify first and second plans of some feasible solution, respectively.

Proof. We prove the theorem by reducing the strongly NP-hard Simple Shortest Path Problem with 1 Must Pass Nodes (SSPP-1-MPN) [21].

Let we are given by an instance I_1 of the SSPP-1-MPN defined by a digraph $G_1 = (V, E, c)$, a pair of specified nodes s and t, and one must-pass node m_1. The corresponding RPPDP instance I_2 is defined by a digraph $G_2 = (V, E, c)$, cluster set $\mathfrak{M} = \mathfrak{M}_0 \cup \mathfrak{M}_1 \cup \mathfrak{M}_2$, where $\mathfrak{M}_0 = \{s\}$, $\mathfrak{M}_1 = \{m_1\}$, and $\mathfrak{M}_2 = \{t\}$, hubs $\mathfrak{H} = \{h : h \in V \setminus \{m_1, s, t\}\}$, and $k = 1$. For any $h \in \mathfrak{H}$ we assign the opening cost $C_h = 0$ and capacity $q_h = 1$. The precedence order is $\mathfrak{M}_0 \prec \mathfrak{M}_1 \prec \mathfrak{M}_2$.

Obviously, an arbitrary feasible plan P consists of two subsequent segments connecting s with m_1 and m_1 with t, respectively and visits any hub node at most once, due to the imposed capacity constraint. Since each hub $h \in \mathfrak{H}$ has zero opening cost, the total cost of P is defined by edge costs exclusively. Therefore, an arbitrary optimal solution of the instance I_1 corresponds to some optimal solution of I_2 and vice versa. Consequently, the RPPDP is strongly NP-hard, as well.

4   MILP Model

In this section, we describe the proposed MILP model. In the sequel, we use it in our branch-and-cut algorithm built on the top of Gurobi optimizer [13].

We introduce assignment variables $x_e^{r,a} \in \{0,1\}$, $(r = \overline{1,k}, a \in A, e \in E)$ that are plan-to-arcs incidence indicators. Namely, $x_e^{r,a} = 1$ if and only if the arc $e \in E$ is employed for routing of the arc $a \in A$ in the production plan r. In addition, we define plan-to-node indicators $y_v^r \in \{0,1\}$, $(r = \overline{1,k}, v \in V)$, where $y_v^r = 1$ if and only if the node $v \in V$ is included to the plan r. Finally, we define real-valued variable $cost$. Our MILP model is represented as follows

$$\min \ cost \tag{1}$$

$$\text{s.t.} \sum_{a \in A, e \in E} c_e x_e^{r,a} + \sum_{h \in \mathfrak{H}} C_h y_h^r - cost \leq 0 \qquad \left(r = \overline{1,k} \right) \tag{2}$$

$$\sum_{u \in \mathfrak{M}_j \cup \mathfrak{H}\,:\,(v,u) \in E} x_{(v,u)}^{r,(i,j)} - y_v^r = 0 \qquad \begin{pmatrix} r = \overline{1,k}, \\ (i,j) \in A, \\ v \in \mathfrak{M}_i \end{pmatrix} \tag{3}$$

$$\sum_{v \in \mathfrak{M}_i \cup \mathfrak{H}\,:\,(v,u) \in E} x_{(v,u)}^{r,(i,j)} - y_u^r = 0 \qquad \begin{pmatrix} r = \overline{1,k}, \\ (i,j) \in A, \\ u \in \mathfrak{M}_j \end{pmatrix} \tag{4}$$

$$\sum_{v \in \mathfrak{M} \cup \mathfrak{H}\,:\,(v,u) \in E} x_{(v,u)}^{r,(i,j)} = 0 \qquad \begin{pmatrix} r = \overline{1,k}, \\ (i,j) \in A, \\ u \in \mathfrak{M} \setminus \mathfrak{M}_j \end{pmatrix} \tag{5}$$

$$\sum_{u \in \mathfrak{M} \cup \mathfrak{H}\,:\,(v,u) \in E} x_{(v,u)}^{r,(i,j)} = 0 \qquad \begin{pmatrix} r = \overline{1,k}, \\ (i,j) \in A, \\ v \in \mathfrak{M} \setminus \mathfrak{M}_i \end{pmatrix} \tag{6}$$

$$\sum_{v \in \mathfrak{M}_i \cup \mathfrak{H}\,:\,(v,h) \in E} x_{(v,h)}^{r,(i,j)} - \sum_{u \in \mathfrak{M}_j \cup \mathfrak{H}\,:\,(h,u) \in E} x_{(h,u)}^{r,(i,j)} = 0 \begin{pmatrix} r = \overline{1,k}, \\ (i,j) \in A, h \in \mathfrak{H} \end{pmatrix} \tag{7}$$

$$\sum_{(i,j \in A)} \sum_{v \in \mathfrak{M}_i \cup \mathfrak{H}\,:\,(v,h) \in E} x_{(v,h)}^{r,(i,j)} - q_h y_h^r \leq 0 \qquad \left(r = \overline{1,k}, h \in \mathfrak{H} \right) \tag{8}$$

$$\sum_{v \in \mathfrak{M}_i} y_v^r = 1 \qquad \left(r = \overline{1,k}, i = \overline{1,m} \right) \tag{9}$$

$$\sum_{r=1}^{k} y_v^r \leq 1 \qquad \left(v \in V \setminus \{s,t\} \right) \tag{10}$$

$$x_e^{r,a} \in \{0,1\}, \ y_v^r \in \{0,1\} \tag{11}$$

In the objective (1), we minimize the maximum cost among the k designed production plans (which is found in (2)). Equations (3) and (4) establish start and finish nodes for a route assigned to any arc (i,j) of the digraph Π and each plan r. Equations (5) and (6) ensure each interim node of such a route (if any) is a hub. Equation (7) is a flow conservation constraint for each r and arc (i, j). Equation (8) establishes a consumption upper bound for any transportation hub. Equation (9) ensures that an arbitrary cluster is visited exactly once. Finally, Eq. (10) ensures node-disjunction of the constructed plans.

Theorem 2. *An arbitrary instance of the RPPDP has a feasible solution if and only if the MILP model defined by Eqs. (1)–(11) is feasible.*

Proof. Suppose that we are given by a set of k plans producing a feasible RPPDP solution. Then, we set $x_e^{r,a} = 1$ for each arc $e \in E$ belonging to the plan r and included into the route connecting clusters $\mathfrak{M}_i$ and $\mathfrak{M}_j$, for the arc $a = (i, j) \in A$ in the order DAG Π. In addition, we set $y_v^r = 1$ for each node $v \in V$ included to the plan r. Then, an arbitrary plan is a set of routes starting from s and finishing at t, therefore, Eqs. (3), (4), and (5) are held. Next, the graph of any feasible solution can be contracted to the digraph isomorphic to the DAG Π, therefore Eqs. (5) and (6) are valid, as well. Every hub from the feasible solution is passed at most q_h times and hence Eq. (8) is also held. Finally, all plans are node-disjoint that makes (10) valid.

On the other hand, let we be given by some feasible solution P of the MILP model. We show that P corresponds to a feasible solution of the initial RPPDP instance. Indeed, from Eqs. (3), (4), (5), (6), and (7) and it follows that every arc $(i, j) = a \in A$ of the graph Π is routed by exactly one path in the graph G departing from the cluster $\mathfrak{M}_i$, traversing some hub nodes and arriving to the cluster $\mathfrak{M}_j$, respectively. By Eq. (8), an arbitrary hub is employed within its capacity. Equation (9) guarantees that every plan visits each cluster in a single node exactly. Finally, Eq. (10) ensures that all plans are pairwise node-disjoint. Hence, the considered feasible solution of the MILP model corresponds to a feasible one of a RPPPD instance.

5   Adaptive Large Neighborhood Search Heuristic for the RPPDP

In this section, we discuss our ALNS-based primal heuristic, which we refer to as reliable production process design ALNS (or RPPD-ALNS) heuristic. The main idea of the classic ALNS [26] appears to be an implementation of the well-known *ruin* and *recreate* principle augmented by online learning over the predefined removal and insertion simple heuristics and simulated annealing stopping criterion. Our approach follows a novel modification of the ALNS proposed recently in [15], which operates over a set of the *local search* and *shaking* procedures. We present its pseudo-code in Algorithm 1.

The proposed heuristic consists of $i_{\max}$ subsequent trials, each of them starts with construction of an initial feasible solution P by applying simple greedy algorithm within a random seed. Then in the internal loop, we employ several '*shaking - local search*' iterations, during each of them this solution is perturbed using a shaking procedure and feed to some local search technique to obtain

Algorithm 1. RPPD-ALNS :: general scheme

Input: an RPPDP instance (G, Π, k)
Parameters: number of trials $i_{\max}$, banks $\mathfrak{Ls}$ and $\mathfrak{Sh}$ of local search and shaking procedures
Output: a feasible solution $\bar{P}$

1: **for** $0 < i \leq i_{max}$ **do**
2:     produce an initial solution P by greedy algorithm;
3:     $P_{best}(i) = P$;
4:     **repeat**
5:         choose sh and ls drawn from current distributions over sets $\mathfrak{Sh}$ and $\mathfrak{Ls}$
6:         **shaking:** using sh, obtain other feasible solution P' from P;
7:         **local search:** until a solution improvement can be made, apply ls and build up a new solution P_{new} from P';
8:         **if** $c(P_{new}) < c(P_{best}(i))$ **then**
9:             $P_{best}(i) = P_{new}$
10:        **end if**
11:        **if** $acceptance\_criterion$ is met for P_{new} and P **then**
12:            $P = P_{new}$
13:            record the improvement made by sh and ls
14:        **end if**
15:     **until** $termination\_criterion$ is met
16:     update the distributions over $\mathfrak{Sh}$ and $\mathfrak{Ls}$ w.r.t. to improvements recorded
17: **end for**
18: **return** $\bar{P} = \arg\min\{P_{best}(i) \colon i \in \{1, \ldots, i_{max}\}\}$

(hopefully) the better feasible solution. Both the shaking and local search techniques are taken from current distributions over the predefined banks $\mathfrak{Sh}$ and $\mathfrak{Ls}$. If the obtained solution meets the simulated annealing acceptance criterion (see, e.g. [31]), we update the current solution and store the statistics of the heuristics usage. Each trial is concluded by the updating of the distributions over $\mathfrak{Sh}$ and $\mathfrak{Ls}$ relying on the collected statistics.

5.1 Basic Heuristics

All of the considered heuristics operate with *routes* each of them departs from some cluster node, visits (optionally) some hub nodes, and arrives to a node belonging to another cluster.

We use six simple techniques to shake the current feasible solution:

- *swapping of hubs* technique switches assignment of two hubs associated initially with a randomly taken couple of plans;
- *swapping of cluster nodes* interchanges the assignment of two nodes from an arbitrary cluster between two randomly taken plans;
- *changing of a cluster node* randomly changes a representative of an arbitrary cluster included to a random plan;
- *swapping of paths* reassigns a whole s-t-paths between two randomly chosen plans;

- *shorten of a route* tries replace a random route of an arbitrary plan with some possible shorter counterpart;
- *reroute of an arc* technique is similar to the previous one except that we try to change routes associated to some randomly chosen arc of the DAG Π in every plan of the current solution.

Generally speaking each proposed shaking technique destroys the current solution. Therefore, all or them are augmented with a simple feasibility restoration procedure.

Obviously, each the aforementioned shaking technique implicitly induced the appropriate neighborhood in the set of feasible plans. That leads us to construct appropriate local search techniques. Namely, we use six local search procedures derived straightforwardly from the proposed shaking techniques.

For instance, we consider the local search procedure induced by *swapping of cluster nodes* technique. In this case, each alternative solution is generated by interchanging a single pair of cluster nodes between two randomly chosen plans (see also Fig. 2). Proceeding with all clusters and possible combinations of the plans, we obtain the desired procedure.

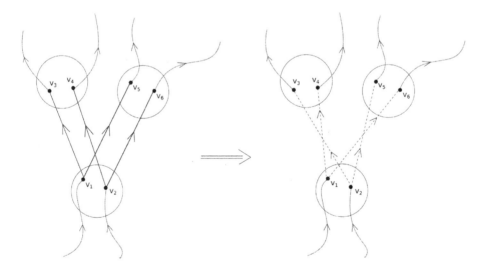

Fig. 2. Local search: changing cluster nodes. We swap nodes v_1 and v_2 between two plans.

6   Numerical Evaluation

We perform the following numerical experiment to check the efficiency of the proposed MILP model and the RPPD-ALNS heuristic.

Instance Generation

As a basis of our test benchmark, we take PCGTSP instances from the well-known library PCGTSPLIB proposed in [28], each of them is transformed as follows:

(i) the graph G together with edge weighting function are inherited from the initial PCGTSP instance

(ii) origin and destination node s and t are obtained from the starting node of the original PCGTSP instance

(iii) all the clusters of size greater or equal to k are inherited as well, together with precedence constraints

(iv) the remaining small clusters are removed, their nodes are regarded as hubs

(v) to any obtained hub h, we assign two integers C_h and q_h taken at random from the range $[1, C]$ and $[1, q]$ respectively, for the predefined parameters C and q.

Each the obtained graph pair (G, Π) is used to synthesize RPPDP instances (G, Π, k), for different number of plans k to design.

Experimental Setup

In this experiment, we take br17.10, ft70.1, ESC47, and p43.1 instances from PCGTSPLIB and generate RPPDP instance as described above, for $k = \overline{2, 6}$. Upper bounds of hub capacity and opening costs are set to 2 and 5, respectively. For an arbitrary combination of graphs and k, we synthesized 100 instances exactly.

Our experiment consists in two stages. In the first exploratory stage, the goal was to assess solvability of the RPPDP problem by out-of-the-box Gurobi MIP-solver within the default settings, equipped with our MILP model. For the second stage, we take RPPDP instances based on p43.1 PCGTSP instance, where approximation gaps observed in the previous stage turned to be high in average. This time, we compare the out-of-the-box Gurobi with the setting, where all the builtin primal heuristics are replaced with the only our RPPD-ALNS heuristic. In both stages, we set time limit for all the algorithms to 5 h.

Results and Discussion

We start with the results of the exploratory stage, summarized in Table 1 in this table, we report names of the original PCGTSP instance, number of their nodes N, clusters M, number of clusters m in the appropriate RPPDP instances,

Table 1. Exploratory stage results

Iname	N	M	m	$|\mathfrak{H}|$	k	Avg. time, sec	Avg. gap(%)	Min time, sec	Min gap, %	Max time, sec,	Max gap, %
br17.10	88	17	8	43	2	1.6	0.0	0.8	0.0	3.36	0.0
					3	3.3	0.0	1.53	0.0	6.2	0.0
					4	9.2	0.0	2.9	0.0	36.1	0.0
					5	469.6	0.01	4.21	0.0	18000	1.47
					6	1458.8	0.04	6.38	0.0	18000	1.47
ft70.1	346	70	15	227	2	75.4	0.0	54.47	0.0	141.33	0.0
					3	133.4	0.0	90.84	0.0	275.5	0.0
					4	340.8	0.0	169.02	0.0	5040.4	0.0
					5	950.0	0.0	233.68	0.0	5856.1	0.0
					6	3173.8	0.0	305.65	0.0	6156.0	0.0
ESC47	244	48	6	213	2	132.8	0.0	97.53	0.0	239.6	0.0
					3	143.9	0.0	105.5	0.0	321.6	0.0
					4	518.8	0.0	165.0	0.0	1821.5	0.0
					5	3892.7	0.02	352.9	0.0	18000	1.69
					6	13052.6	0.23	1854.13	0.0	18000	8.47
p43.1	203	43	9	155	2	1718.2	0.0	956.41	0.0	3202.5	0.0
					3	13598.4	1.8	6420.2	0.0	18000	29.4
					4	17897.1	5.8	12688.5	0.0	18000	22.0
					5	18000	9.7	18000	0.04	18000	24.4
					6	18000	11.3	18000	0.7	18000	26.0

hubs $|\mathfrak{H}|$, and plans k. For each combination of these parameters, we report the minimum, maximum and average running times and obtained gap values.

As it follows from Table 1, out-of-the-box Gurobi optimizer managed to solve all the instances induced by br17.10, ft70.1 and ESC47 within average gap 0.25%. Furthermore, for $k = \overline{2,4}$ all the instances were solved to optimality more than three-times faster then the established time limit. On the other hand, for the instances originated from p43.1, calculation time was significantly bigger in general. Actually, the solver hits the time limit in most cases. Therefore, we select these instances for the subsequent stage.

In the second stage, we compare average approximation gaps reached by the default Gurobi and Gurobi having our RPPD-ALNS as the only primal heuristic to the predefined time stamps. As it follows from Table 2, Gurobi, whose all builtin heuristics are replaced with the RPPD-ALNS, shows significantly better results than its competitor (except the case $k = 4$).

Finally Fig. 3 illustrates gap decay for $k = 2$ and $k = 5$. It can be concluded, that using RPPD-ALNS as the only primal heuristic helps the Gurobi optimizer to obtain approximate solutions with gap less than 1% more faster.

Table 2. Comparing default Gurobi and Gurobi with RPPD-ALNS primary heuristic on p43.1 based instances

k	Time stamp, sec	Avg. gap Gurobi(%)	Avg. gap Gurobi + RPPD-ALNS(%)
2	2000	23.5	**0.9**
	3000	20.9	**0.15**
	5000	20.9	**0.13**
	10000	20.9	**0.09**
	15000	20.9	**0.06**
	18000	0.0	0.0
3	2000	19.9	**18.0**
	3000	19.7	**13.9**
	5000	15.5	**7.7**
	10000	7.0	**2.2**
	15000	**1.8**	1.9
	18000	1.8	**0.2**
4	2000	**15.2**	16.7
	3000	**15.1**	15.9
	5000	**13.9**	15.5
	10000	**9.9**	14.3
	15000	**7.5**	9.7
	18000	5.8	**1.2**
5	2000	17.4	**12.53**
	3000	16.4	**12.4**
	5000	15.4	**12.3**
	10000	13.9	**11.7**
	15000	12.3	**10.2**
	18000	9.7	**0.8**
6	2000	18.6	**10.8**
	3000	16.0	**10.3**
	5000	14.9	**10.3**
	10000	13.9	**10.2**
	15000	13.4	**8.7**
	18000	11.3	**3.1**

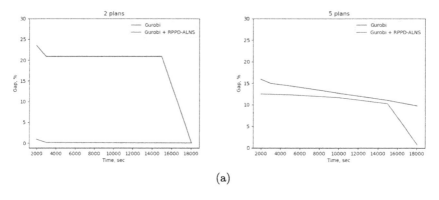

Fig. 3. Comparing gap decreasing for Gurobi and Gurobi + RPPD-ALNS, $k = 2, 5$

7   Summary

In this paper, we introduced the Reliable Production Process Design Problem as a model for construction of resilient supply chains. We proved that this problem is NP-hard in the strong sense. In order to find (sub)optimal solutions of the problem, we proposed the compact MILP model and the ALNS-based heuristic. To prove their efficiency, we carried out numerical evaluation against instances produced from the well-known PCGTSPLIB benchmarking library. The obtained results show, that injection of the proposed RPPD-ALNS heuristic into the state-of-the-art Gurobi MIP-solver helps it to obtain better RPPDP solutions much faster.

As open question we leave more complex numerical evaluation of the proposed heuristic, e.g. for the other possible values of the capacity and hubs opening cost parameters, which we postpone to the forthcoming paper.

Acknowledgements. The work was performed as part of research carried out in the Ural Mathematical Center with the financial support of the Ministry of Science and Higher Education of the Russian Federation (Agreement number 075-02-2023-913).

References

1. de Andrade, R.C.: Elementary shortest-paths visiting a given set of nodes (2013). http://www.din.uem.br/sbpo/sbpo2013/pdf/arq0242.pdf
2. de Andrade, R.C.: New formulations for the elementary shortest-path problem visiting a given set of nodes. Eur. J. Oper. Res. **254**(3), 755–768 (2016). https://doi.org/10.1016/j.ejor.2016.05.008
3. Balas, E., Fischetti, M., Pulleyblank, W.: The precedence-constraint asymmetric traveling salesman polytope. Math. Program. **68**, 241–265 (1995). https://doi.org/10.1007/BF01585767
4. Balas, E., Simonetti, N.: Linear time dynamic-programming algorithms for new classes of restricted TSPs: a computational study. INFORMS J. Comput. **13**(1), 56–75 (2001). https://doi.org/10.1287/ijoc.13.1.56.9748

5. Chentsov, A.G., Khachai, M.Y., Khachai, D.M.: An exact algorithm with linear complexity for a problem of visiting megalopolises. Proc. Steklov Inst. Math. **295**(1), 38–46 (2016). https://doi.org/10.1134/S0081543816090054
6. Deng, C., Xiong, Y., Yang, L., Yang, Y.: A smoothing SAA method for solving a nonconvex multisource supply chain stochastic optimization model. Math. Probl. Eng. **2022** (2022). https://doi.org/10.1155/2022/5617213
7. Dewil, R., Vansteenwegen, P., Cattrysse, D.: A review of cutting path algorithms for laser cutters. Int. J. Adv. Manuf. Technol. **87**(5), 1865–1884 (2016). https://doi.org/10.1007/s00170-016-8609-1
8. Fan, Y., Schwartz, F., Vob, S., Woodruff, D.L.: Catastrophe insurance and flexible planning for supply chain disruption management: a stochastic simulation case study. Int. J. Prod. Res. (2023). https://doi.org/10.1080/00207543.2023.2176179
9. Ferone, D., Festa, P., Guerriero, F.: An efficient exact approach for the constrained shortest path tour problem. Optim. Methods Softw. **35**(1), 1–20 (2020). https://doi.org/10.1080/10556788.2018.1548015
10. Ferone, D., Festa, P., Guerriero, F., Laganà, D.: The constrained shortest path tour problem. Comput. Oper. Res. **74**, 64–77 (2016). https://doi.org/10.1016/j.cor.2016.04.002
11. Gomes, T., Marques, S., Martins, L., Pascoal, M., Tipper, D.: Protected shortest path visiting specified nodes (2015). https://doi.org/10.1109/RNDM.2015.7325218
12. Gomes, T., Martins, L., Ferreira, S., Pascoal, M., Tipper, D.: Algorithms for determining a node-disjoint path pair visiting specified nodes. Opt. Switching Netw. **23** (2017). https://doi.org/10.1016/j.osn.2016.05.002
13. L. Gurobi Optimization: Gurobi optimizer reference manual (2021). https://www.gurobi.com/documentation/9.5/refman/index.html
14. Gutin, G., Punnen, A.P.: The Traveling Salesman Problem and Its Variations. Springer, Boston (2007)
15. Kalateh Ahani, I., Salari, M., Hosseini, S.M., Iori, M.: Solution of minimum spanning forest problems with reliability constraints. Comput. Ind. Eng. **142**, 106365 (2020). https://doi.org/10.1016/j.cie.2020.106365
16. Karuppusamy, N.S., Kang, B.Y.: Minimizing airtime by optimizing tool path in computer numerical control machine tools with application of A^* and genetic algorithms. Adv. Mech. Eng. **9**(12), 1687814017737448 (2017). https://doi.org/10.1177/1687814017737448
17. Khachai, D,, Sadykov, R., Battaia, O., Khachay, M.: Precedence constrained generalized traveling salesman problem: polyhedral study, formulations, and branch-and-cut algorithm. Eur. J. Oper. Res. (2023). https://doi.org/10.1016/j.ejor.2023.01.039
18. Khachai, M.Y., Neznakhina, E.D.: Approximation schemes for the generalized traveling salesman problem. Proc. Steklov Inst. Math. **299**(1), 97–105 (2017). https://doi.org/10.1134/S0081543817090127
19. Khachay, M., Kudriavtsev, A., Petunin, A.: PCGLNS: a heuristic solver for the precedence constrained generalized traveling salesman problem. In: Olenev, N., Evtushenko, Y., Khachay, M., Malkova, V. (eds.) OPTIMA 2020. LNCS, vol. 12422, pp. 196–208. Springer, Cham (2020). https://doi.org/10.1007/978-3-030-62867-3_15
20. Khachay, M., Neznakhina, K.: Complexity and approximability of the Euclidean generalized traveling salesman problem in grid clusters. Ann. Math. Artif. Intell. **88**(1), 53–69 (2020). https://doi.org/10.1007/s10472-019-09626-w

21. Kudriavtsev, A., et al.: The shortest simple path problem with a fixed number of must-pass nodes: a problem-specific branch-and-bound algorithm. In: Simos, D.E., Pardalos, P.M., Kotsireas, I.S. (eds.) LION 2021. LNCS, vol. 12931, pp. 198–210. Springer, Cham (2021). https://doi.org/10.1007/978-3-030-92121-7_17

22. Martin, S., Magnouche, Y., Juvigny, C., Leguay, J.: Constrained shortest path tour problem: branch-and-price algorithm. Comput. Oper. Res. **144**, 105819 (2022). https://doi.org/10.1016/j.cor.2022.105819

23. Morin, T.L., Marsten, R.E.: Branch-and-bound strategies for dynamic programming. Oper. Res. **24**(4), 611–627 (1976)

24. Ogorodnikov, Y., Rudakov, R., Khachay, D., Khachay, M.: A problem-specific branch-and-bound algorithm for the protected shortest simple path problem with must-pass nodes. IFAC-PapersOnLine **55**, 572–577 (2022). https://doi.org/10.1016/j.ifacol.2022.09.455

25. Papadimitriou, C.: Euclidean TSP is NP-complete. Theor. Comput. Sci. **4**, 237–244 (1977)

26. Ropke, S., Pisinger, D.: An adaptive large neighborhood search heuristic for the pickup and delivery problem with time windows. Transp. Sci. **40**, 455–472 (2006). https://doi.org/10.1287/trsc.1050.0135

27. Saksena, J.P., Kumar, S.: The routing problem with 'k' specified nodes. Oper. Res. **14**(5), 909–913 (1966)

28. Salman, R., Carlson, J.S., Ekstedt, F., Spensieri, D., Torstensson, J., Söderberg, R.: An industrially validated CMM inspection process with sequence constraints. Procedia CIRP **44**, 138–143 (2016). https://doi.org/10.1016/j.procir.2016.02.136

29. Salman, R., Ekstedt, F., Damaschke, P.: Branch-and-bound for the precedence constrained generalized traveling salesman problem. Oper. Res. Lett. **48**(2), 163–166 (2020). https://doi.org/10.1016/j.orl.2020.01.009

30. Schilling, L., Seuring, S.: Linking the digital and sustainable transformation with supply chain practices. Int. J. Prod. Res. 1–25 (2023). https://doi.org/10.1080/00207543.2023.2173502

31. Smith, S.L., Imeson, F.: GLNS: an effective large neighborhood search heuristic for the generalized traveling salesman problem. Comput. Oper. Res. **87**, 1–19 (2017). https://doi.org/10.1016/j.cor.2017.05.010

Game Theory and Mathematical Economics

Reciprocal Import Tariffs in the Monopolistic Competition Open Economy

Natalia Aizenberg[1]([✉]) [iD] and Igor Bykadorov[2] [iD]

[1] Melentiev Energy Systems Institute SB RAS, Irkutsk, Russia
ayzenberg.nata@gmail.com
[2] Sobolev Institute of Mathematics SB RAS, Novosibirsk, Russia

Abstract. We use the standard Krugman's one-sector trade model, with unspecified variable-elasticity preferences. We study the impact of reciprocal import tariffs on welfare among symmetric countries (a free-trade agreement). We show that, without transport costs, any ad valorem tariffs or subsidies are always harmful. This is also true under "flatter" demands, satisfying the realistic assumptions of increasingly elastic demand (IED) and decreasingly-elastic utility (DEU).

Keywords: International trade · positive optimal tariffs · monopolistic competition

1 Introduction

Trade liberalization has led to a reduction in reciprocal import tariffs, and in some countries, they have disappeared altogether. This raises the question of the appropriateness of tariffs for countries' welfare. Are these just the result of insufficiently cooperative behavior of governments, or can be supported on welfare grounds, as a common (Pareto) efficient policy for all trading countries?

To give a theoretical answer, we: (1) step aside from the vast theoretical literature on "unilaterally optimal tariffs" towards reciprocal tariffs; (2) choose the classical general-form [1] model as a basic. It includes N symmetric countries, one differentiate good, and one factor of production, is called labor. The "general-form" means that neither constant elasticity of substitution (CES) specification, but we seek restrictions on functional properties like elasticity and curvature of *any additive utility function*, that generates our market and welfare effects. On the supply side, we remain with classic Krugman increasing returns to scale: a fixed cost plus linear variable cost, symmetric firms.

We turn to a utility function with variable elasticity since the CES function ignores important properties. For example, speaking of increasing competition under expanding trade, CES predicts social efficiency, i.e., absent distortion. Instead, we consider the demand functions with two kinds of distortions: (1)

N. Olenev et al. (Eds.): OPTIMA 2023, LNCS 14395, pp. 191–206, 2023.
https://doi.org/10.1007/978-3-031-47859-8_14

excessive entry of firms under *DEU* property under expanding trade – Decreasingly Elastic Utility [2]; (2) incomplete pass-through under Increasingly Elastic Demand – *IED* property.

This paper reveals these two distortions or economic forces' interaction with the demand properties and tariffs – with the help of the classical New Trade model. We consider *ad valorem* tariffs. We assume any elementary utility function suitable for monopolistic competition modeling, notably: increasing, strictly concave, generating elastic demand and strictly concave profit function.

Our results for situations without transportation costs, with ad valorem tariffication, are summarized in Theorems 1, 2. They support the intuitions of a typical trade economist: *any reciprocal positive ad valorem tariff or negative tariff (subsidy) – deteriorates welfare of all trading countries* in this situation. This impact is achieved through *decreasing* firms' output and increasing number (variety of goods), and distortion between domestic/imported consumption, no surprise. In other words, the socially optimal tariff here is zero, irrespectively of any demand properties, the excess-entry distortion cannot be cured.

In the **literature**, there are different conclusions about the benefits from tariffs under perfect or oligopolistic competition (see review in [3,4]), considering a differentiated good and monopolistic competition. We consider New Trade. Usually, unlike us, the dominant hypothesis is the constant elasticity of substitution. In particular, [5] shows the (negative) effects of a tariff war. There is an estimation method for gains/losses from tariff reductions, considering reciprocal tariffs between the two countries. This approach (except for *CES*) is similar to ours as in [6]. Another articles consider a model like ours, but with *CES* preferences and with the second sector (see [7]), and as a variant of the Melitz type model with firms of different sizes [8]. As a result, a one-way tariff may be beneficial, but reciprocal tariffs (tariff war) lead to a welfare loss, which is consistent with our findings under *CES*. An earlier article with *CES* preferences and similar negative conclusions for a bilateral tariff is [9]. [10] and [11] study the impact of trade costs and *CES* preferences on a single country, however, we assume (1) tariffs; (2) *VES* preferences, and; (3) global tariff impact in a K-country world. Our study exploits the parameters of utility consistent with the trade elasticity reported by [12], where the elasticity coefficient is found empirically using gravity equations (elasticity of imports from data on world trade).

Summarizing, several articles have considered the optimal tariff problem under various hypotheses about sectors, preferences, and heterogeneous firms. However, our tariff effects associated with variable elasticity of substitution and tariffs curing excessive variety – have been little studied. Perhaps only in [13], researchers considered such issues. We fill this lacuna in sufficient generality: under any additive utilities using the approaches outlined in [14].

2   Model

We consider the set $\mathcal{K} = \{1, ..., K + 1\}$ of $K + 1$ countries. There is one differentiated good and one aggregate factor of production. It is "labor", inelastically supplied by uniform workers – consumers. Their mass in any country k is L_k (the symmetric case $L_k = L_j$ enable stronger results, but there are some asymmetric results also). Each firm i from country k produces its own variety indexed ik.

Consumers. Consumers' preferences in all countries $j \in \mathcal{K}$ are the same and given by the additively-separable utility function

$$U = U_j = \sum_{k \in \mathcal{K}} \int_0^{N_k} u(x_{ikj}) di,$$

where ikj is the index of a specific variety produced in country k by firm i and sold in country j, where $i \in [0, N_k]$ takes continuous values from the named interval. N_k is the total mass of varieties produced in country k, related consumption x_{ikj} being a function of i.

Assumption 1. Traditionally (see [15,16]), function u must be strictly increasing, strictly concave, thrice differentiable, satisfying the boundary conditions

$$u(0) = 0, \ u'(0) < \infty, \ u''(0) > -\infty, \ u'''(0) \in (-\infty, \infty).$$

The price of a variety is denoted by p_{ikj}. With these notations, utility maximization of every consumer in any country j can be expressed as:

$$\sum_{k \in \mathcal{K}} \int_0^{N_k} u(x_{ikj}) di \to \max_{\{x_{ikj}\}_{i \in [0, N_k], k \in \mathcal{K}}},$$

$$\sum_{k \in \mathcal{K}} \int_0^{N_k} p_{ikj} x_{ikj} di = w_j + T_j. \tag{1}$$

Here income $w_j + T_j$ includes some transfer T_j from the government and the consumer's labor endowment 1, multiplied by wage w_j. When maximizing her utility, each consumer ignores the dependence of transfer T_j on her choice. Using Assumption 1, we derive the consumer's first-order condition, that defines the *inverse demand function* for variety i:

$$p_{ikj} = \frac{u'(x_{ikj})}{\lambda^j}, \quad i \in [0, N_k], k, j \in \mathcal{K}. \tag{2}$$

Here the Lagrange multiplier λ^j is the marginal utility of income.

Demand Properties and Related Notations. From now on, we extensively exploit elasticities, denoting the elasticity of any function g in several forms

$$\mathcal{E}_g(x) \equiv \frac{x g'(x)}{g(x)}.$$

As in [1] and [16], elasticities help to define the Arrow-Pratt measure r_u of concavity of the elementary utility u ("love for variety") and concavity of u':

$$r_u(x) \equiv -\frac{x u''(x)}{u'(x)} > 0, \ r_{u'}(x) \equiv -\frac{x u'''(x)}{u''(x)}. \tag{3}$$

For demand $x(p)$, love for variety expresses the (absolute value of) inverse demand elasticity:

$$r_u(x(p)) \equiv -\frac{1}{\mathcal{E}_x(p)}. \tag{4}$$

Perceiving u' as a price of x, we define now the *elementary revenue function* and express its elasticity through love for variety:

$$R(x) \equiv xu'(x), \quad \mathcal{E}_R(x) \equiv \frac{xR'(x)}{R(x)} = 1 - r_u(x),$$

having in mind that the real per-variety revenue $xu'(x)/\lambda$ will have the same elasticity.

These characteristics are important because when concavity $r_u(\cdot)$ increases (decreases) with consumption, the inverse demand becomes less (more) sensitive, which yields natural (unnatural) market effects. We distinguish two classes of preferences: with increasing elasticity of (inverse) demand (IED): $r'_u(x) > 0$ and with decreasing elasticity (DED): $r'_u(x) < 0$. These two cases respond differently to changes in market parameters, the former looking more realistic economically, but we seek a general theory.

Another characteristic is important for comparing equilibrium and social optimum according to [2], is the utility elasticity:

$$\mathcal{E}_u(x) \equiv \frac{xu'(x)}{u(x)} > 0.$$

For elasticity of utility, we distinguish two cases: increasing elasticity of utility (IEU): $\mathcal{E}'_u(x) > 0$, and decreasing elasticity of utility (DEU): $\mathcal{E}'_u(x) < 0$.

Assumption 2. From now on, utility u should satisfy restrictions:

$$r_u(z) < 1, \quad r_{u'}(z) < 2, \tag{5}$$

$$\lim_{z \to 0}[u'(z)] > 0, \quad \lim_{z \to \infty}[u'(z) + zu''(z)] \leq 0,$$

which are traditional in [16] and other papers on monopolistic competition. The condition $r_u < 1$ means elasticity of demand greater than 1, necessary for monopolistic pricing. The condition $r_{u'} < 2$ makes revenues and profits concave, i.e., provides the second-order condition $2u''(z) + zu'''(z) \leq 0$. The boundary conditions indicate positive marginal utility at zero and saturable demand at infinity.[1]

Producers and Tariffs. Each firm produces a single variety, there is a one-to-one correspondence. Each firm has the same labor requirement $c > 0$ to

---

[1] It seems that these boundary conditions are necessary for equilibria existence. But the questions of the equilibria existence (and uniqueness) are separate problems (often not quite simple), which is not the subject of this study. Let us note that in propositions below, we assume that market equilibrium exists (and, moreover, are unique).

produce one unit and the same fixed cost $F > 0$. Respectively, the cost function $C(q) \equiv (F + cL \cdot q) w$ for per-consumer output q under wage w – exhibits positive economies of scale (gross output is $Q \equiv L \cdot q$ for population L).

In practice, ad valorem tariff denotes $t^{ad} < 1$, which is the share of the revenue $p_{ikj}x_{ikj}$ received by the state from importing any quantity x_{ikj}. When positive, $t^{ad} \geq 0$ mean tariffs, but we shall also consider import subsidies, i.e., $t^{ad} < 0$. Related per capita governmental transfer T_j consists of payments from both components of tariffs:

$$T_j = \sum_{k \in \mathcal{K} \setminus \{j\}} \int_0^{N_k} p_{ikj} t^{ad} x_{ikj} di.$$

Thus, consumers get back the tariffs they have paid, they generally do not gain or lose money, but any positive taxation shifts their the choice towards domestic goods.

We reformulate now such tariffs in the form of *a uniform* iceberg-type trade coefficient

$$\tau \equiv \tau^{ad} = \frac{1}{1 - t^{ad}}.$$

In other words, an ad valorem tariff suppresses with coefficient τ^{ad} the revenue of country k from sales of goods from firm i, country j. Maximizing profit π_{ij} of i-th firm in country j can equivalently be performed in prices or quantities:

$$\pi_{ij} \equiv [p_{ijj} - cw_j] L_j x_{ijj} + \sum_{k \in \mathcal{K} \setminus \{j\}} \frac{p_{ijk}}{\tau} L x_{ijk} - F w_j = \tag{6}$$

$$= \left[\frac{u'(x_{ijj})}{\lambda_j} - cw_j \right] L_j x_{ijj} + \sum_{k \in \mathcal{K} \setminus \{j\}} \frac{u'(x_{ijk})}{\lambda_k \tau} L x_{ijk} - w_j F \to \max_{\mathbf{x}_{ijk}}.$$

Since the profit function is strictly concave under our assumptions (5), each producer i in a country behaves similarly, so, we drop index i. We also introduce the "normalized revenue function"

$$R(z) \equiv zu'(z), \tag{7}$$

extensively exploited further as an alternative "primitive" of the model, instead of $u(\cdot)$.

In these terms, the first-order conditions of profit maximization says that marginal revenue equals marginal costs:

$$\frac{R'(x_{jj})}{\lambda_j} L_j = w_j c L_j, \quad \frac{R'(x_{jk})}{\lambda_k} L_k = \tau \cdot w_j c L_k. \tag{8}$$

Then, taking into account the identity $R'(x_{jk}) \equiv u'(x_{jk}) \cdot (1 - r_u(x_{jk}))$ and prices $p_{jk} = \frac{u'(x_{jk})}{\lambda_k} = \frac{\tau cw_j}{1 - r_u(x_{jk})}$, $p_{jj} = \frac{u'(x_{jj})}{\lambda_j} = \frac{cw_j}{1 - r_u(x_{jj})}$, $p_{kj} = \frac{u'(x_{kj})}{\lambda_j} =$

$\frac{\tau c w_k}{1 - r_u(x_{kj})}$, the conditions on marginal revenue in all countries can be expressed either in relative λ_j or in relative wages:

$$\frac{R'(x_{jj})}{R'(x_{jk})} = \frac{\lambda_j}{\lambda_k} \cdot \frac{1}{\tau}, \tag{9}$$

$$\frac{R'(x_{jj})}{R'(x_{kj})} = \frac{w_j}{w_k} \cdot \frac{1}{\tau}. \tag{10}$$

These equations imply that the ratio $\frac{\lambda_j}{\lambda_k}$ of the marginal utilities of income equals the ratio of domestic/export prices. We proceed so far without using country symmetry, while afterwards we use $\frac{w_j}{w_k} \equiv 1$.

Equilibrium. Firms enter the market whenever profit remains positive, while at equilibrium it must vanish. So, the free entry condition for each country j is

$$\pi_j \equiv \left[\frac{u'(x_{jj})}{\lambda_j} - c w_j \right] L_j x_{jj} + \sum_{k \in \mathcal{K} \setminus \{j\}} \frac{u'(x_{jk})}{\lambda_k \tau} L_k x_{jk} - w_j F = 0.$$

Using $\frac{x_{jj} u'(x_{jj})}{\lambda_j} = \frac{x_{jj} c w_j}{1 - r_u(x_{jj})} = \frac{R(x_{jj}) c w_j}{R'(x_{jj})}$, the free entry condition can be rearranged as

$$L_j \frac{R(x_{jj})}{R'(x_{jj})} = \frac{F}{c} + L_j x_{jj}, \tag{11}$$

As usual in such models, at equilibrium the labor market clearing entails another constraint, which is the consumer's budget for each country, including the tax revenue $T_j = \sum_{k \in \mathcal{K} \setminus \{j\}} N_k p_{kj} t^{ad} x_{kj}$:

$$\sum_{k \in \mathcal{K}} N_k p_{kj} x_{kj} = w_j + \sum_{k \in \mathcal{K} \setminus \{j\}} N_k p_{kj} t^{ad} x_{kj}. \tag{12}$$

Further, we mainly consider symmetric countries: $L_j = L_k = L$ and *the same* reciprocal tariff τ. The symmetric wage of each country can be used as a numeraire:

$$w_k \equiv 1, \quad \forall k \in \mathcal{K}.$$

Symmetric Equilibrium. Under reciprocal tariffs τ on imports, the *symmetric free-entry equilibrium* consumptions $\mathbf{x} \in \mathbf{R}^{(K+1)^2}$, and number of firms $N \in \mathbf{R}^{K+1}$ are those that satisfy labor balances (11), budgets (12) and the following $(K+1)^2 + K$ (firm's FOCs and zero-profit) equations:

$$\frac{R'(x_{jj})}{R'(x_{kj})} = \frac{1}{\tau}, \forall j, k \in \mathcal{K} \tag{13}$$

$$L \frac{R(x_{jj})}{R'(x_{jj})} = \frac{F}{c} + L x_{jj}, \tag{14}$$

whereas equilibrium prices are determined from x_{jk} as

$$p_{jj} = \frac{c}{1 - r_u(x^{jj})}, \quad p_{kj} = \frac{c\tau}{1 - r_u(x^{kj})}. \tag{15}$$

Welfare of a citizen in country j is its (symmetric across varieties in each country) gross utility:

$$W_j = \sum_{k \in \mathcal{K}} N_k u(x_{kj}).$$

To simplify our equations using symmetry, all (the same) imported consumptions is denoted $x_{kj} = x_{il} = y$, while the consumptions of domestic goods is denoted $x_{kk} = x_{jj} = x$. The number of firms is also symmetric: $N^k = N^j = N$ and wages are normalized to $w_k \equiv 1$. Then the equilibrium equations (9), (10), (11) for finding consumption are simplified as follows:

$$\frac{R'(x)}{R'(y)} = \frac{1}{\tau}. \tag{16}$$

$$\frac{R(x)}{R'(x)} - x = \frac{F}{cL} \equiv \frac{f}{c}, \tag{17}$$

where f is the notation for per-consumer investment used further. The budget or labor balance (12) for finding the mass of firms N from (x, y) in each country takes the form:

$$\frac{f}{c} + x + Ky = \frac{1}{cN}. \tag{18}$$

Using the labor balance, the per-consumer welfare function can be simplified as a function of parameters f, c, K and two (x, y) variables only:

$$W(x, y) = \frac{u(x) + Ku(y)}{f + cx + cKy}. \tag{19}$$

3   Impact of Tariffs in the Absence of Transport Costs

Now our task in studying reciprocal tariffs is to find the equilibrium response, including production, consumption, variety – and related welfare consequences. We start with the simple case of K symmetric countries without trade costs.

Totally differentiating our Eqs. (16)–(17) in tariff τ, we come to the following linear equations w.r.t. total derivatives $x'_\tau \equiv dx_\tau/d\tau$, $y'_\tau \equiv dy_\tau/d\tau$:

$$R'(x) + \tau R''(x) x'_\tau = R''(y) y'_\tau$$

$$\left[1 - \frac{R''(x)}{R'(x)}\right] x'_\tau + K \left[1 - \frac{R''(y)}{R'(y)}\right] y'_\tau = x'_\tau + Ky'_\tau.$$

Solving these equations (we recall $R'(x) > 0$, $R''(x) < 0$, see Appendix) we express and estimate the impact of a tariff on consumption as follows:

$$x'_\tau = -\frac{KR(y)}{\tau^2 R''(x)\left(x + Ky + \frac{\ell}{c}\right)} \geq 0, \tag{20}$$

$$y'_\tau = \frac{R(x)}{R''(y)\left(x + Ky + \frac{\ell}{c}\right)} \leq 0, \tag{21}$$

which means *increasing* domestic consumption and *decreasing* imports (naturally).

Combining these expressions (see Appendix), we express firm's per-consumer output in (x, y):

$$q'_\tau = x'_\tau + Ky'_\tau = K \cdot \frac{\tau^2 R(x)R''(x) - R(y)R''(y)}{\tau^2 R''(y)R''(x)\left(x + Ky + \frac{\ell}{cL}\right)}. \tag{22}$$

To find a similar property in a more general situation, observe that $(2u''_x + xu'''_x)$ must be negative because of SOC, but the negativity of the whole sum does not follow from any equilibrium condition. Also, this "uniform demand flatness" does not follow from another kind of "demand flatness". The latter assumption, which is popular in the literature, is increasingly elastic demand (IED), which means

$$r'_u(x) \equiv -\frac{\partial}{\partial x}\frac{xu''(x)}{u'(x)} = -\frac{(xu'''(x) + u''(x))u'(x) - x(u''(x))^2}{(u'(x))^2} > 0, \forall x \quad (IED),$$

while the opposite assumption will be called DED: $r'_u(x) < 0 \ \forall x$.

We define one more characteristic of the utility function—increasing elasticity (IEU) or decreasing elasticity (DEU) and by the related the behavior of the revenue function R:

$$\mathcal{E}_u(x) \equiv -\frac{u'(x)x}{u(x)}, \qquad R(x) \equiv u'(x)x = u(x) \cdot \mathcal{E}_u(x).$$

The sign of the numerator in (22) determines the sign of the derivative q'_τ. Using FOC, we can write (22) as

$$q'_\tau = \frac{K(R'(y))^2}{\tau^2 R''(y)R''(x)\left(x + Ky + \frac{\ell}{c}\right)} \cdot \left(\frac{R(x)R''(x)}{(R'(x))^2} - \frac{R(y)R''(y)}{(R'(y))^2}\right). \tag{23}$$

We also use the following complicated condition on concavity r_u of utility u, explained later on:

$$[(\mathcal{E}_{r'_u}(z) + 2) \cdot (1 - r_u(z)) + 2 \cdot r'_u(z) \cdot z] \cdot r'_u(z) > 0. \tag{24}$$

Using these restrictions, the following theorem describes the general impact of an ad valorem tariff on output, consumption, prices and variety. It discusses the local and global changes over the interval $(1, \tau_a)$ between the free trade point $\tau = 1$ and some finite or infinite autarky -point called $\tau_a : y_{\tau_a} = 0, \ (\tau_a \leq \infty)$.

Theorem 1. *For* $K + 1$ *symmetric countries, an increase in their common (reciprocal) ad valorem import tariff coefficient* τ *modifies the trade equilibrium as follows:*

(i) Domestic individual consumptions increase $(dx_\tau/d\tau > 0 \ \forall \tau \in (0, \tau_a))$ *but imports decrease:* $(dy_\tau/d\tau < 0 \ \forall \tau \in (0, \tau_a))$ *with derivatives (20)–(21), domestic consumption always exceeds import* $(x > y)$ *under positive tariffs* $\tau \in (1, \tau_a)$.

(ii) The firm's output $q_\tau \equiv x_\tau + K y_\tau$ *reaction exhibit derivative (22), with zero impact at the free trade situation* $\tau = 1$*, but negative impact near the autarky point* τ_a *(if* τ_a *is finite). In between, at any* $\tau \in (1, \tau_a)$ *the firm's per-consumer output monotonically decreases in* τ *iff the condition (24) on preferences holds. In particular, negative global monotonicity holds under* IED *preferences satisfying also condition* $\mathcal{E}_{r'_u}(z) + 2 \geq 0$*. By contrast, under some DED preferences output may increase in a positive tariff* $\tau \in (1, \infty)$.

(iii) The equilibrium mass of firms N *always responds to any tariff* τ *inversely to the firm output, i.e.,* $dq_\tau/d\tau \cdot dN_\tau/d\tau \leq 0$*, that implies* $dq_\tau/d\tau < 0 \Leftrightarrow dN_\tau/d\tau > 0$.

(iv) Under IED *preferences, prices always respond inversely to consumptions: domestic price* $p_x = \frac{c}{1 - r_u(x)}$ *decreases with* τ*, while the import price* $p_y = \frac{c\tau}{1 - r_u(y)}$ *increases.*

Proof: Behavior of consumptions/output (20)–(21), (22) has already been proven. For proof of items (ii), (iii) see Appendix, while (iv) is obvious.

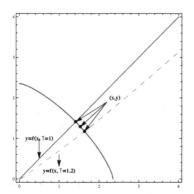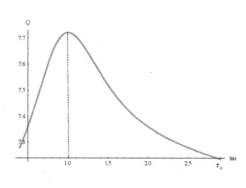

Fig. 1. Ad valorem tariffs' impact on domestic consumption x, import y and output q(under IED). (Color figure online)

For Theorem 1, the geometric intuition behind the changes in consumption can be explained with Fig. 1 (constructed under specific parameters $u(z) = (0.1 + z)^{5.1/6} - 0.1^{5.1/6} - 0.2 \cdot z$, $K = 1$). This figure shows that individual domestic consumption x and export consumption y are positively related to the (producer's) FOC Eq. (16): the increasing red line.

We now formulate such a theorem, using each consumer's welfare function $W(x, y)$ (19), illustrated in Fig. 2. In the formulation, we consider two intervals of changing τ: from some point τ_{x0}, where the domestic consumption disappears $(x_{\tau_{x0}} = 0)$ until free trade, and further from the free trade point $\tau = 1$ to some autarky point $\tau_a \leq \infty$: $y_{\tau_a} = 0$.

Theorem 2. *Symmetric ad valorem tariffs in K symmetric countries have the following impact on welfare:*

(i) Locally, at free trade ($\tau = 1$) welfare has a zero first derivative in tariff τ in any case.

(ii) Under non-strict $IED - DEU$ assumption $(\mathcal{E}'_u(x) \leq 0, r'_u(x) \geq 0)$, or alternative strict $DED - IEU$ assumption $(\mathcal{E}'_u(x) > 0, r'_u(x) < 0)$, welfare W has a strict arg-maximum in a free trade situation $\tau = 1$ and monotonically decreases in tariff τ on the right wing $\tau \in (1, \tau_a]$, in subsidy $(-\tau)$ on the left wing ($\tau \in (0, 1)$) of our interval.

Proof: See Appendix.

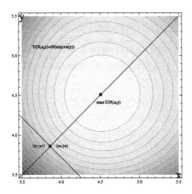

Fig. 2. The indifference curves of welfare $W(x, y)$, the equilibrium consumptions $(x_\tau, y_\tau) \forall \tau$, and the (free-trade) socially optimal equilibrium (x_1, y_1), under DEU and ad valorem tariffs.

Figure 2 explains the geometric intuition for the effects described in Theorem 2 under parameters $(K = 1, L = 1, u = (0.1 + x)^{5.1/6} - 0.1^{5.1/6} - 0.2 \cdot x)$. We display various levels of welfare $W_\tau(\cdot)$ as a function of consumption x, y (the lighter the higher) and observe quasi-concave welfare. An important DEU case is presented, this welfare function has its arg-maximum (x^o, y^o) (social optimum) *above* all equilibria, which means a socially insufficient equilibrium output $Q = L \cdot (x_\tau + y_\tau)$ under any τ. Here all equilibria $((x_\tau, y_\tau) \forall \tau)$ are displayed with the black downward-slope curve (x_τ, y_τ) of equilibria responses to τ. At free trade $((x_1, y_1), \tau = 1)$ the 45-degree line $x + Ky = const = x_1 + Ky_1$ is tangent to the curve of all equilibria $((x_\tau, y_\tau) \forall \tau)$ and also tangent to some indifference curve of welfare $W_\tau(x, y) = W_\tau(x_1, y_1)$, thereby separating all lower outputs (all equilibria) from all situations with higher welfare. Hence, any introduction of tariffs (subsidies) deteriorates welfare. Similarly, the right panel displays the $DED - IEU$ case, only the zone of better welfare is now separated from *higher* outputs.

4   Conclusion

We study *reciprocal* ad valorem import tariffs in the general-form Krugman's model of international trade among several symmetric countries (complimenting the theory of *unilateral* tariffs under monopolistic competition).

It is shown that under any preferences with variable elasticity of substitution – $DEU - CES - IEU$ – any ad valorem tariffs or subsidies are harmful, because they induce product distortion, whereas variety is socially optimal without regulation. The mechanism of such an impact of tariffs on trade is as follows. Such tariffs shrink the firm size, but under DEU, for example, firm size was already too small; the mass of firms in equilibrium exceeds the socially-optimal one [2]. Essentially, the introduction of low import tariffs attracts new businesses and exacerbates non-optimality of equilibrium, adding up to the distortion of demand, expressed in the insufficient consumption of imports. Under "flatter" demands, satisfying the realistic assumptions of increasingly elastic demand (IED) and decreasingly-elastic utility (DEU), any ad valorem tariff deteriorates welfare, when transportation costs are zero.

Acknowledgements. This work was carried out under the State Assignment Project (no. FWEU- 2021-0001) of the Fundamental Research Program of Russian Federation 2021–2030 and under the State contract of the Sobolev Institute of Mathematics (project no. FWNF-2022-0019).

Appendix

Proof of Theorem 1. *i*) Reaction of **consumption**. To find how consumptions in $K + 1$ symmetric countries respond to tariff τ, we denote total derivatives as $x'_\tau \equiv \frac{dx}{d\tau}$, $y'_\tau \equiv \frac{dy}{d\tau}$ and $R'_z \equiv R'(z)$. Our symmetric-equilibrium equations are:
Free entry:

$$\pi(x, y) \cdot \lambda \equiv R(x) + K\frac{R(y)}{\tau} - c \cdot (x + Ky)\lambda - f\lambda = 0,$$

FOC:

$$R'(x) = c\lambda, \quad R'(y) = c\tau\lambda.$$

Thus,

$$\frac{R(x)}{R'(x)} + K\frac{R(y)}{R'(y)} = x + Ky + \frac{f}{c}.$$

Totally differentiating the latter equations in τ (and applying $\frac{d\pi}{dx} = 0$, $\frac{d\pi}{dy} = 0$ or Envelope Theorem to the third equation) we get

$$R''(x)x'_\tau = c\lambda'_\tau, \quad R''(y)y'_\tau = c\lambda + c\tau\lambda'_\tau$$

and

$$\lambda'_\tau = -\frac{KR(y)\left(R'\left(x\right)\right)^2}{\left(R'\left(y\right)\right)^2 \left(c\left(x + Ky\right) + f\right)}.$$

It follows that the total derivatives of consumptions are (20) and (21).

(ii) **Reaction of output.** Under *general* tariff, to find its impact on sales (output), we combine the changes in x and y:

$$q'_\tau = x'_\tau + Ky'_\tau = \frac{K\left(\tau^2 R(x)R''(x) - R(y)R''(y)\right)}{\tau^2 R''(y)R''(x)\left(x + Ky + \frac{f}{c}\right)}. \tag{25}$$

We would like to know the sign of this derivative. For linear demand $R''(x) = R''(y) = constant$, so, the sign is clear: output decreases in τ on the whole interval.

To find the derivative of output $q'_\tau = x'_\tau + Ky'_\tau$ we substitute (20) and (21) into $q'_\tau = x'_\tau + Ky'_\tau$ and

$$q'_\tau \mid_{\tau=1} = 0.$$

To find the derivative of sales at **autarky** $(\tau_a : y(\tau_a) = 0)$ we just plug $y(\tau_a) = 0$ into our formulate and obtain

$$x'_{\tau_a} = -\frac{K \cdot R(0)}{\tau^2 \cdot R''(x)\left(x + \frac{f}{c}\right)} = 0,$$

$$y'_{\tau_a} = \frac{K \cdot R(x)}{R''(0)\left(x + \frac{f}{c}\right)} < 0,$$

$$q'_\tau < 0.$$

To find the derivative of sales at **global point** $(\tau_{x0} : x(\tau_{x0}) = 0)$ we just plug $x(\tau_{x0}) = 0$ into our formula and obtain

$$x'_{\tau_{x0}} = -\frac{KR(y)}{\tau^2 \cdot R''(0)\left(y + \frac{f}{c}\right)} > 0,$$

$$y'_{\tau_{x0}} = \frac{K\tau \cdot R(0)}{R''(y)\left(y + \frac{f}{c}\right)} = 0,$$

$$q'_{\tau_{x0}} > 0.$$

So it decreases when the subsidy grows.

Global impact q'_τ of τ. We use that $\tau = \frac{R'(y)}{R'(x)}$. Substitute τ into (25):

$$q'_\tau = \frac{K\left(R'(y)\right)^2}{\tau^2 R''(y) R''(x)\left(x + Ky + \frac{f}{c}\right)} \cdot \left(\frac{R(x)R''(x)}{\left(R'(x)\right)^2} - \frac{R(y)R''(y)}{\left(R'(y)\right)^2}\right). \quad (26)$$

The sign of the bracket determines the sign of the derivative. Let us introduce the function $\phi(z) \equiv \frac{R(z)R''(z)}{(R'(z))^2}$. If function $\phi(\cdot)$ is decreasing then the derivative q'_τ of the total output is negative under a positive tariff $\tau > 1$.

For IED:

$$\frac{R(z)R''(z)}{\left(R'(z)\right)^2} \equiv \frac{r'_u(z) \cdot z + r_u(z) - \left(r_u(z)\right)^2}{\left(1 - r_u(z)\right)^2}. \quad (27)$$

Find derivative:

$$-\left(\frac{r'_u(z) \cdot z + r_u(z) - \left(r_u(z)\right)^2}{\left(1 - r_u(z)\right)^2}\right)'$$

$$= -\frac{\left(\mathcal{E}_{r'_u}(z) + 2\right) \cdot \left(1 - r_u(z)\right) + 2 \cdot r'_u(z) \cdot z}{\left(1 - r_u(z)\right)^3} \cdot r'_u(z).$$

Proof of Theorem 2 (about welfare)

Recall that any consumer's welfare to be studied is expressed through consumptions as

$$W_\tau(x, y) = \frac{u(x) + Ku(y)}{f + c(x + Ky)}.$$

Using notations $C_q \equiv C(x + Ky) \equiv f + c(x + Ky)$, we estimate the welfare total derivative W'_τ w.r.t. tariff τ:

$$W'_\tau(x, y) = \frac{u'(x) x'_\tau + Ku'(y) y'_\tau}{(f + c(x + Ky))} - \frac{u(x) + Ku(y)}{(f + c(x + Ky))} \cdot \frac{c(x'_\tau + Ky'_\tau)}{(f + c(x + Ky))}.$$

Multiplying everything by $\frac{C_q}{U}$ we come to

$$\frac{C_q}{U} \cdot W'_\tau(x, y) = \frac{u'(x) x'_\tau + Ku'(y) y'_\tau}{u(x) + Ku(y)} - \frac{c(x'_\tau + Ky'_\tau)}{(f + c(x + Ky))},$$

which can be expressed in elasticities as

$$\frac{C_q}{U} \cdot W'_\tau = \frac{x'_\tau}{x} \cdot \left[\mathcal{E}_{U|x} - \mathcal{E}_{C|x}\right] + K\frac{y'_\tau}{y} \cdot \left[\mathcal{E}_{U|y} - \mathcal{E}_{C|y}\right], \quad (28)$$

Unchanging Welfare at the Point of Free Trade. Using elasticities, at free trade we plug $\tau = 1$, $x = y$ into expression (28), substitute $\mathcal{E}_{C|y} = \mathcal{E}_{C|x}$ (which is true everywhere, not only at $x = y$) and obtain

$$\frac{C_q}{U} \cdot W'_\tau = \left[\mathcal{E}_{U|x} - \mathcal{E}_{C|x}\right]\left(\frac{x'_\tau}{x} + K\frac{y'_\tau}{x}\right) = 0$$

because of zero change $q'_\tau = x'_\tau + Ky'_\tau = 0$ at $\tau = 1$, by Theorem 1.

The global welfare change under $IED - DEU$

Using (4) and let $x'_\tau \cdot [A] + Ky'_\tau \cdot [B] \equiv \frac{C_q}{U} \cdot W'_\tau$ we would like to prove to be negative everywhere, except the point of free trade. At free trade, this $\frac{C_q}{U} \cdot W'_\tau$ is zero, because the first bracket is equal to the second one ($A = B$), while $q'_\tau = x'_\tau + Ky'_\tau = 0$ at $\tau = 1$. Both brackets at free trade ($\tau = 1$, $x = y = z$) are positive under DEU ($\mathcal{E}'_u(z) < 0$) because

$$[B]_{\tau=1} = \frac{u'(z)}{2u(z)} - \frac{c}{f + c(1 + K)z} = -\frac{\mathcal{E}'_u(z)}{2\mathcal{E}_u(z)} > 0,$$

and identity $z \cdot \mathcal{E}'_u(z) \equiv \mathcal{E}_u(z) \cdot (1 - \mathcal{E}_u(z) - r_u(z))$ that can be easily derived for any function u.

Further, under positive tariff, the second bracket $[B]$ should *increase* (and remain positive) when τ increases and thereby y decreases. Indeed, differentiating $[B]$ we get

$$[B]'_\tau = \frac{u''(y)\,y'_\tau}{u(x) + Ku(y)} - \frac{u'(y)\,(u'(x)\,x'_\tau + Ku'(y)\,y'_\tau)}{(u(x) + Ku(y))^2} + \frac{c^2\,(x'_\tau + Ky'_\tau)}{(f + c(x + Ky))^2} > 0$$

under DEU. First summands here is positive due to $u''(y)\,y'_\tau > 0$, see part (i) of the theorem. Two of the remaining amount are

$$y'_\tau \nu_y + x'_\tau \nu_x \equiv y'_\tau\left[-\frac{(u'(y))^2}{(u(x) + Ku(y))^2} + \frac{1}{\left(\frac{f}{c} + (x + Ky)\right)^2}\right]$$

$$+x'_\tau\left[-\frac{u''(y)\,u'(x)}{(u(x) + Ku(y))^2} + \frac{1}{\left(\frac{f}{c} + (x + Ky)\right)^2}\right].$$

Let's compare two of the remaining amount ν_y and ν_x. If $-\frac{(u'(y))^2}{(u(x)+Ku(y))^2} + \frac{1}{\left(\frac{f}{c}+(x+Ky)\right)^2} < 0$ then we have $y'_\tau \nu_y > -x'_\tau \nu_x$ (using $\frac{(u'(y))^2}{(u(x)+Ku(y))^2} > \frac{(u'(x))^2}{(u(x)+Ku(y))^2}$, i.e., $u'(z)$ is decreasing function). Then we get that positive summand x'_τ weighted with small positive or negative multiplier ν_x, whereas the negative summand y'_τ is weighted with bigger negative multiplier ν_y, while without multipliers $x'_\tau + Ky'_\tau < 0$ under IED by Theorem 1. Then $y'_\tau \nu_y + x'_\tau \nu_x > 0$.

We prove that $-\frac{\left(u'(y)\right)^2}{(u(x)+Ku(y))^2} + \frac{1}{\left(\frac{L}{c}+(x+Ky)\right)^2} < 0$. So, bracket $[B]$ increases in τ, remaining positive, whereas its multiplier Ky'_τ is negative. At the same time, for all positive tariffs $[A] < [B]$, because $u'(x) < u'(y)$, other parts of these expressions being similar. Further, consider the sum $x'_\tau \cdot [A] + Ky'_\tau \cdot [B]$, where the first positive summand is weighted with a smaller multiplier $[A]$ (positive or negative), then the negative summand Ky'_τ. So, the sum remains negative (provided it was negative: $x'_\tau + Ky'_\tau < 0$ without any multipliers).

One can extend exactly the same reasoning to the case of subsidies, where $\tau = 1 - s < 1$, here welfare decreases in subsidy.

The global welfare change under $DED - IEU$ – just exactly mirror the proof for the previous $IED - DEU$ case. Only both signs change in the derivation: $x'_\tau + Ky'_\tau > 0$ for $\tau \in (1, \tau_a)$ under IED and $\mathcal{E}'_u(z) > 0$ under DEU in assessing the sign of term $[A]$, instead of term $[B]$, both of them starting from a negative value, and $[A]$ decreasing.

References

1. Krugman, P.R.: Increasing returns, monopolistic competition, and international trade. J. Int. Econ. **9**, 469–479 (1979)
2. Dixit, A.K., Stiglitz, J.E.: Monopolistic competition and optimum product diversity. Am. Econ. Rev. **67**, 297–308 (1977)
3. Krugman, P.R., Obstfeld, M.: International Economics: Theory and Policy. Harper Collins College Publishers (1994)
4. Helpman, E.: Trade Understanding Global Trade. Harvard University Press, Cambridge (2011)
5. Gros, D.: A note on the optimal tariff, retaliation and the welfare loss from tariff wars in a framework with intra-industry trade. J. Int. Econ. **23**(3–4), 357–367 (1987)
6. Ossa, R.: A new trade theory of GATT/WTO negotiations. J. Polit. Econ. **119**(1), 122–152 (2011)
7. Pfluger, M., Suedekum, J.: Subsidizing firm entry in open economies. IZA Discussion Paper No. 4384, 41 p. (2012)
8. Bagwell, K., Lee, S.H.: Trade policy under monopolistic competition with firm selection. J. Int. Econ. **127**, Article no. 103379 (2020)
9. Jorgensen, J.G., Schreder, P.J.: Effects of tariffication: tariffs and quotas under monopolistic competition. Open Econ. Rev. **18**, 479–498 (2007)
10. Arkolakis, C., Costinot, A., Rodríguez-Clare, A.: New trade models, same old gains? Am. Econ. Rev. **102**(1), 94–130 (2012)
11. Arkolakis, C., Costinot, A., Donaldson, D., Rodríguez-Clare, A.: The elusive pro-competitive effects of trade. Rev. Econ. Stud. **86**(1), 46–80 (2019)
12. Anderson, J.E., Van Wincoop, E.: Trade costs. J. Econ. Lit. **42**(3), 691–751 (2004)
13. Morgan, J., Tumlinson, J., Vardy, F.: Bad trade: the loss of variety. Available at SSRN 3529246 (2020)
14. Kokovin, S., Molchanov, P., Bykadorov, I.: Increasing returns, monopolistic competition, and international trade: revisiting gains from trade. J. Int. Econ. **137**, Article no. 103595 (2022)

15. Mrázová, M., Neary, J.P.: Not so demanding: demand structure and firm behavior. Am. Econ. Rev. **107**(12), 3835–3874 (2017)
16. Zhelobodko, E., Kokovin, S., Parenti, M., Thisse, J.-F.: Monopolistic competition in general equilibrium: beyond the CES. Econometrica **80**(6), 2765–2784 (2012)

Numerical Modelling of Mean-Field Game Epidemic

Andrei Neverov[1]([✉])[iD] and Olga Krivorotko[2][iD]

[1] Institute of Computational Mathematics and Mathematical Geophysics SB RAS, Novosibirsk, Russia
a.neverov@g.nsu.ru
[2] Sobolev Institute of Mathematics SB RAS, Novosibirsk, Russia
https://math.nsc.ru
https://icmmg.nsc.ru

Abstract. The mean-field game model of infectious disease local propagation is formulated and solved numerically considering social behavior of modelled population. The numerical algorithm based on collocation method is proposed. As a result of numerical modelling with specific assumptions about population, its movement cost, knowledge about infected group, initial distribution and its optimal behavior is acquired and discussed.

Keywords: Mean field game · Kolmogorov-Fokker-Planck · Hamilton-Jacobi-Bellman · Collocation method · SIR model

1 Introduction

The mean-field game model of infectious disease local propagation is formulated and solved numerically, considering the social behavior of the modeled population. The numerical algorithm based on the collocation method is proposed. As a result of numerical modelling with specific assumptions about the population, its movement cost, and knowledge about the infected group's initial distribution, its optimal behavior is acquired and discussed.

The first epidemic models described a population divided into several compartments dedicated to different stages of disease [1]. Such models are named SIR models and are characterized by system of ordinary differential equations (ODEs). SIR model, based on three equations, describes the dynamics of susceptible (S), infectious (I) and removed (R) people. The addition of a new compartment (for example, exposed group (E)) leads to the SEIR model described by 4 ODEs. For modeling the nonuniform epidemic in regions, the spatial distribution of the population is considered in papers [2,3]. It could be described by systems of partial differential equations of parabolic type with an optimal

---

Supported by the Mathematical Center in Akademgorodok.

N. Olenev et al. (Eds.): OPTIMA 2023, LNCS 14395, pp. 207–217, 2023.
https://doi.org/10.1007/978-3-031-47859-8_15

control part that characterized economic or social influence on the epidemic process. These models are characterized by their parameters. For better forecasting the epidemic spread in the population and the influence of economic and social processes on it, the parameters of models should be identified (as solution of inverse problems).

This work considers the problem of epidemic modeling using the social behavior of the population. This is achieved by modeling the population as a group of agents, that are distributed in the space of strategies and that are minimizing their cost functional [4]. We assume that the dynamic of agents is described by a stochastic differential equation (SDE).

$$dx_t = \mu_\alpha(x,t)x_t dt + \sigma dB_t,$$

where x is the trajectory of the representative agent, μ_α is an average speed, σ is a dispersion of Brownian motion B_t. Thus, we may approximate all agents with their distribution function, and their dynamics are described by the Kolmogorov-Fokker-Plank (KFP) equation:

$$\frac{\partial \rho}{\partial t} - \sigma \nabla^2 \rho + \mathbf{min}(\rho \partial_q H(\alpha, \rho)) = 0, \tag{1}$$

where ρ – density function of agents, and $H(q,p) = \alpha(-q \cdot \mu_\alpha - L)$ – Hamiltonian of the system with initial and boundary conditions:

$$\begin{cases} \rho|_{x \in \nabla^2 \Omega} = \rho_1(x,t) \\ \rho(x,0) = \rho_0(x,0) \end{cases} \tag{2}$$

The movement of agents in the space of strategies is motivated by the minimization of cost functional for all agents:

$$J(\rho, \alpha) = E\left[\int_0^T L(x_t, \alpha_t, \rho_t)dt + g(x_T, \rho)\right] \to \mathbf{min}$$

where L – Lagrangian of the system, g – terminal cost, α_t – control functional parameter. The minimization problem of J where ρ is a solution of the problem for KFP (1)–(2) is called the mean-field game (MFG) problem [4]. For the existing and uniqueness of the solution of MFG, we require g to be Lipschitz and ρ_0 to be uniformly continuous [1]. This minimization problem could be rewritten into a system of equations with adjoint differential operators. Such a problem is ill-posed and has different time boundary conditions then (1)–(2), which makes a numerical solution more difficult. We present an attempt to overcome this difficulty by using the collocation method with overdetermination and weights [9]. The paper is organized as follows: in Sect. 2, we formulate the MFG problem for the SIR model that describes the epidemic outbreak in a closed population divided into three groups and based on the mass balance law [1,3]; the numerical algorithm based on the collocation method for solving the MFG problem is presented in Sect. 3; the numerical experiments for the synthetic population and the social behavior are discussed in Sect. 4.

2   Problem Statement

Let us consider the derivation of the MFG system for the basic epidemiological SIR model [1] based on the system of ODEs:

$$\begin{cases} \dfrac{dS}{dt} = -\dfrac{\beta I S}{N}, & S(0) = S_0, \\[2mm] \dfrac{dI}{dt} = \dfrac{\beta I S}{N} - \gamma I, & I(0) = I_0, \\[2mm] \dfrac{dR}{dt} = \gamma I, & R(0) = 0. \end{cases} \tag{3}$$

where S, I, R are groups of susceptible, infectious, and removed people in a closed population, $N = S(t) + I(t) + R(t)$ is a constant population size at any time (in people), β is the infection rate, and γ is the recovery or death rate.

Suppose that people are distributed locally in some territory Ω, for example, around some city.

$$\int_\Omega \begin{pmatrix} s(x,t) \\ i(x,t) \\ r(x,t) \end{pmatrix} dx = \begin{pmatrix} S(t) \\ I(t) \\ R(t) \end{pmatrix} \tag{4}$$

Including diffusion part $\sigma_{\{s,i,r\}} > 0$ (people random walk) and velocity field $\alpha_{\{s,i,r\}} \in L^2(\Omega, [0,T])$ (people deliberate movement), we acquire a system of equations of KFP type:

$$\begin{cases} \dfrac{\partial s}{\partial t} + \dfrac{\beta i s}{N} - \sigma_s \nabla^2 s + \mathrm{div}(-\alpha_s s) = 0, & s(x,0) = s_0(x), \\[2mm] \dfrac{\partial i}{\partial t} - \dfrac{\beta i s}{N} + \gamma i - \sigma_i \nabla^2 i + \mathrm{div}(-\alpha_i i) = 0, & i(x,0) = i_0(x), \\[2mm] \dfrac{\partial r}{\partial t} - \gamma i - \sigma_r \nabla^2 r + \mathrm{div}(-\alpha_r r) = 0, & r(x,0) = 0, \end{cases} \tag{5}$$

that could be rewritten as

$$\begin{cases} \mathcal{L}_1(s,i,r) = 0, & s(x,0) = s_0(x), \\ \mathcal{L}_2(s,i,r) = 0, & i(x,0) = i_0(x), \\ \mathcal{L}_3(s,i,r) = 0, & r(x,0) = 0. \end{cases} \tag{6}$$

Assume that people do not want to be infected by the end of the modelling period and movement is associated with quadratic costs:

$$J = \int_\Omega \int_0^T |\alpha(x,t)|^2 dt dx + i^2(x,T) dx$$

We minimize this functional with restrictions on s, i, r from the Eqs. (6) with Lagrangian approach. This results in a modified functional:

$$J^* = \int\limits_{\Omega} \int\limits_0^T \left[\varphi_s \mathcal{L}_1(s,i,r) + \varphi_i \mathcal{L}_2(s,i,r) + \varphi_r \mathcal{L}_3(s,i,r) + |\alpha(x,t)|^2 \right] dt dx$$
$$+ i^2(x,T) dx$$

For example, a derivative with respect to i results in

$$0 = \frac{\partial J^*}{\partial i} = \int\limits_{\Omega} \int\limits_0^T [\varphi_s \left(\frac{\beta s}{N} \right) + \frac{\partial}{\partial i} \varphi_i \left(\frac{\partial i}{\partial t} - \beta \frac{is}{N} + \gamma i - \sigma_i \nabla^2 i + \mathbf{div}(-\alpha_i i) \right) - $$
$$- \varphi_r \gamma] dt dx + 2i(x,T) dx.$$

Using integration by parts, we move derivatives from i to φ_i, and get an adjoint equation

$$\frac{\partial}{\partial i} \int\limits_{\Omega} \int\limits_0^T \varphi_s L_1 dt dx = \int\limits_{\Omega} \int\limits_0^T \frac{\partial}{\partial i} \varphi_s \left(\frac{\partial s}{\partial t} + \frac{\beta i s}{N} - \sigma_s \nabla^2 s + \text{div}(-\alpha_s s) \right) = $$

$$= \int\limits_{\Omega} \int\limits_0^T \varphi_s \frac{\beta s}{N} dt dx$$

$$\frac{\partial}{\partial i} \int\limits_{\Omega} \int\limits_0^T \varphi_i L_2 dt dx = \int\limits_{\Omega} \int\limits_0^T \frac{\partial}{\partial i} \varphi_i \left(\frac{\partial i}{\partial t} - \frac{\beta i s}{N} + \gamma i - \sigma_i \nabla^2 i + \text{div}(-\alpha_i i) \right) = $$

$$= \int\limits_{\Omega} \int\limits_0^T \left(-\frac{\partial \varphi_i}{\partial t} - \varphi_i \left(\frac{\beta s}{N} + \gamma \right) - \sigma_i \nabla^2 \varphi_i - \nabla \varphi_i \cdot \alpha_i \right) dt dx + $$

$$+ \frac{\partial}{\partial i} \int\limits_{\Omega} \varphi_i (\sigma_i \frac{\partial i}{\partial x} + i \alpha_i)|_{t=0}^T dx + \frac{\partial}{\partial i} \int\limits_0^T i \varphi_i |_{x=x_0}^{x_1} dt$$

$$\frac{\partial}{\partial i} \int\limits_{\Omega} \int\limits_0^T \varphi_r L_3 dt dx = \int\limits_{\Omega} \int\limits_0^T \frac{\partial}{\partial i} \varphi_r \left(\frac{\partial r}{\partial t} - \gamma i - \sigma_r \nabla^2 r + \text{div}(-\alpha_r r) \right) = $$

$$= \int\limits_{\Omega} \int\limits_0^T -\gamma \varphi_r dt dx$$

$$\frac{\partial}{\partial i} \int_\Omega \int_0^T \nabla^2 (t-T) i(x,t)^2 dt dx = \int_\Omega \int_0^T 2\nabla^2 (t-T) i(x,t) dt dx =$$

$$= \int_\Omega 2i(x,T) dt dx$$

$$-\frac{\partial \varphi_i}{\partial t} + \beta s \left(\varphi_s - \varphi_i\right)/N + \gamma(\varphi_i - \varphi_r) - \sigma_i \nabla^2 \varphi_i - \nabla\varphi \cdot \alpha_i = 0$$

with terminal and boundary conditions

$$\varphi_i \left(x, T\right) = 2i \left(x, T\right),$$
$$\varphi_i \left(x, t\right)|_{x \in \nabla^2 \Omega} = 0.$$

And taking the derivative by α_i we get

$$0 = \frac{\partial J^*}{\partial \alpha_i} = \int_\Omega \int_0^T \left[\frac{\partial}{\partial \alpha_i}\varphi_i \left(\mathbf{div}\left(-\alpha_i i\right)\right) + 2\alpha_i\right].$$

Thus

$$\alpha_i = -\frac{\nabla\varphi_i}{2i}.$$

Likewise, other derivatives with respect to $(s, i, r, \varphi_s, \varphi_i, \varphi_r, \alpha_s, \alpha_i, \alpha_r,)$ result in acquiring a system of Eqs. (3) with $\alpha_p = -\frac{\nabla\varphi_p}{2p}$, where $p \in (s, i, r)$ called Hamilton-Jacobi-Bellman (HJB):

$$\begin{cases} -\frac{\partial \varphi_s}{\partial t} + \beta i \left(\varphi_s - \varphi_i\right) N - \sigma_s \nabla^2 \varphi_s + \frac{\nabla\varphi_s^2}{2} = 0, & \varphi_s(x, T) = 0, \\ -\frac{\partial \varphi_i}{\partial t} - \beta s \left(\varphi_s - \varphi_i\right) N - \sigma_i \nabla^2 \varphi_i + \frac{\nabla\varphi_i^2}{2} = 0, & \varphi_i(x, T) = 2i(x, T), \\ -\frac{\partial \varphi_r}{\partial t} - \sigma_r \nabla^2 \varphi_r + \frac{\nabla\varphi_r^2}{2} = 0, & \varphi_r(x, T) = 0, \\ \varphi_s(x, T) = 0 \\ \varphi_i(x, T) = 2i(x, T) \\ \varphi_r(x, T) = 0 \end{cases} \tag{7}$$

with uniform boundary conditions

$$\varphi_p \left(x, t\right)|_{x \in \nabla^2 \Omega} = 0, p \in (s, i, r).$$

Combining the KFP (5) and HJB (7) systems, we acquired the MFG system for the SIR model.

3   Numerical Method

For numerical modelling, we used a modified collocation method with overdetermination. Converging results were acquired in works [5–7]. This method was chosen because of its versatility in application to unusual boundary problems. The MFG system has initial conditions for KFP and terminal conditions for HJB equations, which makes it difficult to solve without forward-backward iterations in time, while these iterations result in converging to the local minimum of cost functional J, what is known as the turnpike effect [8]. Let us describe the method for a one-dimensional case. At the start, we split the region of computation into subregions.

$$A = (x_0, x_{N_x}) \times (t_0, t_{N_t}), C_{(m,n)} = (x_m, x_{m+1}) \times (t_n, t_{n+1}),$$
$$x_0 < x_1 < \cdots < x_m < \cdots < x_{N_x},$$
$$t_0 < t_1 < \cdots < t_n < \cdots < t_{N_t}.$$

Then we consider the solution to be the sum of functions defined on subregions, i.e., the solution is approximated in the form of a polynomial spline.

$$u(x,t) = \sum_{m=0}^{N_t} \sum_{n=0}^{N_x} u_{m,n}(x,t) I((x,t) \in C_{m,n})$$

$$u_{m,n}(x,t) = \sum_{k=0}^{P_t} \sum_{l=0}^{P_x} c_{m,n,k,l} t^k x^l$$

When we substitute this approximation from an arbitrary subregion (m, n) into the KFP equation, we acquire the following equation:

$$\sum_{i=0}^{P_t} \sum_{l=0}^{P_x} c_{k,l} \left(\frac{\partial}{\partial t} t^k x^l - \sigma \nabla^2 t^k x^l - \mathbf{div}(t^k x^l \cdot \varphi(x,t)) \right) = 0 \tag{8}$$

As soon as this equation cannot be satisfied in every point of the subregion, we choose a set of collocation points, where this equation must be satisfied. This gives us a system of nonlinear algebraic equations that is linearized with the Newton method. The same way we obtain connection and boundary conditions at the boundaries of subregions with weights:

$$\begin{cases} w_0 u_{m,n}(x_{m+1}, t) = w_0 u_{m+1,n}(x_{m+1}, t) \\ w_1 \frac{\partial}{\partial x} u_{m,n}(x_{m+1}, t) = w_1 \frac{\partial}{\partial x} u_{m+1,n}(x_{m+1}, t) \\ w_0 u_{m,n}(x, t_{n+1}) = w_0 u_{m,n+1}(x, t_{n+1}) \\ w_{0b} u_{0,n}(x_0, t) = w_{0b} f_0(t) \\ w_{1b} u_{N_x,n}(x_{N_x}, t) = w_{1b} f_1(t) \\ w_b u_{m,0}(x, t_0) = w_b u_0(x) \end{cases} \tag{9}$$

This way, we may satisfy both initial for KFP and terminal for HJB conditions.

The algorithm of numerical solving of systems (5)–(7) based on the collocation method is following:

1. Set discretization of computational area by setting size of subregions, number and places of collocation, connection, and boundary points. Set basis of solution.
2. By substitution of solution in Eqs. (5), (7) and evaluating basis functions in chosen points, acquire system of non-linear Eqs. (8).
3. Linearize system with Newton approach.
4. Solve linearized system of equations until relaxation.

4   Numerical Results

Suppose that the modelled domain could be parameterized in space by the parameter $x \in [0, 1]$. In numerical experiments for the SIR model in terms of the MFG model, we choose the initial conditions of populations as:

$$\begin{cases} s(x,0) = 0.95 \cdot 6\left(x - \frac{1}{3}\right)^2 \left(\frac{2}{3} - x\right), \\ i(x,0) = 0.05 \cdot 6\left(x - \frac{1}{3}\right)^2 \left(\frac{2}{3} - x\right), & , x \in [1/3, 2/3] \\ r(x,0) = 0. \end{cases} \tag{10}$$

and equals zero elsewhere.

This means that the density of susceptible s and infectious i individuals is concentrated in regions $1/3 < x < 2/3$ and with amplitudes $S(0) = 0.95 \cdot 0.6$ and $I(0) = 0.05 \cdot 0.6$ respectively. We consider the start of the epidemic outbreak while there are no immune people.

In Fig. 1 the inaccurate fulfillment of the boundary conditions is demonstrated, as well as a violation of the continuity and smoothness conditions for the solution between the cells for the functions φ_s and φ_r, the values of which are comparable with the reached residual error of the system. The smallness of these values indicates the inactivity of the susceptible and recovered groups. These violations are connected with overdetermination of the linear system of equations and solving it in the least squares sense. On the other hand, a group of infected people "leaves" the crowded place in the center of the computational area, thus showing that the self-isolation strategy is optimal under the existing assumptions.

The numerical method caused additional complexity due to the poor conditionality of the problem, which had to be dealt with by manually selecting the weight parameters in (9) [9–11].

To achieve more stable results we enforce border and smoothness conditions with multiplying them by larger weights. This resulted in more accurate results (Fig. 2).

In Fig. 2 boundary conditions are well met, for all functions. For ψ_s and ψ_r we see the same situation as in Fig. 1, as values of this functions are order of error

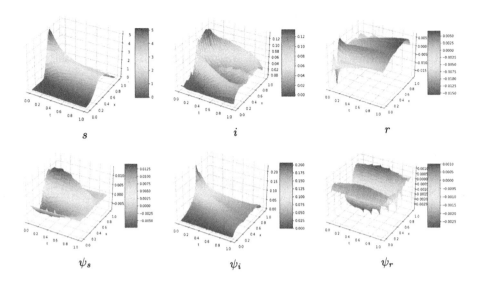

Fig. 1. Results of the modelling for SIR-MFG system with initial conditions 10.

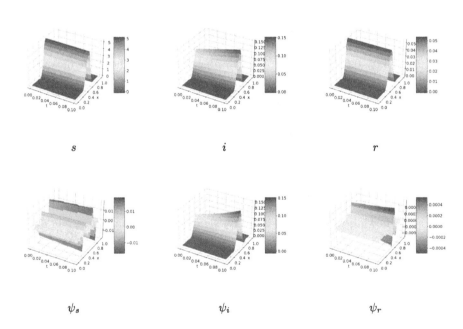

Fig. 2. Results of the modelling for SIR-MFG system with initial conditions (10) and enforced border and smoothness conditions.

values. Smooth solutions do not achieve interesting situation with segregation of s and i parts of population.

To illustrate effects of changes in cost functional J we introduce weighted variant:

$$J_d = \int_\Omega \int_0^T \left(d_0|\alpha|^2 + d_1(-s^2 + i^2) \right) dt dx + d_2 i(T, x)^2 dx \qquad (11)$$

where part with d_0 coefficient represents transport cost, part with d_1 represents cost of falling ill, and d_2 represents cost of being ill at terminal time.

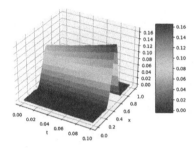

Fig. 3. i compartment in results of modelling SIR-MFG system, if no control is applied (i.e. $d_1 = d_2 = 0$)

We consider coefficient $d_0 = 1$ to see effects of ratios of d_1/d_2 on i as s and r compartments do not change significantly on short period of time, and controls on this compartments are almost 0 (Fig. 4).

In Fig. 4 we see, that d_2 coefficient has much greater influence on system, as it forces $i(T)$ to be lower much faster, thus control with terminal cost is more efficient, than control throughout time.

5   Conclusion

An algorithm for the numerical solution of the MFG system is obtained, which allows solving a pair of KFP and HJB equations without iterations in time as well as for an arbitrary number of discrete states of agents. The disadvantage of this algorithm is the inaccurate fulfillment of the boundary conditions. Additionally, there is the possibility of setting matching conditions between grid cells, for example, fulfilling the conservation law, or matching solution derivatives with different weight coefficients to reduce the conditionality of the resulting system; however, there is no algorithm for selecting these coefficients yet.

The operation of the algorithm is demonstrated by the example of an epidemiological SIR model written in terms of MFG. The interpretation of the results showed the adequacy of the model.

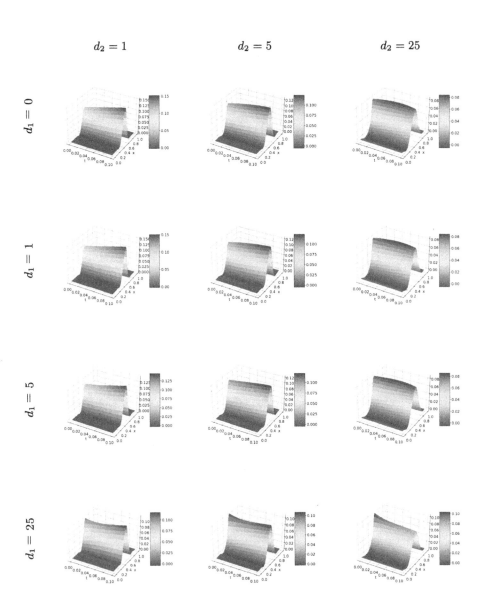

Fig. 4. i compartment dynamics comparison in results of the modelling for SIR-MFG system with initial conditions (10) and different weights d_1 and d_2 in cost functional. In rows, $d1$ changes in $\{0,1,5,25\}$

Acknowledgements. The work is supported by the Russian Science Foundation, project No. 23-71-10068.

References

1. Kermack, W.O., McKendrick, A.G.: A contribution to the mathematical theory of epidemics. Proc. R. Soc. **115**, 700–721 (1927)
2. Bognanni, M., Hanley, D., Kolliner, D., Mitman, K.: Economics and Epidemics: Evidenceliman Estimated Spatial Econ-SIR Model. Finance and Economics Discussion Series (2020)
3. Petrakova, V., Krivorotko, O.: Mean field game for modeling of COVID-19 spread. J. Math. Anal. Appl. **514**, 126271 (2022)
4. Lasry, J.-M., Lions, P.-L.: Mean field games. Jpn. J. Math. **2**(1), 229–260 (2007)
5. Houstis, E.: A collocation method for systems of nonlinear ordinary differential equations. J. Math. Anal. Appl. **62**, 24–37 (1978)
6. Ascher, U., Christiansen, J., Russell, R.D.: A collocation solver for mixed order systems of boundary value problems. Math. Comput. **33**(146), 659–679 (1979)
7. Cerutti, J.: Collocation for Systems of Ordinary Differential Equations. Computer Sciences Technical Report 230. University of Wisconsin-Madison (1974)
8. Trusov, N.V.: Numerical solution of mean field games problems with turnpike effect. Lobachevskii J. Math. **41**(4), 561–576 (2020). https://doi.org/10.1134/S1995080220040253
9. Belyaev, V., Bryndin, L., Golushko, S., Semisalov, B., Shapeev, V.: H-, P-, and HP-versions of the least-squares collocation method for solving boundary value problems for biharmonic equation in irregular domains and their applications. Comput. Math. Math. Phys. **62**, 517–537 (2022)
10. Belyaev, V.: Solving a Poisson equation with singularities by the least-squares collocation method. Numer. Anal. Appl. **13**, 207–218 (2020)
11. Shapeev, V., Golushko, S., Belyaev, V., Bryndin, L., Kirillov, P.: New versions of the least-squares collocation method for solving differential and integral equations. In: Journal of Physics: Conference Series, vol. 1715, no. 1, p. 012031 (2021)

Models of Decision-Making in a Game with Nature Under Conditions of Probabilistic Uncertainty

Victor Gorelik[1,2] and Tatiana Zolotova[3(✉)]

[1] FRC CSC RAS, Vavilova Str. 40, 119333 Moscow, Russia
[2] Moscow Pedagogical State University, M. Pirogovskaya Str. 1/1, 119991 Moscow, Russia
[3] Financial University under the Government of RF, Leningradsky Prospekt 49, 125993 Moscow, Russia
tgold11@mail.ru

Abstract. A game with nature for known state probabilities is considered. An optimality principle is proposed for decision-making for games with nature, based on efficiency and risk estimates. In contrast to the traditional approach to the definition of a mixed strategy in game theory, this paper considers the possibility of correlation dependence of random payoff values for initial alternatives. Two variants of the implementation of the two-criteria approach to the definition of the optimality principle are suggested. The first option is to minimize the variance as a risk estimate with a lower threshold on the mathematical expectation of the payoff. The second option is to maximize the mathematical expectation of the payoff with an upper threshold on the variance. Analytical solutions of both problems are obtained. The application of the obtained results on the example of the process of investing in the stock market is considered. An investor, as a rule, does not form a portfolio all at once, but as a sequential process of purchasing one or another financial asset. In this case, the mixed strategy can be implemented in its immanent sense, i.e. purchases are made randomly with a distribution determined by the previously found optimal solution. If this process is long enough, then the portfolio structure will approximately correspond to the type of mixed strategy. This approach of using the game with nature, taking into account the correlation dependence of random payoff of pure strategies, can also be applied to decision-making problems in other areas of risk management.

Keywords: Risk management · Optimality principle · Two-criteria approach · Mathematical expectation · Standard deviation

1 Introduction

Decision theory describes and explains the behavior of a complex system consisting of human and information resources. In this case, the decision maker

© The Author(s), under exclusive license to Springer Nature Switzerland AG 2023
N. Olenev et al. (Eds.): OPTIMA 2023, LNCS 14395, pp. 218–231, 2023.
https://doi.org/10.1007/978-3-031-47859-8_16

makes an informed choice between several options, each of which is considered achievable. This selection is based on available information. The result of a combination of the preferences of the decision maker and various decision options is the identification of a subjective decision that best meets the decision criteria [1,2].

When modeling decision-making processes, the game-theoretic approach is widely used [3–13]. Control processes in complex systems are characterized by incomplete information about the state of the system and the environment. If one participant is explicitly distinguished, then a game with nature can be used as a mathematical model for making decisions in such situations. One of the players is a person or an institution acting as a decision-maker. The other player is nature and can affect the outcome of the game to various degrees. Nature constitutes a set of conditions affecting the results of taken decisions. To resolve a game of nature, it is required to apply certain decision-making criteria indicating the choice of the optimal decision which is to be made under the conditions of uncertainty concerning the future states of nature. In Wald's criterion [14], based on the loss function, the optimal decision corresponds to the lowest value of the maximum loss. On the other hand, when using the effectiveness function, the decision that maximizes the lowest value of the effectiveness function is the optimal one. In the case of Hurwicz's criterion [15], the parameter α is adopted to determine the coefficient of pessimism (expectations as to the realization of a given state of nature) about the possible future states of nature. This criterion determines the optimal decision, which maximizes the average value of the lowest and highest decision efficiency function with the weights α and $1 - \alpha$ respectively. For a specific loss function, Hurwicz's criterion determines the optimal decision which minimizes the average value of the highest and the lowest loss function with the weights α and $1 - \alpha$ respectively. Savage's criterion [16], which is based on Wald's criterion, refers to the minimum regret function (alternative loss function) resulting from wrong decisions for particular states of nature. The minimum regret function is formulated based on the decision effectiveness function or the loss function. According to Savage's criterion, it is first necessary to find the relative loss matrix (regret matrix). A loss is defined as the difference between the largest win possible in a particular state of nature, and the win corresponding to the decision currently under investigation.

When building a model and setting an optimization problem, the question arises about the availability of information concerning the states of nature. The definition of the concept of optimality or, as is sometimes said, the principle of optimality, depends on this. In this paper, it is assumed that the decision maker has information about the probabilities of the states of nature, i.e. the case of probabilistic uncertainty is considered (or, as it is fashionable to say, we are talking about risk management).

A large number of works are devoted to the application of mathematical methods in risk-based decision making (see, for example, [17–23]). In the paper [20], a two-criteria approach "efficiency-risk" was proposed to determine the principle of optimality when making decisions in stochastic conditions. The mathematical expectation of the gain was used as an efficiency assessment, and the VAR function was used as a risk assessment. As it is known, the VAR function and variance are the most widely used quantities as a risk assessment (see, for example, [24–26]).

The paper [21] outlined the two-criteria approach "efficiency-risk" to the definition of the principle of optimality in decision-making under stochastic conditions, using the mathematical expectation of the payoff as an efficiency estimate and the standard deviation as a risk estimate. Note that if, under known probabilities of states of nature, we are talking about maximizing the mathematical expectation of the payoff, then using a mixed strategy does not make sense. The situation is different with the two-criterion approach, namely, the optimal mixed strategy, generally speaking, gives a greater gain than any pure strategy.

The main difference of this paper from the traditional approach to the definition of a mixed strategy in game theory is that it takes into account the possibility of correlation dependence of random payoff values of the original alternatives (pure strategies). It should be noted that it is in the two-criteria approach that taking into account correlation becomes essential. Usually, in games with nature, either the mathematical expectation of the payoff or the risk according to Savage is considered as a criterion. In this case, the possible correlation of random payoffs under different pure strategies does not play any role. If there are two criteria, one of which is the standard deviation, taking into account the correlation significantly affects the formulation of the problem and the method of its solution. '

Here we consider two problems: the first is to minimize the variance as a risk criterion with a lower threshold on the mathematical expectation of the payoff; the second is to maximize the mathematical expectation of the payoff with an upper threshold on the variance. Analytical and algorithmic results will be obtained concerning the solution of these problems taking into account the correlation of random payoffs of each pair of pure strategies. These results are illustrated by the example of the investment process using real statistical data.

2   The Problem of Minimizing the Variance Under the Efficiency Constraint

So, we consider the situation when the decision maker can choose one of the strategies (alternatives) $i = 1, \ldots, n$, with a known set of possible options for the states of the environment (nature) $j = 1, \ldots, m$. The gain from the i-th decision in the j-th state of the environment is a_{ij}. The payoff matrix from the implementation of possible solutions is $A = \|a_{ij}\|$. The probabilities of states of nature q_j will be considered known. The decision maker needs to choose the strategy that will lead, if possible, to a greater gain, but at the same time, possible losses due to the ambiguity of the outcome will be as small as possible.

As an estimation of the effectiveness of a pure strategy i we take the mathematical expectation of a payoff $\bar{a}_i = \sum_{j=1}^{m} a_{ij} q_j$, and as a risk estimate - the standard deviation $\sigma_i = \left(\sum_{j=1}^{m} \left(a_{ij} - \bar{a}_i \right)^2 q_j \right)^{0.5}$.

When using a mixed strategy, value $\bar{a}_i$ is a conditional mathematical expectation of payoff under realization of the pure strategy i. We denote by p_i the probability of choosing the pure strategy i. Then the mathematical expectation of payoff when using the strategy $p = (p_1, \ldots, p_n)$ is $\sum_{i=1}^{n} \bar{a}_i p_i$.

Let σ_{ik} be the covariance moments of random values of payoff for pure strategies i and k, which are determined by the formula

$\sigma_{ik} = \sum_{j=1}^{m} (a_{ij} - \bar{a}_i) (a_{kj} - \bar{a}_k) q_j$.

We denote the covariance matrix $D = \|\sigma_{ik}\|$. As known, the covariance matrix is always non-negative definite. In what follows, we will assume a little more, namely, that it is positive definite.

The standard deviation of the random value of payoff for the strategy $p = (p_1, \ldots, p_n)$ in the case of correlation is determined, obviously, by the formula $\sigma = (\sum_{i=1}^{n} \sum_{k=1}^{n} \sigma_{ik} p_i p_k)^{0.5}$ or in the matrix-vector form $\sigma = \langle p, Dp \rangle^{0.5}$, where $\langle \cdot, \cdot \rangle$ denotes the scalar product of vectors.

It is convenient to present all the data in the form of Table 1.

Table 1. Model Data.

	q_1	q_2	$\cdots$	q_m	$\bar{a}_i$	1	2	$\cdots$	n
1	a_{11}	a_{12}	$\cdots$	a_{1m}	$\bar{a}_1$	σ_{11}	σ_{12}	$\cdots$	σ_{1n}
2	a_{21}	a_{22}	$\cdots$	a_{2m}	$\bar{a}_2$	σ_{21}	σ_{22}	$\cdots$	σ_{2n}
$\cdots$	$\cdots$	$\cdots$	$\cdots$	$\cdots$	$\cdots$	$\cdots$	$\cdots$	$\cdots$	$\cdots$
n	a_{n1}	a_{n2}	$\cdots$	a_{nm}	$\bar{a}_n$	σ_{n1}	σ_{n2}	$\cdots$	σ_{nn}

The first m columns of the table are the initial data imported from external sources, and the last $n + 1$ columns are the calculated data.

We introduce n-dimensional vectors $\bar{a} = (\bar{a}_1, \ldots, \bar{a}_n)$ and $e = (1, \ldots, 1)$.

Let us formulate a problem for the minimum variance under a lower bound on the mathematical expectation of the payoff:

$$\min_{p \in P} \langle p, Dp \rangle, \quad P = \{p | \langle \bar{a}, p \rangle \geq a_0, \quad \langle p, e \rangle = 1, \ p \geq 0\}. \tag{1}$$

The set P is non-empty, closed, bounded if the threshold value a_0 is not greater than the maximum of the values $\bar{a}_i$. Hence, for

$$a_0 \leq \max_{i=1,\ldots,n} \bar{a}_i, \tag{2}$$

problem (1) has a solution.

Let us find the left boundary a^* of the range of values a_0, at which the first constraint in problem (1) becomes significant. To do this, consider an auxiliary problem of quadratic programming:

$$d_0 = \min_{p \in P_0} \langle p, Dp \rangle, \quad P_0 = \{p \mid \langle p, e \rangle = 1, \ p \geq 0\}. \tag{3}$$

Problem (3) has a unique solution p^*. Obviously, $a^* = \langle \overline{a}, p^* \rangle$. Denote by $\hat{D}$ an arbitrary square submatrix of the matrix D of dimension $k \times k$, obtained by deleting rows and columns with the same numbers, I_1 - the set of not deleted row and column numbers, I_2 - the set of deleted row and column numbers, $\hat{D}^+$ - additional submatrix obtained from D by deleting rows with numbers from I_1 and columns with numbers from I_2, $\hat{e}$ – part of the vector e of dimension k, $\hat{e}^+$ – part of the vector e of dimension $n - k$, $\hat{a}$ – part of the vector $\overline{a}$ with components from I_1. The following lemma gives a formula for finding a^*.

Lemma 1. *There is a unique matrix $\hat{D}$ such that $\hat{D}^+\hat{p} - \left\langle \hat{D}^{-1}\hat{e}, \hat{e} \right\rangle^{-1} \hat{e}^+ \geq 0$, where*

$$\hat{p} = \left\langle \hat{D}^{-1}\hat{e}, \hat{e} \right\rangle^{-1} \hat{D}^{-1}\hat{e}. \tag{4}$$

wherein

$$a^* = \left\langle \hat{D}^{-1}\hat{e}, \hat{e} \right\rangle^{-1} \left\langle \hat{a}, \hat{D}^{-1}\hat{e} \right\rangle. \tag{5}$$

Proof. Compose the Lagrange function $L_0(p, \mu) = 0.5 \langle p, Dp \rangle + \mu(1 - \langle p, e \rangle)$.

The Karush-Kuhn-Tucker (KKT) extremum conditions for problem (3): $\frac{\partial L_0(p,\mu)}{\partial p_i} = 0, i \in I, \frac{\partial L_0(p,\mu)}{\partial p_i} \geq 0, i \notin I$, where I – the set of indices corresponding to nonzero p_i. For problem (2) these conditions are necessary and sufficient, and since the solution of problem (3) p^* is unique, they are satisfied only for the given vector. For nonzero components of the vector p^*, the first part of the KKT conditions gives the system of equations: $\hat{D}\hat{p} - \mu\hat{e} = 0$. The square submatrices of the positive-definite matrix D are also positive-definite and hence non-degenerate. Therefore $\hat{p} = \mu\hat{D}^{-1}\hat{e}$ and from the restriction we have $\mu \left\langle \hat{D}^{-1}\hat{e}, \hat{e} \right\rangle = 1$. The matrix $\hat{D}^{-1}$ is also positive definite, so $\mu = \left\langle \hat{D}^{-1}\hat{e}, \hat{e} \right\rangle^{-1}$ and $\hat{p} = \left\langle \hat{D}^{-1}\hat{e}, \hat{e} \right\rangle^{-1} \hat{D}^{-1}\hat{e}$, i.e. we get (4). The second part of the KKT conditions leads to the inequality $\hat{D}^+\hat{p} - \left\langle \hat{D}^{-1}\hat{e}, \hat{e} \right\rangle^{-1} \hat{e}^+ \geq 0$.

Multiply the vector (4) by the vector $\hat{a}$:

$$\langle \hat{a}, \hat{p} \rangle = \left\langle \hat{a}, \left\langle \hat{D}^{-1}\hat{e}, \hat{e} \right\rangle^{-1} \hat{D}^{-1}\hat{e} \right\rangle = \left\langle \hat{D}^{-1}\hat{e}, \hat{e} \right\rangle^{-1} \left\langle \hat{a}, \hat{D}^{-1}\hat{e} \right\rangle.$$

Hence $a^* = \left\langle \hat{D}^{-1}\hat{e}, \hat{e} \right\rangle^{-1} \left\langle \hat{a}, \hat{D}^{-1}\hat{e} \right\rangle$, i.e. we get (5). The lemma is proven.

Note that for this case, the KKT conditions are necessary and sufficient. Therefore, if $\hat{p} \geq 0$ and the rest of the KKT conditions are satisfied, namely, the nonnegativity of the derivatives of the Lagrange function with respect to p_i with numbers corresponding to zero components, then the vector $\tilde{p}$, padded with zeros in the corresponding places, is a solution to problem (3).

Thus, the method for solving the problem (3) is reduced to enumerating the square submatrices of the matrix D, solving the systems of equations based on them using the obtained formulas, and checking the conditions for non-negativity of the components of the obtained vectors and the corresponding derivatives of the Lagrange function. Moreover, since the conditions of the KKT are necessary and sufficient, the enumeration stops as soon as a vector satisfying them is found.

In what follows, we will assume that all $\bar{a}_i$ are distinct. We will need this purely technical assumption to formulate a theorem on the method for solving problem (1). It allows us to exclude trivial cases when the optimal solution is a pure strategy. But this assumption is quite natural and does not violate the generality of the consideration.

The following theorem substantiates a method for finding optimal truly mixed (containing at least two nonzero components) strategies.

Theorem 1. *If*

$$a^* < a_0 < \max_{i=1,\ldots,n} \bar{a}_i,$$

all $\bar{a}_i$ are distinct, matrix $D = \|\sigma_{ik}\|$ is positive definite, then problem (1) has a unique solution p^0 and true mixed optimal strategy can be represented as

$$\tilde{p}^0 = \tilde{D}^{-1}\left(\lambda^0 \tilde{a} + \mu^0 \tilde{e}\right), \tag{6}$$

$$\lambda^0 = \frac{\max\left\{a_0 \langle \tilde{e}, \tilde{D}^{-1}\tilde{e}\rangle - \langle \tilde{a}, \tilde{D}^{-1}\tilde{e}\rangle, 0\right\}}{\langle \tilde{a}, \tilde{D}^{-1}\tilde{a}\rangle\langle \tilde{e}, \tilde{D}^{-1}\tilde{e}\rangle - \langle \tilde{a}, \tilde{D}^{-1}\tilde{e}\rangle^2},$$

$$\mu^0 = \frac{\langle \tilde{a}, \tilde{D}^{-1}\tilde{a}\rangle - a_0 \langle \tilde{a}, \tilde{D}^{-1}\tilde{e}\rangle}{\langle \tilde{a}, \tilde{D}^{-1}\tilde{a}\rangle\langle \tilde{e}, \tilde{D}^{-1}\tilde{e}\rangle - \langle \tilde{a}, \tilde{D}^{-1}\tilde{e}\rangle^2}, \tag{7}$$

$\tilde{D}$ *is some (unique) square submatrix of the matrix D obtained by deleting rows and columns with the same numbers, $\tilde{p}^0$ is a vector of nonzero components of the vector p^0, $\tilde{a}$ is a vector of the part of the components of the vector $\bar{a}$, $\tilde{e}$ is a vector from a part of the components of the vector e, obtained by deleting the components with numbers corresponding to the zero components of the vector p^0.*

Proof. If condition (2) is satisfied, the set P is not empty, closed, and bounded; therefore, the convex programming problem (1) has a solution, and it is unique, because the objective function is strictly convex. The KKT conditions for it are necessary and sufficient (in a problem with linear constraints, the Slater regularity condition is not required). The Lagrange function has the form $L_1(p, \lambda, \mu) = 0.5\langle p, Dp\rangle + \lambda(a_0 - \langle \bar{a}, p\rangle) + \mu(1 - \langle p, e\rangle)$, $\lambda \geq 0$. Let I be the set of indices corresponding to non-zero p_i. The KKT extremum conditions for problem (1) have the form $\frac{\partial L_1(p, \lambda, \mu)}{\partial p_i} = 0$, $i \in I$, $\frac{\partial L_1(p, \lambda, \mu)}{\partial p_i} \geq 0$, $i \notin I$.

For the non-zero components of the vector p, we have the system of equations: $\tilde{D}\tilde{p} - \lambda\tilde{a} - \mu\tilde{e} = 0$, where $\tilde{D}$ is a square submatrix of the matrix D obtained by deleting rows and columns with numbers corresponding to the zero components of the vector p, $\tilde{p}$ is a vector of non-zero components of the vector p, $\tilde{a}$ is a vector from the part of the components of the vector $\bar{a}$, $\tilde{e}$ is a vector from the part of the components of the vector e, obtained by deleting the components with numbers corresponding to the zero components of the vector p.

Suppose first that $\lambda > 0$, then the first constraint in (1) is active. As mentioned above, the square submatrices of the positive-definite matrix D are also positive-definite and, therefore, non-degenerate. Therefore, we have $\tilde{p} = \tilde{D}^{-1}(\lambda\tilde{a} + \mu\tilde{e})$. We

substitute this expression into the constraints of problem (1):
$\left\langle \tilde{a}, \tilde{D}^{-1}(\lambda\tilde{a}+\mu\tilde{e})\right\rangle =a_0, \left\langle \tilde{D}^{-1}(\lambda\tilde{a}+\mu\tilde{e}),\tilde{e}\right\rangle = 1$. We transform the first equality to
the form $\lambda\left\langle \tilde{a}, \tilde{D}^{-1}\tilde{a}\right\rangle +\mu\left\langle \tilde{a}, \tilde{D}^{-1}\tilde{e}\right\rangle =a_0$. From the second equality, we express
$\mu= (1-\lambda\left\langle \tilde{e}, \tilde{D}^{-1}\tilde{a}\right\rangle)\left\langle \tilde{e}, \tilde{D}^{-1}\tilde{e}\right\rangle^{-1}$ and substitute into the first:
$\left\langle \tilde{a}, \tilde{D}^{-1}\tilde{a}\right\rangle +(1-\lambda\left\langle \tilde{e}, \tilde{D}^{-1}\tilde{a}\right\rangle)\left\langle \tilde{e}, \tilde{D}^{-1}\tilde{e}\right\rangle^{-1}\left\langle \tilde{a}, \tilde{D}^{-1}\tilde{e}\right\rangle =a_0.$

Thus, taking into account the fact that the matrix $\tilde{D}^{-1}$ is symmetric, we obtain an expression for λ:

$$\lambda=\frac{a_0-\left\langle \tilde{e}, \tilde{D}^{-1}\tilde{e}\right\rangle^{-1}\left\langle \tilde{a}, \tilde{D}^{-1}\tilde{e}\right\rangle}{\left\langle \tilde{a}, \tilde{D}^{-1}\tilde{a}\right\rangle -\left\langle \tilde{e}, \tilde{D}^{-1}\tilde{e}\right\rangle^{-1}\left\langle \tilde{a}, \tilde{D}^{-1}\tilde{e}\right\rangle^2}, \tag{8}$$

or, after transformation $\lambda=\frac{a_0\left\langle \tilde{e},\tilde{D}^{-1}\tilde{e}\right\rangle-\left\langle \tilde{a},\tilde{D}^{-1}\tilde{e}\right\rangle}{\left\langle \tilde{a},\tilde{D}^{-1}\tilde{a}\right\rangle\left\langle \tilde{e},\tilde{D}^{-1}\tilde{e}\right\rangle-\left\langle \tilde{a},\tilde{D}^{-1}\tilde{e}\right\rangle^2}$. Let us show that the
denominator is positive, i.e. there is an inequality

$$\left\langle \tilde{a}, \tilde{D}^{-1}\tilde{a}\right\rangle \left\langle \tilde{e}, \tilde{D}^{-1}\tilde{e}\right\rangle -\left\langle \tilde{a}, \tilde{D}^{-1}\tilde{e}\right\rangle^2>0. \tag{9}$$

Indeed, since $\tilde{D}^{-1}$ is a positive definite matrix, there exists a nondegenerate matrix B such that $\tilde{D}^{-1}=B^T B$. Substituting this decomposition of the matrix into the left-hand side of the inequality (9), we have
$\left\langle \tilde{a}, B^T B\tilde{a}\right\rangle \left\langle \tilde{e}, B^T B\tilde{e}\right\rangle -\left\langle \tilde{e}, B^T B\tilde{a}\right\rangle^2= \left\langle B\tilde{a}, B\tilde{a}\right\rangle \left\langle B\tilde{e}, B\tilde{e}\right\rangle -\left\langle B\tilde{e}, B\tilde{a}\right\rangle^2.$
We apply the Cauchy-Bunyakovsky inequality: $\left\langle x,y\right\rangle^2 \leq \|x\|^2 \cdot \|y\|^2$, setting
$x=B\tilde{a}, y=B\tilde{e}$. In the Cauchy-Bunyakovsky inequality equality holds only if the vectors x and y are collinear. But the vectors $B\tilde{a}$ and $B\tilde{e}$ cannot be collinear, since otherwise, when they are multiplied by the matrix B^{-1}, the vectors $\tilde{a}$ and $\tilde{e}$ are also collinear. This contradicts the condition of the theorem, since by assumption, all $\bar{a}_i$ are distinct, and all components of the vector e are equal to ones. Therefore, if these vectors have at least two components, (9) holds.

The numerator in (8) is non-negative, because otherwise for the submatrix $\tilde{D}$ the threshold value a_0 is less than the mathematical expectation of the payoff corresponding to the minimum of the variance (it follows from the lemma, see formula (5)).

Substituting λ into the expression for μ we have
$$\mu= \left(1-\frac{a_0-\left\langle \tilde{e},\tilde{D}^{-1}\tilde{e}\right\rangle^{-1}\left\langle \tilde{a},\tilde{D}^{-1}\tilde{e}\right\rangle}{\left\langle \tilde{a},\tilde{D}^{-1}\tilde{a}\right\rangle-\left\langle \tilde{e},\tilde{D}^{-1}\tilde{e}\right\rangle^{-1}\left\langle \tilde{a},\tilde{D}^{-1}\tilde{e}\right\rangle^2}\left\langle \tilde{e}, \tilde{D}^{-1}\tilde{a}\right\rangle\right)\left\langle \tilde{e}, \tilde{D}^{-1}\tilde{e}\right\rangle^{-1}=$$
$$=\frac{1}{\left\langle \tilde{e},\tilde{D}^{-1}\tilde{e}\right\rangle}-\frac{a_0\left\langle \tilde{a},\tilde{D}^{-1}\tilde{e}\right\rangle-\left\langle \tilde{e},\tilde{D}^{-1}\tilde{e}\right\rangle^{-1}\left\langle \tilde{a},\tilde{D}^{-1}\tilde{e}\right\rangle^2}{\left\langle \tilde{a},\tilde{D}^{-1}\tilde{a}\right\rangle\left\langle \tilde{e},\tilde{D}^{-1}\tilde{e}\right\rangle-\left\langle \tilde{a},\tilde{D}^{-1}\tilde{e}\right\rangle^2}=\frac{\left\langle \tilde{a},\tilde{D}^{-1}\tilde{a}\right\rangle-a_0\left\langle \tilde{a},\tilde{D}^{-1}\tilde{e}\right\rangle}{\left\langle \tilde{a},\tilde{D}^{-1}\tilde{a}\right\rangle\left\langle \tilde{e},\tilde{D}^{-1}\tilde{e}\right\rangle-\left\langle \tilde{a},\tilde{D}^{-1}\tilde{e}\right\rangle^2}.$$

If $\tilde{p} \geq 0$ and the rest of the KKT conditions are satisfied, namely, the non-negativity of the derivatives of the Lagrange function with respect to p_i with numbers corresponding to zero components, then the vector $\tilde{p}$, padded with zeros at the appropriate places, is a solution to problem (1).

Let now $\lambda = 0$, then we have $\tilde{D}\tilde{p} - \mu\tilde{e} = 0$. Combining both cases, we obtain formulas (6), (7). The theorem has been proven.

Note: If formula (8) gives $\lambda < 0$, i.e. numerator $a_0 - \left\langle \tilde{e}, \tilde{D}^{-1}\tilde{e} \right\rangle^{-1}$ $\left\langle \tilde{a}, \tilde{D}^{-1}\tilde{e} \right\rangle < 0$, then this means that for the given submatrix $\tilde{D}$ the first constraint of problem (1) for a given a_0 cannot be active and the case $\lambda = 0$ takes place. The algorithm for finding a solution to problem (1) includes enumeration of sets of nonzero components I. Since for the convex programming problem (2) the KKT optimality conditions are also sufficient, if a solution satisfies them, then the enumeration process ends.

3   The Problem on Maximum Efficiency with a Restriction to Risk

The problem for the maximum of mathematical expectation of payoff under an upper bound on the standard deviation has the form:

$$\max_{p \in P} \langle \overline{a}, p \rangle \ , \quad P = \{p | \langle p, Dp \rangle^{0.5} \leq \sigma_0, \ \langle p, e \rangle = 1, \ p \geq 0\}. \tag{10}$$

The set P is not empty if the threshold value σ_0 is not less than the minimum value of the standard deviation on the set $P_0 = \{p \mid \langle p, e \rangle = 1, \ p \geq 0\}$. To find this value, you need to solve an auxiliary quadratic programming problem (2). Substituting this vector $\hat{p} = \left\langle \hat{D}^{-1}\hat{e}, \hat{e} \right\rangle^{-1} \hat{D}^{-1}\hat{e}$ into the objective function (2), we obtain the value $d_0 = \left\langle \hat{D}^{-1}\hat{e}, \hat{e} \right\rangle^{-1}$.

In what follows, we will assume again that all $\overline{a}_i$ are different. This does not violate the generality of the consideration, since if two pure strategies have the same mathematical expectation of payoff and the standard deviations are also equal, then such strategies are equivalent within the framework of this approach and one of them can be excluded. If one of these strategies has a larger standard deviation than the other, then, within the framework of this approach, such a pure strategy cannot be included in the optimal mixed strategy with a nonzero weight.

The following theorem substantiates a method for finding the optimal truly mixed (containing at least two nonzero components) strategies.

Theorem 2. *If $\sigma_0 > d_0^{0.5}$, all $\overline{a}_i$ are different, the matrix $D = \|\sigma_{ik}\|$ is positive definite, then the problem (10) has a solution p^0 and the truly mixed optimal strategy can be represented as*

$$\tilde{p}^0 = \lambda^{0^{-1}} \tilde{D}^{-1} \left(\tilde{a} - \mu^0 \tilde{e} \right), \tag{11}$$

$$\lambda^0 = \sqrt{\frac{\left\langle \tilde{a}, \tilde{D}^{-1}\tilde{a} \right\rangle \left\langle \tilde{e}, \tilde{D}^{-1}\tilde{e} \right\rangle - \left\langle \tilde{e}, \tilde{D}^{-1}\tilde{a} \right\rangle^2}{\sigma_0^2 \left\langle \tilde{e}, \tilde{D}^{-1}\tilde{e} \right\rangle - 1}}, \quad \mu^0 = \frac{\left\langle \tilde{e}, \tilde{D}^{-1}\tilde{a} \right\rangle - \lambda^0}{\left\langle \tilde{e}, \tilde{D}^{-1}\tilde{e} \right\rangle}, \tag{12}$$

$\tilde{D}$ is some square submatrix of matrix D obtained by deleting rows and columns with the same numbers, $\tilde{p}^0$ is a vector from nonzero components of the vector p^0, $\tilde{a}$ is a vector from a part of the components of the vector $\bar{a}$, $\tilde{e}$ is a vector from parts of the components of the vector e obtained by deleting the components with numbers corresponding to the zero components of the vector p^0.

Proof. For $\sigma_0 > d_0^{0.5}$ the set P is not empty, closed and bounded; therefore, convex programming problem (10) has a solution and satisfies the Slater condition, and the KKT conditions for it are necessary and sufficient. In problem (10), to apply the extremum conditions, it is more convenient to square the first constraint. Then the Lagrange function has the form

$L_2(p, \lambda, \mu) = \langle \bar{a}, p \rangle + \frac{1}{2}\lambda \left(\sigma_0^2 - \langle p, Dp \rangle \right) + \langle \mu, 1 - \langle p, e \rangle \rangle, \ \lambda \geq 0.$

Let, as before, I be the set of indices corresponding to nonzero p_i. The conditions KKT of extremum for the problem (10) have the form $\frac{\partial L_2(p, \lambda, \mu)}{\partial p_i} = 0, i \in I$, $\frac{\partial L_2(p, \lambda, \mu)}{\partial p_i} \leq 0, \ i \notin I$.

For nonzero components of the vector p, we have a system of equations: $\tilde{a} - \lambda \tilde{D}\tilde{p} - \mu \tilde{e} = 0$, where $\tilde{D}$ is a square submatrix of the matrix D obtained by deleting rows and columns with numbers corresponding to the zero components of the vector p, $\tilde{p}$ is a vector of nonzero components of the vector p, $\tilde{a}$ is a vector from a part of the components of the vector $\bar{a}$, $\tilde{e}$ is a vector from a part of the components of the vector e obtained by deleting the components with numbers corresponding to the zero components of the vector p.

If $\lambda = 0$, then we have $\tilde{a} - \mu \tilde{e} = 0$. But by virtue of the assumption of the theorem that all $\bar{a}_i$ are different, this equality is possible only for one index, so in this case, the optimality conditions can be satisfied only for the set I containing one index. If the quadratic constraint in the problem (10) is not active, then $\lambda = 0$. Therefore, for a truly mixed optimal strategy with at least two components different from zero, the quadratic constraint in (10) must be active and $\lambda > 0$.

As mentioned above, the square submatrices of the positive definite matrix D are also positive definite and, therefore, nondegenerate. Therefore, we have $\tilde{p} = \lambda^{-1}\tilde{D}^{-1}(\tilde{a} - \mu\tilde{e})$. We substitute this expression into the constraints of problem (10): $\left\langle \tilde{D}^{-1}(\tilde{a} - \mu\tilde{e}), (\tilde{a} - \mu\tilde{e}) \right\rangle = \lambda^2 \sigma_0^2, \quad \lambda^{-1} \left\langle \tilde{D}^{-1}(\tilde{a} - \mu\tilde{e}), \tilde{e} \right\rangle = 1.$

We transform the first equality to the form: $\left\langle \tilde{a}, \tilde{D}^{-1}\tilde{a} \right\rangle + \mu^2 \left\langle \tilde{e}, \tilde{D}^{-1}\tilde{e} \right\rangle - 2\mu \left\langle \tilde{e}, \tilde{D}^{-1}\tilde{a} \right\rangle = \lambda^2 \sigma_0^2$. From the second equality, we express μ:

$\mu = \left(\left\langle \tilde{e}, \tilde{D}^{-1}\tilde{a} \right\rangle - \lambda \right) \left\langle \tilde{e}, \tilde{D}^{-1}\tilde{e} \right\rangle^{-1}$, and substitute into the first equality

$\left\langle \tilde{a}, \tilde{D}^{-1}\tilde{a} \right\rangle \left\langle \tilde{e}, \tilde{D}^{-1}\tilde{e} \right\rangle + \left\langle \tilde{e}, \tilde{D}^{-1}\tilde{a} \right\rangle^2 - 2\lambda \left\langle \tilde{e}, \tilde{D}^{-1}\tilde{a} \right\rangle + \lambda^2 -$

$-2 \left(\left\langle \tilde{e}, \tilde{D}^{-1}\tilde{a} \right\rangle^2 - \lambda \left\langle \tilde{e}, \tilde{D}^{-1}\tilde{a} \right\rangle \right) = \lambda^2 \sigma_0^2 \left\langle \tilde{e}, \tilde{D}^{-1}\tilde{e} \right\rangle.$

In this expression, the coefficient at λ is zero. Thus, we obtain a quadratic equation for λ: $\lambda^2 \left(\sigma_0^2 \left\langle \tilde{e}, \tilde{D}^{-1}\tilde{e} \right\rangle - 1 \right) = \left\langle \tilde{a}, \tilde{D}^{-1}\tilde{a} \right\rangle \left\langle \tilde{e}, \tilde{D}^{-1}\tilde{e} \right\rangle - \left\langle \tilde{e}, \tilde{D}^{-1}\tilde{a} \right\rangle^2.$

As it was shown the free term in the last equation is positive (see the inequality

(9)). Let us show that the coefficient at λ^2 is also positive. To do this, we will use the form of solution of the problem (3) obtained above. If we solve a similar problem of minimizing the variance with the covariance matrix $\tilde{D}$, corresponding to the solution of the problem (10), we obtain the minimum value of the variance $\left\langle \tilde{D}^{-1}\tilde{e}, \tilde{e} \right\rangle^{-1}$. By the assumption of the theorem, σ_0^2 is greater than this value, i.e. $\sigma_0^2 > \left\langle \tilde{e}, \tilde{D}^{-1}\tilde{e} \right\rangle^{-1}$. Considering, that $\lambda > 0$, λ is a solution with a plus sign in front of the radical.

If $\tilde{p} \geq 0$ and the rest of the KKT conditions are satisfied, namely, the nonpositiveness of the derivatives of the Lagrange function with respect to p_i with numbers corresponding to zero components, then the vector $\tilde{p}$, padded with zeros in the corresponding places, is a solution to the problem (10).

As a result, we obtain formulas (11) and (12). Q.E.D.

4   Practical Interpretation of the Model on the Example of Investment Management

Let us consider the application of the obtained results on the example of the process of investing in the stock market. Usually, a mixed strategy is interpreted as a vector of shares of financial instruments in a portfolio. Without excluding such an interpretation, we will offer a slightly different point of view. An investor, as a rule, does not form a portfolio all at once, but as a sequential process of purchasing one or another financial asset. In this case, the mixed strategy can be implemented in its immanent sense, i.e. purchases are made randomly with a distribution determined by the previously found optimal solution. If this process is long enough, then the portfolio structure will approximately correspond to the type of mixed strategy. Within the framework of this model, as a game with nature, when applied to the stock market, short sales are unacceptable, because the solution is mixed strategies, the components of which, in principle, cannot be negative.

We will conduct a technical analysis and find the optimal investment strategy using real data on stock quotes of Russian companies for the period from 02/01/2021 to 05/01/2021. This period was chosen because the later data period characterizes the fall of market indices and is associated not so much with economic as with political reasons.

Three relatively successful companies were selected, namely VTB Bank (VTBR), Gazprom (SAGP), Sberbank of Russia (SBER). Based on data on daily closing prices, the daily value of company returns, average returns, variance, and covariance for a given period were calculated (data taken from the site of FINAM Investment Company [27]).

Strategy 1 − investment in shares of VTB Bank, strategy 2 − investment in shares of Gazprom, strategy 3 − investment in shares of Sberbank of Russia. In this case, the average values of returns are equal to $\overline{a}_1 = 0.00548$ (0.548%),

$\bar{a}_2 = 0.00127$ (0.127%), $\bar{a}_3 = 0.002$ (0.2%), the covariance matrix has the form

$$D = \begin{pmatrix} 0.00034 & 0.00010 & 0.000095 \\ 0.00010 & 0.00016 & 0.000094 \\ 0.000095 & 0.000094 & 0.00017 \end{pmatrix}.$$

At first, we will solve the problem (1) for the minimum variance with a constraint on the mathematical expectation of the payoff. According to the condition of Theorem 1, we calculate the left and right ends of the interval

$$\left\langle \hat{D}^{-1}\hat{e}, \hat{e} \right\rangle^{-1} \left\langle \hat{a}, \hat{D}^{-1}\hat{e} \right\rangle < a_0 < \max_{i=1,\dots,n} \bar{a}_i.$$

The solution of problem (3) gives a full-size portfolio $p = (0.11532, 0.44388, 0.44079)$, therefore, for the initial matrix D and the initial vector of expected payoffs $\bar{a} = (0.00548, 0.00127, 0.002)$ we have

$$D^{-1} = \begin{pmatrix} 3815.458 & -1597.97754 & -1296.22457 \\ -1597.97754 & 9840.62394 & -4696.6592 \\ -1296.22457 & -4696.6592 & 9514.1783 \end{pmatrix}, \langle e, D^{-1}e \rangle = 7988.538,$$

$\langle \bar{a}, D^{-1}e \rangle = 16.60911$. Then we get $0.00208 < a_0 < 0.00548$.

Let us solve the problem (1) with the threshold value of the mathematical expectation of the payoff $a_0 = 0.003$. For clarity, we present a detailed procedure for solving this problem using formulas (6), (7).

Let us take $I = \{1, 2, 3\}$, i.e. we use the original vector of expected payoffs $\bar{a} = (0.00548, 0.00127, 0.002)$ and the original covariance matrix D, then we get $\langle \bar{a}, D^{-1}\bar{a} \rangle = 0.093993$. By formulas (7) we have $\lambda = 0.01549$, $\mu = 0.00009$. Using formula (6), we have $p = (0.33775, 0.24212, 0.42013)$.

Let us now solve the problem (1) with the threshold value of the mathematical expectation of the payoff $a_0 = 0.0045$. Let's take $I = \{1, 2, 3\}$, then similarly by formulas (7) we have $\lambda = 0.04071$, $\mu = 0.00004$. Using formula (6), we have $p = (0.70007, -0.08654, 0.38648)$. The non-negativity condition $p \geq 0$ is not satisfied in this case, which means that this vector p is not a solution. Since p_2 is negative in this case, we can assume that the optimal mixed strategy contains a second zero component.

So let us take $I = \{1, 3\}$, then $\tilde{a} = (0.00548, 0.002)$,

$$\tilde{D} = \begin{pmatrix} 0.00034 & 0.000095 \\ 0.000095 & 0.00017 \end{pmatrix}, \tilde{D}^{-1} = \begin{pmatrix} 3555.96955 & -2058.89531 \\ -2058.89531 & 7272.59202 \end{pmatrix},$$

$\langle \tilde{a}, \tilde{D}^{-1}\tilde{a} \rangle = 0.090743$, $\langle \tilde{e}, \tilde{D}^{-1}\tilde{e} \rangle = 6710.771$, $\langle \tilde{e}, \tilde{D}^{-1}\tilde{a} \rangle = 18.64708$ and by formulas (7) we obtain $\lambda = 0.04422$, $\mu = 0.00003$. Using formula (6), we have a vector of nonzero components $\tilde{p} = (0.71830, 0.28170)$.

Let us check the fulfillment of the KKT condition for the crossed-out number $i = 2$. The derivative of the Lagrange function with respect to p_2 is $\frac{\partial L_1(p, \lambda, \mu)}{\partial p_2} = \sum_{k=1}^{3} \sigma_{2k} p_k - \lambda \bar{a}_2 - \mu$. When substituting the vector $(0.71830, 0, 0.28170)$ and the Lagrange multipliers $\lambda = 0.04422$ and $\mu = 0.00003$, it is equal $\frac{\partial L_1(p, \lambda, \mu)}{\partial p_2} = 0.00002$. This means that all KKT conditions are satisfied and the optimal solution to problem (1) has the form $p^0 = (0.71830, 0, 0.28170)$.

Now let us solve problem (10) for the maximum mathematical expectation of the payoff with a restriction on the variance. For the original matrix D, the solution of the problem (3) is the strategy $p = (0.11532, 0.44388, 0.44079)$ and the corresponding minimum value of the objective function is $d_0 = 0.00013$.

Let us solve the problem (10) at the threshold value of the standard deviation $\sigma_0 = 0.014$ (or the variance $\sigma_0^2 = 0.0002$).

Take $I = \{1, 2, 3\}$, i.e. we will use the original vector of expected returns $\bar{a} = (0.00548, 0.00127, 0.002)$ and the original covariance matrix D, then we get by formulas (11), (12)
$\lambda = 28.1906, \mu = -0.00145, p = (0.62479, -0.01826, 0.39346)$. The non-negativity condition $p \geq 0$ is not satisfied in this case, which means that this vector p is not a solution. Since p_2 is negative in this case, it can be assumed that the optimal mixed strategy contains the second zero component.

Therefore, we take $I = \{1, 3\}$, then $\tilde{a} = (0.00548, 0.002)$,
$$\tilde{D} = \begin{pmatrix} 0.00034 & 0.000095 \\ 0.000095 & 0.00017 \end{pmatrix},$$ and we get by (12) $\lambda = 27.63183$, $\mu = -0.00134$.
Using (11), we have the vector of nonzero components $\tilde{p} = (0.62840, 0.37161)$.

Let us check the fulfillment of the KKT condition for the crossed-out number $i = 2$. The derivative of the Lagrange function with respect to p_2 is $\frac{\partial L_2(p,\lambda,\mu)}{\partial p_2} = \bar{a}_2 - \lambda \sum_{k=1}^{3} \sigma_{2k} p_k - \mu$. When substituting the vector $(0.62840, 0, 0.37161)$ and Lagrange multipliers $\lambda = 27.63183$ and $\mu = -0.00134$, we have $\frac{\partial L_2(p,\lambda,\mu)}{\partial p_2} = -0.00009$.

This means that all the KKT conditions are satisfied and the optimal solution of the problem (10) has the form $p^0 = (0.62840, 0, 0.37161)$.

In [21], an example of investing in shares of Russian companies for the period from 10/01/2019 to 12/31/2019 was considered. Analysis of statistical data showed that the values of the covariance of the returns of the companies under consideration were an order of magnitude less than the values of their variances. So covariances practically did not affect the calculation results, and it was legitimate to assume that they could be neglected.

In the above example with data on stock quotes of Russian companies for the period from 02/01/2021 to 05/01/2021 the covariances and variances have approximately the same order, and the covariances are positive. It can be assumed that this is due to recovery growth after the peak of the pandemic.

If we neglect the covariances in this example, then we have the following results.

Having solved problem (10) with covariances equal to zero and the same threshold value of the standard deviation, we obtain a solution of the problem (10) $p = (0.75124, 0, 0.24876)$. As you can see, the structure of the strategy has not changed, but the values of the first and third components differ significantly from these values of the vector $p^0 = (0.62840, 0, 0.37161)$.

Thus, the idea of mixed strategy calculations without neglecting the covariance of random payoffs of different pure strategies in games with nature is founded. Of course, this idea is not new in portfolio analysis, but games with nature can be models for other management tasks.

5 Conclusion

The purpose of this work is to develop a new approach in game theory, specifically in games with nature, related to the consideration of the correlation of random payoffs for each pair of pure strategies. The obtained theoretical results, in our opinion, can find applications in various decision-making problems. The considered example of stock investment is an illustration of the practical application of the results obtained. At the same time, we note that in general theoretical terms, we are talking about finding an optimal mixed strategy for which the condition of non-negativity of the components is mandatory (which, by the way, significantly complicates the search for a solution). Therefore, when applying this approach to stock investing, short selling is excluded. However, for the stock markets, restrictions on short sales, up to their complete ban, are not so rare.

References

1. Simon, H. A.: Administrative behavior. Simon and Schuster, United States (2013)
2. Wald, A.: Sequential analysis. Courier Corporation, United States (2004)
3. Samuelson, L.: Game theory in economics and beyond. Voprosy Ekonomiki **13**, 89–115 (2017). https://doi.org/10.32609/0042-8736-2017-5-89-115
4. Allen, F., Morris, S. E.: Game theory models in finance. In: International Series in Operations Research and Management Science, pp. 17–41. Springer, New York LLC (2014). https://doi.org/10.1007/978-1-4614-7095-3_2
5. Breton, M.: Dynamic games in finance. In: Başar, T., Zaccour, G. (eds.) Handbook of Dynamic Game Theory, pp. 827–863. Springer, Cham (2018). https://doi.org/10.1007/978-3-319-44374-4_23
6. Askari, G., Gordji, M.E., Park, C.: The behavioral model and game theory. Palgrave Commun. **5**(15), 1–8 (2019). https://doi.org/10.1057/s41599-019-0265-2
7. Fox, W.P., Burks, R.: Game theory. In: Applications of Operations Research and Management Science for Military Decision Making. ISORMS, vol. 283, pp. 251–329. Springer, Cham (2019). https://doi.org/10.1007/978-3-030-20569-0_6
8. Allayarov, S.: Game theory and its optimum application for solving economic problems. Turkish J. Comput. Math. Educ. **12**(9), 3432–3441 (2021)
9. Jiang, Y., Zhou, K., Lu, X., Yang, S.: Electricity trading pricing among prosumers with game theory-based model in energy blockchain environment. Appl. Energy **271**, 115239 (2020). https://doi.org/10.1016/j.apenergy.2020.115239
10. Lee, S., Kim, S., Choi, K., Shon, T.: Game theory-based security vulnerability quantification for social internet of things. Futur. Gener. Comput. Syst. **82**, 752–760 (2018). https://doi.org/10.1016/j.future.2017.09.032
11. Hafezalkotob, A., Mahmoudi, R., Hajisami, E., Wee, H.M.: Wholesale-retail pricing strategies under market risk and uncertain demand in supply chain using evolutionary game theory. Kybernetes **47**(8), 1178–1201 (2018). https://doi.org/10.1108/K-02-2017-0053
12. Piraveenan, M.: Applications of game theory in project management: a structured review and analysis. Mathematics **7**(13), 858 (2019). https://doi.org/10.3390/math7090858

13. Shafer, G., Vovk, V.: Game-Theoretic Foundations for Probability and Finance, p. 455, John Wiley and Sons, United States (2019)
14. Wald, A.: Statistical Decision Functions. Wiley, Oxford UK (1950). https://doi.org/10.2307/2333373
15. Hurwicz, L.: Optimality criteria for decision making under ignorance. Statistics 370 (1951)
16. Savage, L.J.: The theory of statistical decision. J. Am. Stat. Assoc. **46**, 55–67 (1951)
17. Harman, R., Prus, M.: Computing optimal experimental designs with respect to a compound Bayes risk criterion. Statist. Probab. Lett. **137**, 135–141 (2018). https://doi.org/10.1016/j.spl.2018.01.017
18. Kuzmics, C.: Abraham Wald's complete class theorem and Knightian uncertainty. Games Econom. Behav. **104**, 666–673 (2017). https://doi.org/10.1016/j.geb.2017.06.012
19. Radner, R.: Decision and choice: bounded rationality. International Encyclopedia of the Social and Behavioral Sciences (Second Edition). Elsevier, Florida United States, pp. 879–885 (2015)
20. Gorelik, V.A., Zolotova, T.V.: The optimality principle mathematical expectation-var and its application in the Russian stock market. In: 12th International Conference Management of Large-Scale System Development Proceedings, pp. 1–4. IEEE Conference Publications, Moscow Russia (2019). https://doi.org/10.1109/MLSD.2019.8911018
21. Gorelik, V.A., Zolotova, T.V.: Risk management in stochastic problems of stock investment and its application in the russian stock market. In: 13th International Conference Management of Large-Scale System Development Proceedings, pp. 1–5. IEEE Conference Publications, Moscow Russia (2020). https://doi.org/10.1109/MLSD49919.2020.9247801
22. Zhukovsky, V.I., Kirichenko, M.M.: Risks and outcomes in a multicriteria problem with uncertainty. Risk Manage. **2**, 17–25 (2016)
23. Labsker, L. G. : The property of synthesizing the Wald-Savage criterion and its economic application. Econ. Math. Methods **55**(4), 89–103 (2019). https://doi.org/10.31857/S042473880006775-1
24. García, F., González-Bueno, J.A., Oliver, J.: Mean-variance investment strategy applied in emerging financial markets: evidence from the Colombian stock market. Intellect. Econ. **9**(15), 22–29 (2015). https://doi.org/10.1016/j.intele.2015.09.003
25. Xu, Y., Xiao, J., Zhang, L.: Global predictive power of the upside and downside variances of the U.S. equity market. Econ. Model. **93**, 605–619 (2020). https://doi.org/10.1016/j.econmod.2020.09.006
26. Sharpe, W. F., Alexander, G. J., Bailey, J. V.: Investments. Prentice-Hall, Englewood Cliffs (1999). https://doi.org/10.4236/ajps.2019.103030
27. Investment company FINAM. https://www.finam.ru/. Accessed 5 Mar 2021

Optimization in Economics and Finance

Application of Optimization Methods in Solving the Problem of Optimal Control of Assets and Liabilities by a Bank

Alexey Chernov[1,2] , Anna Flerova[2] , and Aleksandra Zhukova[2(✉)]

[1] Moscow Institute of Physics and Technology, Dolgoprudny, Institute lane, 9, 141701 Moscow, Russian Federation
[2] Federal Research Center "Computer Science and Control" of the Russian Academy of Sciences, Vavilov st, 44, building 2, 119333 Moscow, Russian Federation
sasha.mymail@gmail.com

Abstract. In this work the possibilities of applying optimization methods to solving the optimal control problem are studied. We consider a model of interaction between a bank and its depositors, borrowers and owners in the form of an optimal control problem, where the control parameters are interest rates on loans and deposits, and the dividend payments to the owners. The model has the form of optimal control problem with phase constraints, which makes it difficult to solve analytically. The optimality conditions in the form of the Pontryagin maximum principle lead to a system of differential equations, for the numerical solution of which the shooting method is used. With a large number of model parameters, optimization methods are used to speed up calculations.

Keywords: Optimal control · Asset and liability management · Gradient descent

1 Introduction

Financial systems of many countries experienced turbulence of varying scale as systemically significant banks got on the verge of bankruptcy. Two important features specific to the current banking systems were exposed in these events. Firstly, that the regulatory environment plays a crucial role in the behavior of banks and their planning. Secondly, the market for savings and loans is not perfectly competitive and demonstrates more features of what is called imperfect capital markets. In order to analyze the participants on such markets, evaluate

The research by Aleksandra Zhukova in Sect. 4.1-4.2 was supported by RSCF grant No. 22-21-00746 "Models, methods and software to support the modeling of socio-economic processes with the possibility of forecasting and scenario calculations".
The research by Alexey Chernov in Sect. 2.3 was supported by RSCF grant No. 21-71-30005, https://rscf.ru/en/project/21-71-30005/.

N. Olenev et al. (Eds.): OPTIMA 2023, LNCS 14395, pp. 235–250, 2023.
https://doi.org/10.1007/978-3-031-47859-8_17

the effectiveness of regulation measures in this area, we propose a general mathematical model of a bank that manages assets and liabilities using interest rates on loans and deposits. The framework is based on the approach to modelling the banking sector in [6–9].

Our model has some specifics. The bank controls the demand for its products by controlling the interest rates on issued loans and attracted deposits issued. The existence of major banks contradicts the typical assumption of competitive markets for capital in general equilibrium models. Rather, the system contains several price-makers and the rest of the players follows the leaders. Another detail that is important for our description and was not analyzed thoroughly so far is the system of constraints that limits the planning of crucial indicators, such as own capital, liquidity, assets. Some of these constraints come from regulation (capital adequacy), some are internal limitations that might come from the risk management or internal policy (liquidity constraints), other might be defined by the owners of the bank as the target performance indicators for the bank's management (e.g. growth of the own capital by 10% over the two-years' term).

With these assumption we formulate the model of asset and liability management by the bank, formalize it as the optimal control problem. We then formulate the optimality conditions and discuss the optimization problems that appear when we solve it numerically. These optimization problems are built in the algorithm that is presented in the Sect. 4. We present the results of application of this algorithm. The results are discussed and possible developments are outlined in the Sect. 5. The last section makes the concluding remarks.

2 Related Literature

2.1 Asset-Liability Management

"Asset-Liability Management (ALM) is a field that maximizes the wealth of shareholders and tries to increase the efficiency of banks and reduce the risks in banks" [1]. The behavior by a bank on capital markets typically is modelled in the context of a partial or general equilibrium. The standard framework is the model of a competitive market that defines the equilibrium interest rate from the interaction of small price-taking lenders and price-taking borrowers. For example, in the paper [6] that describes the national banking sector, the bank is a small price taking entity that accepted the interest rates as defined by the equilibrium on the market. The authors admit the existence of major players, but concentrate on the banking sector as a block of an applied general equilibrium model of a national economy and consider the interest rates as information variables.

There exist models of imperfect capital markets [9] where the authors model the environment with substantial discrepancy in the interest rates on loans and deposits. However, they do not assume any financial entity directly controls the interests or uses information about the demand for loans and deposits as the control mechanism. In order to analyze the participants on such markets, evaluate the effectiveness of regulation measures in this area, we propose a general mathematical model of a bank that manages assets and liabilities using interest rates on loans and deposits. The framework relies on the approach to modelling the banking sector in [6] and assumes discrepancy in the interest on loans and deposits as in [9].

Our model has some specifics. In contrast to [6], the bank controls the demand for its products by controlling the interest rates on issued loans and attracted deposits issued. The similarity to [6] is the liquidity constraints [6] and the targets by the owners of the bank regarding the performance indicators for the bank's management (the growth of the own capital by 10% over the two-years' term). One should note that the authors in [6] derived the optimality condition from another approach and due to technical challenges had to simplify the optimality conditions and in fact, solve a different system of equations.

2.2   Optimal Control Problems with Phase Constraints

When proving the classical maximum principle for problems without phase constraints, arbitrarily small neighborhoods of optimal trajectories are used. The case when phase constraints form a closed region and the optimal trajectory lies on this region in whole or in part is more complex and requires the fulfillment of regularity conditions [3, 4].

Consequently, an obvious question arises about such necessary optimality conditions in problems with state constraints that would extend the existing optimality conditions. Such extended optimality conditions have been proposed. In the extended maximum principle, the multiplier of the function that defines the constraint is a measure, as well as the joint state function, might not be finite. However, at the end points of time they have poles and satisfy a certain growth condition. In [5], using a set of Gamkrelidze optimality conditions, it is shown that in the special case it is possible to guarantee that the joint state is bounded, while the measure must still have poles of order dictated by a special condition. Continuity of the multiplier measurement is important both theoretically and for the numerical implementation of indirect computational methods. if the regularity conditions used in [3] are met, the multiplier measure is continuous.

2.3   Numerical Analysis of Optimal Control Problems Using Optimization Methods

In our study we consider the following loss function f_{loss} to qualify the optimality of the trajectory:

$$f_{loss} = (x_3(T) - \eta x_3(0))^2 + p_2(T)^3 + p_3(T)^2 \tag{1}$$

Note that any element of the loss function (1) is the function of the initial parameters (x_0, p_0), where values of x_0 are calculated from the problem parameters while the values $p_0 \in R^3$ identify the exact trajectory. Of course the problem itself is smooth and one can expect the smoothness of the loss function (1) as well. And this happens in some unknown $G \subset R^3$ while outside this subset loss function is not defined. But even if the set G is known, the calculation of the loss function values for each point is very expensive that does not allow us to use full search approach applied in the other works [11,12]. Thus, in order to find the optimal trajectory we study it as the solution of the optimization problem:

$$f_{loss}(p) \rightarrow \min_{p \in G} \tag{2}$$

We assume that the formulated problem has the following properties:

1. If p^* produces optimal trajectory then $f_{loss}(p^*) = 0$ and vice versa.
2. The solution p^* (where $f_{loss}(p^*) = 0$) exists.
3. If $p \in G$ then f_{loss} is smooth in p.

However we are not able to directly use common known optimization methods like gradient descent, etc. due to the following:

1. There is no explicit form of the derivative of the objective function $f_{loss}(p)$.
2. The objective function $f_{loss}(p)$ is not convex as shown by our experiments[1].
3. The set G is unknown and the function $f_{loss}(p)$ is not defined if $p \notin G$.

Let us consider each point above separately and build problem's specific numerical optimization method.

The item 1 requires the usage of the zero-order optimization. There is a well-known Nelder-Mead method [15], however it looks quite complex to use it for solution and there is now any guarantee that is solves the problem. On the other hand we can use produce zero-order method on the basis of the gradient approximation applied to the first-order method [16,17]. In that case there is a known theory that method converges to the local minimum. We concentrated on the usage of this method. In that case we have the following challenges: calculate the approximation of the gradient value, calculate the step size. Since no data for the problem's properties is known and one dimensional search of the optimal step size is quite expensive, we use the constant step size.

There are several ways to calculate the approximation of the gradient: (1) approximate partial derivatives, (2) stochastic approach. The first approach requires computing three additional function values for each partial derivative i:

$$\frac{\partial f}{\partial p_i}(p) = \frac{f(p + he_i) - f(p)}{h}, \tag{3}$$

where $e_i \in R^3$ is the i-th ort vector, h is the method's parameter that defines the approximation quality. Of course there are many other schemes to calculate the derivative but we use the simplest one in this study. The second stochastic approach leads to the following schema

$$f'(p, e) = \frac{f(p + he) - f(p)}{h}, \tag{4}$$

---

[1] These experiments are not included to the article due to its size limitations.

where $e \in R^3$, $\|e\|_2 = 1$ is a uniformly distributed random vector. In the second case the scheme of the classical gradient method will be transformed to the following one:

$$p^k = p^{k-1} - \alpha f'(p^{k-1}, e)e. \tag{5}$$

It is well known that for the suggested scheme $E_e\{f'(p^{k-1}, e)e\} = \nabla f(p^{k-1})$.

The next point 2 requires an additional extension that corresponds to the selection of the initial point. This extension is very similar to the well known approach when initial set is divided into the set of the subsets - 3d cubes of the same size. Initial point is selected as a center of such a cube and method is started from this point. If for any reason the method will later generate the point that belongs to another cube then we exclude this another cube from the analysis as well. The method is repeated for all cubes. Number of the cubes can be selected less than infinity if one selects some reasonable boundaries for initial conditions p, for example, $-\infty < a_i \leq p_i \leq b_i < \infty$ $i = 1, 2, 3$ where a_i, b_i, $i = 1, 2, 3$ are some constants.

The next point 3 requires some criteria that allows us to calculate objective function $f_{loss}(p)$ when $p \notin G$. In order to do it we extend $f_{loss}(p)$ on the basis of the smoothness of the trajectory via the following criteria:

- if trajectory is not smooth than $p \notin G$;
- if trajectory is not smooth than related dual multiplier is not change smoothly. Its derivation from the smoothness we consider as the value of the loss function and try to reduce via the same optimization method described above.

3 The Model of Asset-Liability Management by a Bank

We consider a basic model of a bank that has sufficient market power to control the demand for its services by varying the interest rates for deposits $r_s(t)$ and loans $r_l(t)$. The outstanding debt to depositors $S(t)$ grows with the interest with coefficient γ_s and declines as depositors withdraw the deposits at the rate $\beta_s(t)$ which we assume to be constant in the basic version of the model. Generally, this indicator varies and might be computed from the available data reported by banks [6]. The interest on loans work in the opposite direction. The outstanding assets $L(t)$ grow at a lower pace or even decline if the interest is high. The parameter that measures sensitivity is γ_l.

In order to take into account the whole balance sheet of the bank, one has to control the flow of cash balances $A(t)$, that involve the expenditures $OC(t)$, payment of dividends to the owners $Z(t)$, flow of income from interest on loans and flow of expenditures on interest on deposits. Reserve requirements on deposits are characterised by the coefficient n_s.

$$\int_0^T (Z(t)^{1-\rho}e^{-\delta t} - \epsilon_1 r_l^2(t) - \epsilon_1 r_s^2(t))dt \to max_{r_l(\cdot), r_s(\cdot), Z(\cdot)} \tag{6}$$

$$\frac{d}{dt}L(t) = (\alpha_l - \gamma_l r_l(t) - \beta_l)L(t), \tag{7}$$

$$\frac{d}{dt}S(t) = (\alpha_s + \gamma_s r_s(t) - \beta_s)S(t), \tag{8}$$

$$\frac{d}{dt}A(t) = -OC - Z(t) + ((\gamma_l + 1)r_l(t) + \beta_l - \alpha_l)L(t)+ \tag{9}$$

$$+((-1 + (1 - n_s)\gamma_s)r_s(t) + (1 - n_s)(\alpha_s - \beta_s))S(t) \tag{10}$$

with constraints on the controls

$$0 \leqslant r_l(t) \leqslant R_l, 0 \leqslant r_s(t) \leqslant R_s, 0 \leqslant Z(t) \leqslant M, \tag{11}$$

phase constraints on the bank's liquidity and the capital adequacy

$$A(t) - \tau_s S(t) - \tau_l L(t) \geqslant 0, \tag{12}$$

$$L(t)(1 - K_A w_l) + (1 - K_A)A(t) - S(t)(1 - n_s) \geqslant 0. \tag{13}$$

and the terminal constraint that formalizes the management's goal on the growth (by the factor of η of the own's capital of the bank

$$A(0) + L(0) - (1 - n_s)S(0) \leqslant \eta(A(T) + L(T) - (1 - n_s)S(T)). \tag{14}$$

The initial conditions are known and assumed to satisfy all the constraints $L(0) = L_0, S(0) = S_0, A(0) = A_0$.

The goal of the management of a commercial bank in this model is to maximize the wealth of its owners, formalized by the integral discounted utility of the flow of dividends

$$\int_0^T Z(t)^{1-\rho}e^{-\delta t}dt. \tag{15}$$

This functional is strictly concave with respect to the control variable $Z(t)$, but in order to regularize it, we add the terms $-\epsilon_1 r_l^2(t) - \epsilon_1 r_s^2(t)$. The parameters ϵ are assumed to be small.

The capital adequacy ratio (13) is a modified form of the ratio of the own capital $A(t) + L(t) + n_s S(t) - S(t)$ to the weighted liabilities $w_l L(t)$, which should not exceed the value K_A prescribed by the regulator (the Central Bank). This phase constraint limits the liabilities from above if the coefficient $(1 - K_A w_l)$ is negative and from below if it is positive.

The constraint (12) means the bank keeps a portion of its assets and liabilities in cash for possible requests from its clients. This constraint is rather informal and subject to the internal policy of the bank. The parameters τ_s, τ_l are identified for individual banks from statistics.

There is a challenge in solving the optimal control problem with two constraints of this sort. Our paper is aimed at applying the indirect method of solving the optimal control problem with phase constraints by the approach in [11]. This approach involves analysis of regularity conditions and for two constraints these conditions become quite complicated to verify, especially in the numerical analysis, where the regularity conditions should be verified at each step as the trajectory reaches the border of the constraint. We plan to extend the analysis on two constraint in our future works. In the present work we omit the liquidity constraint (12) and leave only the capital adequacy constraint.

4   Numerical Analysis of the Optimal ALM by a Bank

4.1   The Optimal Control Problem

By changing the variables

$$x(t) = (x_1(t), x_2(t), x_3(t))^T = (S(t), L(t), A(t) + L(t) - (1 - n_s)S(t))^T, \quad (16)$$
$$u(t) = (u_1(t), u_2(t), u_3(t))^T = (r_s(t), r_l(t), Z(t))^T. \quad (17)$$

we formalize the model (10)-14) in the form:

$$\int_0^T (u_3(t)^{1-\rho}e^{-\delta t} - \epsilon_1 u_2(t)^2 - \epsilon_2 u_1(t)^2)dt \to max \quad (18)$$
$$\dot{x}(t) = B_1\left(x\left(t\right)\right) \cdot u(t) + B \cdot x(t) + R,$$
$$C^T x(t) \le 0, \quad (19)$$
$$u \in U = \{u(t)|\ 0 \le u_1(t) \le R_s,\ 0 \le u_2(t) \le R_l,\ 0 \le u_3(t) \cdot \le M\} \quad (20)$$
$$\eta x_3(0) \le x_3(T), x_2(0) = L_0, \quad x_1(0) = S_0, \quad x_3(0) = W_0, \quad (21)$$

As we discussed above, the indirect method we apply in this paper requires substantially more complex analysis for several phase constraints, so in this first attempt we limit the inequality constraints (19) to one:

$$C = (K_A(1 - n_s),\ K_A(w_l - 1),\ (1 - K_A))^T. \quad (22)$$

For solving this bank's ALM optimal control problem, we used the Pontryagin's maximum principle according to the paper [12]. In their approach, the necessary optimality conditions are formulated using the extended Hamilton-Pontryagin function for the problem:

$$H(\lambda_0, t, x, p, u, \mu) = \lambda_0 \left((u_3(t))^{1-\rho}e^{-\delta t} - \epsilon_1\left(u_2\left(t\right)\right)^2 - \epsilon_2\left(u_1\left(t\right)\right)^2\right) + \quad (23)$$
$$+p_1(t)\left(\gamma_s u_1(t) + \alpha_s - \beta_s\right)x_1(t) + p_2(t)\left(\gamma_l u_2(t) + \alpha_l - \beta_l\right)x_2(t) +$$
$$+p_3(t)\left(-x_1(t)u_1\left(t\right) + x_2(t)u_2(t) - u_3(t)\right) +$$
$$+\mu(t)\,C^T\left(B_1\left(x\left(t\right)\right) \cdot u(t) + B \cdot x(t) + R\right).$$

4.2   Optimality Conditions

We assume that the regularity condition [12] holds. This condition is essential for the proof of the Pontryagin's maximum principle. If they do not hold, the trajectory and the corresponding control still might be optimal [3]. In the case of the current problem, the regularity conditions have the following form. As we show by the expression (26), the coefficients to the control variables $u_i(t)$ are nonzero, therefore it is easy to check that the regularity conditions might be violated only at the vertices of the set of admissible controls U.

Regularity condition. The problem (18)-(21) is regular with respect to the state constraints if for all $x \in R^3$ and $u \in R^3$, such that $C^T x = 0$,

$$\Gamma(x, u) = C^T (B_1(x(t)) \cdot u(t) + B \cdot x(t) + R) = 0,$$

and $h(u) = 0$, the set of vectors $\frac{\partial \Gamma}{\partial u}(x, u)$ and $\nabla h(u)$ is linearly independent. Here $h(u)$ is the vector function that describes the set of admissible controls (20) so that $U = \{u \in R^3 \mid h(u) \leq 0\}$.

Under the regularity condition, for an optimal process (x^*, u^*), the maximum principle ensures the existence of Lagrange multipliers: a scalar $\lambda_0 \geq 0$, an absolutely continuous dual vector-function $p(t) \in W_{1\infty}([0, T]; R^3)$, and a scalar continuous monotone function $\mu \in C([0, T]; R)$, such that the optimality conditions below hold. Since the Lagrange multipliers are defined up to a constant multiplier, we may set $\lambda_0 = 1$.

The dual system for the dual variables $p(t)$ is

$$\dot{p}(t) = -p(t) \cdot B_1'(x(t)) \cdot u(t) - Bp(t) - \mu(t) (C \cdot B_1'(x(t)) \cdot u(t) + C \cdot B). \quad (24)$$

The initial conditions for this system are not defined by the optimality conditions.

The optimal control is found by maximizing the Hamilton-Pontryagin function (23). It might be represented by the piece-wise expressions

$$u^*(t) = \begin{bmatrix} \max\left(0, \min\left(R_s, \frac{(p_1(t)\gamma_s - p_3(t) + \mu(t)(k_A(1-n_s)\gamma_s + (1-k_A))x_1(t)}{2\epsilon_1}\right)\right) \\ \max\left(0, \min\left(R_l, \frac{(p_3(t) - p_2(t)\gamma_l + \mu(t)(k_A(1-w_l)\gamma_l - (1-k_A))x_2(t)}{2\epsilon_1}\right)\right) \\ \max\left(0, \min\left(M, \left(\frac{p_3(t) - \mu(t)(1-k_A)}{(1-\rho 1)}\right)^{-\frac{1}{\rho}} e^{-\frac{\delta t}{\rho}}\right)\right) \end{bmatrix}. \quad (25)$$

It is demonstrated in [11] that the dual variable $\mu(t)$ might be set equal to zero at $t = 0$, then stay constant as the trajectory $x^*(t)$ stays outside the boundary of the phase constraint, where $C^T x^* < 0$ and $\mu(t)$ increases, when the trajectory reaches the boundary of the phase constraint, i.e. $C^T x^* = 0$. The function $\mu(t)$ must be continuous [11]. The equation for determining μ while on the boundary of the phase constraint for the optimal trajectory is found from the $\dot{x} = f(x, u)$ and optimal control $u^*(x^*(t), p(t), \mu(t))$ by the condition

$$\Gamma(x^*, u^*) = C^T f(x^*, u^*) = (k_A(1 - n_s)\gamma_s + (1 - k_A))x_1^*(t)u_1^*(t) + \quad (26)$$
$$+ (k_A(1 - w_l)\gamma_l - (1 - k_A))x_2^*(t)u_2^*(t) + (1 - k_A)u_3^*(t) +$$
$$k_A(1 - n_s)(\alpha_s - \beta_s)x_1^*(t) + k_A(w_l - 1)(\alpha_l - \beta_l)x_2^*(t) + (1 - k_A)C_O = 0.$$

This expression depends on $\mu(t)$ in a complex way due to (25). Therefore, there appears an optimization problem

$$\max_{u:h(u)\leq 0} H(1,t,x^*(t),p(t),u,\mu(t)) \ \ w.r.t. \ \ \Gamma(x^*,u) = C^T f(x^*,u) = 0 \qquad (27)$$

that might define the value of $\mu(t)$ explicitly by applying the Lagrange multipliers method. In problems such as [11,12] the authors used the explicit form of $\mu(t)$, which substantially simplified the numerical procedure. However, in our case, due to cubic form of the region U for possible controls and therefore the controls of the form (25), one cannot assume any smoothness or even continuity, nothing to say about solving (26) explicitly as was in the model in the previous works. Therefore, numerical optimization methods are used in order to solve this equation at every point in time.

The transversality conditions require that

$$p_1(T) = p_2(T) = 0, x_3(T) = \eta x_3(0). \qquad (28)$$

In the numerical analysis these conditions define the target criterion for the optimization procedure to find the initial conditions for the dual variables $p(0) = (p_1(0), p_2(0), p_3(0))^T$ such that the necessary optimality conditions hold and the transversality conditions (28) are satisfied with the highest possible precision

$$\rho(p(0)) = (p_1(T))^2 + (p_2(T))^2 + (x_3(T) - \eta x_3(0))^2 \to \min_{p(0)}. \qquad (29)$$

4.3   The Algorithm

The algorithm for numerical analysis contains main components:

- The procedure named **RK4(DYNSYS,CURRSTATE,timestep)** to make a step of the dynamical system DYNSYS for $x_k, p_k, \mu_k \to x_{k+1}, p_{k+1}, \mu_{k+1}$ from the state $CURRSTATE = \{x_k, p_k, \mu_k\}$ for the trajectory inside the constrained region (when $C^T x < 0$ and μ is constant) or along the boundary $C^T x = 0$ when $\mu(t)$ increases. This procedure was taken according to the Runge-Kutta mid-point method. The parameter *timestep* defines the step in time of the dynamical system.
- The procedure named **FindStep(C, DYNSYS, CURRSTATE, timestep, ϵ)** to find the precise step Δ_t as the dynamical system for $x_k, p_k, \mu_k \to x_{k+1}, p_{k+1}, \mu_{k+1}$ crosses the boundary $C^T x = 0$;
- The procedure for search for $\mu(t)$ from the condition $\Gamma(x^*, u^*) = C^T f(x^*, u^*) = 0$. It might involve the use of the regularity conditions to find the value of $\mu(t)$ explicitly;
- The procedure **FindRho(p(0))** to compute the deviation of the final point of the computed trajectory from the terminal conditions (29). This procedure should account for the possibility the trajectory is not computed till the endpoint if it violates the optimality conditions at some point in the middle of computation. In this case we propose to compute the residual (29) by using the last computed point $x(t), p(t)$ instead of $x(T), p(T)$ to penalize the unfinished trajectories.

– The procedure for finding the stationary trajectories, that satisfy the Maximum principle, by choosing the initial $p(0)$. We use the random search approach presented in Algorithm 3. We present the basic version where the algorithm makes at most 100 steps just for compactness of presentation. In practice, the limiting factor is either $\Delta\rho \to 0$ or $\alpha_k \to 0$.

Algorithm 1: FindStep(C, DYNSYS, CURRSTATE, Δt, ϵ)

The Procedure to Find the Step to the border $C^T x = 0$.
1. compute the state of the system before and after the step
    $x_{Left} \leftarrow x_k$ from the $CURRSTATE$;

$$\langle t_k + \Delta t, x_{Right}, p_{Right}, \mu(t_k) \rangle \leftarrow RK4(DYNSYS, CURRSTATE, \Delta t);$$

2. if $C^T x_{Left} < -\epsilon$ and $C^T x_{right} > \epsilon$ then
    (a) compute the step of time $\tilde{\Delta} t \leftarrow \dfrac{C^T x_{Left}}{C^T x_{Left} - C^T x_{Right}} \Delta t$
    (b) make the step $MIDPX \leftarrow RK4(DYNSYS, CURRSTATE, \tilde{\Delta} t)$
    (c) return $(MIDPX, hb = \tilde{\Delta} t)$
3. else return $(CURRSTATE, hb = 0)$.

The procedure Search comes from the DirectSearch package in Maple, uses universal derivative-free searching methods to find the minimum of the function under inequality constraints. It appears to be more stable and efficient than the standard Optimization package built in Maple [13,14].

The results of numerical experiments for the set of parameters $\alpha_l = .5, \alpha_s = .6, \beta_s = .5, \beta_l = .3, \delta = .1, \rho = .5, \gamma_l = .2, \gamma_s = .2, n_s = 0.09, \epsilon_l = 0.01, \epsilon_s = 0.01, x_1(0) = 5, x_2(0) = 2, x_3(0) = 7, R_s = 3, R_l = 3, M = 20, T = 5.0, C0 = .3, w_l = 1.5, \epsilon = 0.1e - 6, \eta = 1.1, \lambda_0 = 1.$ are presented on Figs. (1), (2), (3).

5   Discussion

The numerical method to find the optimal control of the economic agent's problem presented in this work seems to be different to what our group has been using in the ECOMOD system for economic modelling. This example revealed several challenges:

– an efficient method to search for the value of $\mu(t)$ from

$$\Gamma(x^*, u^*) = C^T f(x^*, u^*) = 0$$

which delivers continuous μ;
– an efficient method to solve the boundary value problem in the system of optimality conditions;

Algorithm 2: FindRho(PKLIST). Part 1.

The procedure to compute the residual $\rho(p(0))$ for the initial values of
$p(0) = (p_1(0), p_2(0), p_3(0))^T$ in $PKLIST$.
Set the constants. 10 is added to NMAX for the case of additional steps near
the border $C^T x = 0$;
$TMAX \leftarrow 5;$, $TIMESTEP \leftarrow 10^{-4}$; $NMAX \leftarrow [\frac{TMAX}{NMAX}] + 10$;
Initialize
$CURXP \leftarrow \langle t = 0, x_0 = x^0, p_0 = p^0, \mu_0 = 0 \rangle$;
$CURMU \leftarrow \mu(t) = 0$; current time $whattime \leftarrow 0$.
for i from 1 to NMAX while $whattime \leq TMAX - TIMESTEP$ **do**
1. decide if the trajectory is inside the constrained region and make a step
   $CXCUR \leftarrow C^T x_i$;
   $NEWXP \leftarrow RK4(DYNSYS(CURMU), CURXP, TIMESTEP)$;
   $whattime \leftarrow rhs(NEWXP[1])$;
2. if $CXCUR < -\varepsilon$ then the current state x_i is inside the constraint
   (a) evaluate the $C^T x$ at the x_{i+1}
       $CXNEW \leftarrow C^T x_{i+1}$, where x_{i+1} is from $NEWXP$;
   (b) if $CXNEW > \varepsilon$ then the step takes x_{i+1} beyond the region $C^T x \leq 0$, and
       the step should be smaller for x_{i+1} to stay admissible.
       i. Make a smaller step to the state named $MIDXP$
          $MIDXP \leftarrow$
          $FindStep(C, DYNSYS(CURMU), CURXP, TIMESTEP, \varepsilon)$;
       ii. The new state x_{i+1} is on the border $C^T x_{i+1} = 0$.
           $NEWXP \leftarrow MIDXP[1.. - 2]$;
           $CXCUR \leftarrow C^T x_{i+1}$, where x_{i+1} is from $NEWXP$;
       iii. find the μ_{i+1}^- that solves $\Gamma(x_{i+1}, u(t_{i+1}, x_{i+1}, p_{i+1}, \mu_{i+1})) = 0$ in **(26)**.
            Since the stationary trajectory should have the $\mu(t)$ continuous and
            monotone, μ_{i+1} must be close enough to μ_i (differ at most by μ_{tol})
            $GAMUR \leftarrow l.h.s.((\textbf{26}))^2$ at $NEWXP$ and $\mu_{i+1} = \mu$ as the unknown
            $\langle GAMUR(\mu_{i+1}), \mu_{i+1} \rangle \leftarrow Search(GAMUR, \mu \geq \mu_i - \varepsilon, \mu \leq \mu_i + \mu_{tol})$;
       iv. if the equation (26) is solved with satisfactory precision
           $GAMUR(\mu_{i+1}) < \varepsilon$ then the algorithm continues to compute the
           trajectory with the found μ_{i+1}
           $CURMU \leftarrow \mu(t) = \mu_{i+1}$ from the solution.
           end if;
       end if;
   (c) move to the new state
       $CURXP \leftarrow NEWXP$; $whattime \leftarrow r.h.s.(NEWXP[1])$;
       end if;

Algorithm 2: FindRho(PKLIST). Part 2. Continued

---

3. if $CXCUR \geq -\varepsilon$ and $CXCUR \leq \varepsilon$ then
   the current point x_i of the trajectory is on the border $C^T x_i = 0$. There are two
   options for the state x_{i+1} stored in NEWXP. Either the trajectory stays on the
   border or leaves it. Only in the case of $C^T x_{i+1} = 0$ the dual μ_{i+1} is updated
   using $\Gamma(x_{i+1}, u_{i+1}) = 0$.
   (a) $GAMUR \leftarrow l.h.s.((\mathbf{26}))^2$ at $NEWXP$ and $\mu_{i+1} = \mu$ as the unknown;
       find the μ so that $\mu(t)$ is continuous
       $\langle GAMUR(\mu_{i+1}), \mu_{i+1} \rangle \leftarrow Search(GAMUR, \mu \geq \mu_i - \varepsilon, \mu \leq \mu_i + \mu_{tol})$;
   (b) **if the equation (26) is solved with satisfactory precision**
       $GAMUR(\mu_{i+1}) < \varepsilon$ **then**
       the algorithm continues to compute the trajectory with the found μ_{i+1}
       $CURMU \leftarrow mu(t) = \mu_{i+1}$
       end if;
   (c) move to the new state $CURXP \leftarrow NEWXP$;
   end if;
4. if the time is in the close vicinity of TMAX, so that only one step remains **if**
   $whattime > TMAX - TIMESTEP$ **and** $whattime <= TMAX$) **then**
   Make the step over the remaining time interval $TMAX - whattime$
   $LASTXP \leftarrow RK4(DYNSYS(CURMU), CURXP, TMAX - whattime)$
   $whattime := rhs(LASTXP[1])$ Compute the residual
   $NEVAK \leftarrow (p_1(T))^2 + (p_2(T))^2 + (x_3(T) - \eta x_3(0))^2$ evaluated at $LASTXP$.
   else if $whattime > TMAX$ **then**
   print("whattime is above TMAX", whattime, "and we stop here")
   end if;
   in the remaining contingency, the computation stopped before the final moment
   of time and the residual is computed just to deliver some definite value, by
   evaluation the residual at the last computed point
   $NEVAK \leftarrow (p_1(t))^2 + (p_2(t))^2 + (x_3(t) - \eta x_3(0))^2$
   end if;
   end do;

---

return NEVAK.

Algorithm 3: Main algorithm to compute the $p(0)$ that defines the solution $x(t), p(t)$ with minimum deviation of the terminal state from the optimal terminal values (29).

Initialize $p_k \leftarrow (1,1,1)$; $s_k \leftarrow 1/1000$; $\alpha_k \leftarrow 1/100000$;
$\rho_k \leftarrow FindRho(p_k)$;
$\rho_k^e \leftarrow \rho_k$;
for k from 1 to 100 do Attempt a random direction but take the step only
   when it improves the criterion ρ_k
   for i from 1 to 200 while $\rho_k^e \geq \rho_k - \varepsilon$ **do**
      Generate a random vector with elements in the range $-1..1$
      $E \leftarrow random(e_1, e_2, e_3)$;
      $p_k^e \leftarrow p_k + E\, s_k$;
      $\rho_k^e := FindRho(p_k^e)$;
   end do;
   $\Delta\rho \leftarrow -\alpha_k(\rho_k^e - \rho_k)/s_k$;
   $p_k^{NEW} \leftarrow p_k + E\Delta\rho$
   $p_k \leftarrow p_k^{NEW};$ $\rho_k \leftarrow FindRho(p_k)$;
   $\rho_k^e := \rho_k$;
end do;

– to keep the balance between the time grid and error accumulaion, especially along the trajectory on the boundary of the constraint. In our case $C^T x = 0$.

Implications for the bank's ALM: we found that the bank may reach the limit in the capital adequacy constraint, see Fig. 3, but we found no evidence that, once on the constraint, it is trapped on it. Additional finding, which might be the outcome of the regularization in (6) that penalizes the interest rates, is almost

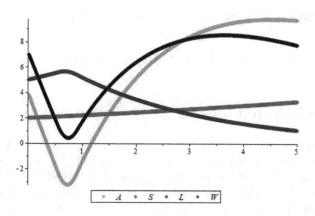

Fig. 1. The dynamics of the bank's deposits (S), loans (L), cash balances (A), own capital (W).

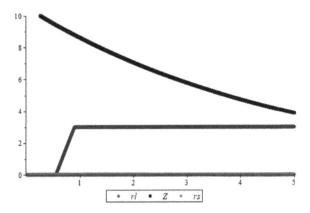

Fig. 2. The dynamics of the interest on deposits (r_s), interest on loans (r_l), dividends to owners (Z).

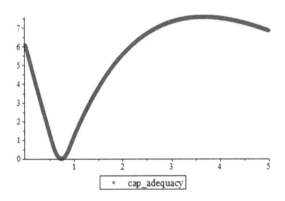

Fig. 3. Capital adequacy ratio is satisfied when above zero.

zero values of the interest on deposits. As one may show, for some parameters it might become positive.

6   Concluding Remarks

For the future directions of the research, we may suggest several options. To extend the analysis on two constraints, including liquidity constraints or solvency constraints. To add the automated procedure to solve for the dual variable $\mu(t)$ along the border of the constraint in explicit form using the regularity conditions and the Lagrange's multipliers methods. To omit the regularization terms $\epsilon_l r_l(t) + \epsilon_s r_s(t)$ in the target function of the ALM management problem. In this case there might appear ambiguity in the optimal controls $r_l(t)$ and $r_s(t)$ since the optimality conditions $max_u H(t, x, u, p, \mu)$ would define the maximum of a linear function with respect to u. In order to speed up the shooting method, it seems to be productive to smooth the problem. This would enable finding the region for refined search of the minimum.

Acknowledgements. The authors are very grateful to D.Yu. Karamzin and R.A. Chertovskih for discussions and advises.

References

1. Peykani, P., Sargolzaei, M., Botshekan, M.H., Oprean-Stan, C., Takaloo, A.: Optimization of asset and liability management of banks with minimum possible changes. Mathematics **11**(12), 2761 (2023)
2. Chunxiang, A., Shen, Y., Zeng, Y.: Dynamic asset-liability management problem in a continuous-time model with delay. Int. J. Control **95**(5), 1315–1336 (2022)
3. Pontryagin, L.S., Boltyansky, V.G., Gamkrelidze, R.V., Mishchenko, E.F.: Mathematical Theory of Optimal Processes. Moscow: Fizmatgiz, p. 391 (1969)
4. Arutyunov, A.V., Karamzin, D.: Review of regularity conditions for optimal control problems with state constraints and the principle of non-degenerate maximum. J. Optim. Theory Appl. **184**, 697–723 (2020)
5. Karamzin, D., Pereira, F.L.: On several issues concerning the study of optimal control problems with state constraints. J. Optim. Theory Appl. **180**, 235–255 (2019)
6. Pilnik, N., Radionov, S., Yazykov, A.: Model of optimal behavior of the modern Russian banking system. Econ. J. HSE **22**(3), 418–447 (2018)
7. Pilnik, N., Radionov, S., Yazikov, A.: The model of the Russian banking system with indicators nominated in rubles and in foreign currency. In: Evtushenko, Y., Jaćimović, M., Khachay, M., Kochetov, Y., Malkova, V., Posypkin, M. (eds.) The model of the Russian banking system with indicators nominated in rubles and in foreign currency. CCIS, vol. 974, pp. 427–438. Springer, Cham (2019). https://doi.org/10.1007/978-3-030-10934-9_30
8. Chernov, A., Zhukova, A.: Numerical analysis of the model of optimal savings and borrowing. In: Olenev, N., Evtushenko, Y., Jaćimović, M., Khachay, M., Malkova, V., Pospelov, I. (eds) Optimization and Applications. OPTIMA 2022. Lecture Notes in Computer Science, vol. 13781, pp. 165–176. Springer, Cham (2022). https://doi.org/10.1007/978-3-031-22543-7_12
9. Shananin, A.A.: Mathematical modeling of investments in an imperfect capital market. In: Proceedings of the Steklov Institute of Mathematics, vol. 313, pp. S175–S184 (2021)
10. Shananin, A.A., Tarasenko, M.V., Trusov, N.V.: Mathematical modeling of household economy in Russia. Comp. Math. Math. Phys. **61**, 1030–1051 (2021)
11. Chertovskih, R., Karamzin, D., Khalil, N.T., Pereira, F.L.: An indirect method for regular state-constrained optimal control problems in flow fields. IEEE Trans. Autom. Control **66**(2), 787–793 (2020)
12. Chertovskih, R., Karamzin, D., Khalil, N.T., Pereira, F.L.: Regular path-constrained time-optimal control problems in three-dimensional flow fields. Eur. J. Control. **56**, 98–106 (2020)
13. Moiseev, S.: Universal derivative-free optimization method with quadratic convergence //arXiv preprint arXiv:1102.1347 (2011)
14. https://www.maplesoft.com/Applications/Detail.aspx?id=87637

15. Nelder, J.A., Mead, R.: A simplex method for function minimization. Comput. J. **7**(4), 308–313 (1965). https://doi.org/10.1093/comjnl/7.4.308
16. Nesterov, Y., Spokoiny, V.: Random gradient-free minimization of convex functions. Found. Comput. Math. **17**, 527–566 (2017)
17. Gasnikov, A., Dvinskikh, D., Dvurechensky, P., Gorbunov, E., Beznosikov, A., Lobanov, A.: Randomized gradient-free methods in convex optimization. arXiv preprint arXiv:2211.13566 (2022)

An Endogenous Production Function in the Green Solow Model

Nicholas Olenev[✉][iD]

Federal Research Center "Computer Science and Control" of the Russian Academy of Science, Moscow, Russia
nolenev@mail.ru

Abstract. The paper amends the Green Solow model to replace the neoclassical production function by an endogenous production function. This endogenous function incorporates the growth rate, the rate of degradation of production capacities, and their age limit. All these internal parameters of the economic system make the analysis of environmental problems more interesting.

Keywords: Solow model · Endogenous production function · Vintage capacity model · Age limit · Balanced growth path

1 Introduction

An endogenous production function was constructed in [1,2] on the basis of the initial microeconomic description of the dynamics of production capacities distributed by technology. It considers the possibility of structural changes in the economy. The parameters of this production function can be found for different countries indirectly by comparing the calculations of macroeconomic time series on the original microeconomic model of production with their statistical data (see, for example, [1,3–5]). These parameters depend significantly on the country under study, on the established structure of its economic system, and on the structure of distribution of production capacities by technology.

The production factors of this endogenous production function are labor and total capacity. In addition, this production function reflects the dependence on external technological parameters, so its use allows us to take a new look at the classical problems of mathematical economics, as well as at the newly emerging topical problems related to the interaction between the economy and the environment.

In particular, with this endogenous production function in [6], the classical problem of Phelps's golden rule of accumulation [7] in the Solow model [8,9] was considered.

The publication has been prepared with the support of the "Research and development program using the infrastructure of the Shared Center of FRC CSC RAS".

N. Olenev et al. (Eds.): OPTIMA 2023, LNCS 14395, pp. 251–262, 2023.
https://doi.org/10.1007/978-3-031-47859-8_18

In the same time and independently, Robert Solow and Trevor Swan considered the simplest dynamic model of the economy with a classical production function, which served and still serves as a starting point for further study in research area of economic growth. However, empirical verification of a number of the model's statements showed that they are not confirmed in practice. In particular, this is due to the fact that the classical production function may not correspond to reality, and its parameters are set outside from the model.

The Solow model is based on an exogenous savings rate and a neoclassical aggregate production function with the possibility of technical change [10]. If we take an endogenous production function [1] that also accounts for technical change, then Solow's problem on the optimal rate of economic growth becomes more interesting. This is due to the fact that this endogenous production function contains among its parameters the rate of economic growth, as well as the fact that it is possible to determine the endogenous savings rate [6].

In work [11] it was found that economic growth brings an initial phase of environmental quality deterioration followed by a subsequent phase of it's improvement. They proposed to use an inverted U-shaped curve to describe the relationship between the level of environmental pollution and economic growth. According to the environmental Kuznets curve (EKC) hypothesis, it is assumed that as GDP per capita rises to a certain level, the amount of pollution per capita first increases and then decreases.

In [12] it is shown that the environmental Kuznetz curve can be explained by technological progress in pollution control and diminishing returns, just as the Solow model explains economic growth. In addition, there found the parameters of their model using data from 173 countries, assuming that all countries have the same technological performance.

In this paper, we propose to use the endogenous production function in the green Solow model, and to find indirectly the parameters of this model in each country from its own data by comparing the time series of macro-indicators calculated by the model with the corresponding statistical data.

Section 2 presents problem statement on the base of Green Solow model.

Section 3 gives statistical time series of macroindices for the Russian Federation and presents criteria of closeness of calculations and statistics.

Section 4 describes the Green Solow model in values of production capacities [1] and presents the Houthakker-Johansen model of production.

Section 5 provides conclusions and future plans for the study.

2   Problem Statement

Paper [12] considers the standard one sector Solow model with a fixed savings rate s, a constant return to scale and strictly concave neoclassical production function. Effective labor $B(t)L(t)$ and capital $K(t)$ are the production factors of a production function F at time t, they produce output $Y(t)$. Capital $K(t)$ accumulates via savings $sY(t)$ and depreciates at rate $\delta > 0$.

$$Y = F(K, BL), \dot{K} = sY - \delta K, \dot{L} = nL, \dot{B} = g_B B, \tag{1}$$

where $B(t)$ represents labor augmenting technological progress, n is population growth, and g_B is a rate of labor augmenting technological progress.

To describe the impact of pollution papers [12,13] assume that pollution $E(t)$ is a joint product of output $Y(t)$, $E(t) = \Omega Y(t)$. An abatement A removes ΩA units of total pollution. If the economy's efforts on the abatement is $Y^A(t)$, than

$$E = \Omega(Y - A(Y, Y^A)) = \Omega Y(1 - A(1, Y^A/Y) = \Omega Y a(\theta), \qquad (2)$$

where $\theta = Y^A/Y$, the intensive abatement function $a(\theta) = 1 - A(1, Y^A/Y)$, and $a(0) = 1, a'(\theta) < 0, a''(\theta) > 0$. In special cases we will use the form [12] $a(\theta) = (1 - \theta)^\kappa$ where $\kappa > 1$.

Taking into account the abatement $Y^A(t) = \theta Y(t)$ in a closed economy of the Solow model we have

$$(1 - \theta)Y = bJ + C, \qquad (3)$$

where $C = C(t)$ is a total consumption, and $bJ(t) = (1 - \theta)sY(t)$ is a total production investment. Here $b > 0$ is an incremental capital intensity [1] of a new production capacity $J(t)$.

The paper [12] supposes an exogenous technological progress in abatement lowering $\Omega(t)$ at rate $g_A > 0$, and transforms the measures of output $Y(t)$, capital $K(t)$ and pollution $E(t)$ into intensive units: $y(t) = Y(t)/(B(t)L(t)), k(t) = K(t)/(B(t)L(t)), e(t) = E(t)/(B(t)L(t))$, and $\varphi(k) = F(k,1)$.

$$y = \varphi(k)(1 - \theta), \qquad (4)$$

$$\dot{k} = s\varphi(k)(1 - \theta) - (\delta + n + g_B)k, \qquad (5)$$

$$e = \varphi(k)\Omega a(\theta), \dot{\Omega} = -g_A\Omega. \qquad (6)$$

This model was used [12] to test the Environmental Kuznets Curve hypothesis for the Cobb-Douglas production function $\varphi(k) = k^{1-\beta}$ with a constant capital share $1 - \beta$, or a constant effective labor share β, upon $0 < \beta < 1$.

The paper [12] built the Environmental Kuznets Curve for aggregate emissions on a balabced growth path (see Fig. 1). Here on the Fig. 1 we use the next parameters of the Russian economy partly found in [1,14]: $\mu = 0.03155$, $\theta = 0.1$, $s = 0.19$, $n = 0.011$, $g_B = \alpha * \varepsilon = 0.11 * 0.3465 = 0.038115$, and $\beta = 0.75$.

Here we'll write this model in the variables of the Houthacker-Johansen model [1] and try to identify the parameters of this model from the statistical data of the Russian economy.

3 Data

3.1 Statistical Time Series of Macro Indices

Table 1 gives statistical data of Russian Federation which are used here to identify the model parameters.

As a composite indicator of pollution P, we can take a difference between the production and consumption waste generation W and the waste utilization

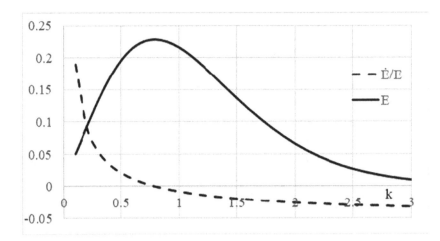

Fig. 1. Rate of change of aggregate emissions $\dot{E}/E$ and emissions E by capital on effective worker k.

and neutralization U, so that $P = W - U$, mln tons. Source for W and U is Rosprirodnadzor data [15].

Source for Gross Domestic Product (GDP) Y, and Gross capital formation bJ at constant 2015 prices in National currency is [16].

The economy's efforts on the abatement $Y^A(t)$ is my estimation by GDP deflator [16] and dynamics of investments aimed at environmental protection from harmful impact of production and consumption waste in the Russian Federation, 2012–2021, mln rub [15].

Table 1. Statistical data of Russia.

year	P, mln tons	Y^A, bln rub 2015	Y, trillion rub 2015	L, mln people	bJ, trillion rub 2015
2012	2659.9	9.036	60.726	71.545	23.484
2013	3109.4	8.629	63.053	71.391	22.274
2014	2810.8	8.241	63.650	71.539	20.851
2015	2374.9	12.732	58.531	71.324	18.403
2016	2197.3	8.190	57.629	71.393	18.285
2017	2956.4	10.099	59.589	71.316	19.457
2018	3447.6	12.771	61.656	72.500	19.148
2019	3869.1	11.157	63.757	71.900	19.596
2020	3527.0	8.882	60.545	70.600	18.802
2021	4511.8	6.989	65.024	71.700	20.479

3.2 Reconciliation of Proximity Criteria for Calculation and Statistics

To compare the closeness of time series for macroeconomic indices, it is common to use Theil's inequality index [1,17], a generalization of standard deviation for an economy where indices often grow exponentially. Theil's inequality index for index X is

$$T_X = \sqrt{\frac{\sum_{t=t_0}^{t_n} (X(t) - X_S(t))^2}{\sum_{t=t_0}^{t_n} (X^2(t) + X_S^2(t))}} \to \min_a, \tag{7}$$

where X_S is a statistical counterpart of index X, vector $\boldsymbol{a}$ is is a set of sought parameters $\boldsymbol{a} = (a_1, ..., a_n)$. The closer the Theil index T_X is to zero, the more the estimated $X(t)$ and statistical $X_{stat}(t)$ time series on the interval $t = t_0, ..., t_n$ coincide.

When using parallel computing on a cluster supercomputer for parameter identification, it is important to reduce the number of time-consuming computations at each step, so it is advantageous to get rid of the root and consider [18] the following analog $N_X = 1 - T_X^2$ of the Theil index to reduce the computation time.

$$N_X = \frac{2 \sum_{t=t_0}^{t_n} X(t)X_S(t)}{\sum_{t=t_0}^{t_n} (X^2(t) + X_S^2(t))} \to \max_a. \tag{8}$$

If the calculation matched the statistics, then $N_X = 1$. Here we also take into account the possibility of using this criterion in multiplicative convolution of criteria, where the difference from zero of each multiplier is important.

In our problem, we want to approximate simultaneously the calculation and statistical measures for output Y, employment L and waste emission rate P, then the convolution of the criteria can be, for example, their product.

$$O = N_Y N_L N_P \to \max_a. \tag{9}$$

The parameter identification problem (9) is solved numerically using parallel calculations and preliminary computational results for a Houthakker-Johansen model of endogenous production function are given in this paper.

4   The Houthakker-Johansen Model

In the Houthakker-Johansen model [1] with the abatement $Y^A(t)$ the total capacity $M(t)$ [19,20] is used instead of capital $K(t)$, a reciprocal of the lowest labor intensity $1/\nu(t)$ [1] is used as $B(t)$. So that, from (1)

$$\dot{\nu}/\nu = -\dot{B}/B = -g_B. \tag{10}$$

In the model the output of production $Y(t) = F(M(t), L(t)/\nu(t))$ is determined by the capacity utilization function $f(x) = F(1, x) \leq 1$ [1]:

$$Y(t) = M(t)f(x(t)), \tag{11}$$

where

$$x = \frac{L(t)}{\nu(t)M(t)} \tag{12}$$

is the value of the average labor intensity of capacities $L(t)/M(t)$ in relative units to the lowest labor intensity $\nu(t)$, or by other words, x is the average amount of effective labor $L(t)/\nu(t)$ per unit of the production capacity $M(t)$.

If a share of new production capacities in the total capacity

$$\alpha = J(t)/M(t) = (1 - \theta)f(x)s/b, \tag{13}$$

than [1] the rate of labor augmenting technological progress (10)

$$g_B = \varepsilon\alpha = \varepsilon(1 - \theta)f(x)s/b, \tag{14}$$

and the rate of exogenous technological progress in abatement lowering

$$g_A = \eta\alpha = \eta(1 - \theta)f(x)s/b. \tag{15}$$

The dynamics of the total production capacity, labor and the lowest labor intensity are determined by the equations

$$\dot{M} = \frac{s}{b}(1 - \theta)Y - \mu M, \dot{L} = nL, \dot{\nu} = -\varepsilon\frac{s}{b}(1 - \theta)f(x)\nu, \tag{16}$$

where $\mu > 0$ is a depreciation rate of total production capacity $M(t)$.

Assuming as in [12] that exogenous technological progress in abatement lowering at a rate $g_A = \eta\alpha > 0$. Converting our values of output $Y(t) = M(t)f(x)$, production capacity $M(t)$ and pollution $E(t) = \Omega(t)Y(t)a(\theta)$ into intensive units, we have an analogue of the Green Solow model in terms of the endogenous production function of the Houthakker-Johansen model [1]:

$$\dot{x} = \left[\mu + n - (1 - \varepsilon)(1 - \theta)\frac{s}{b}f(x)\right]x, \tag{17}$$

$$y = \frac{f(x)}{x}(1 - \theta), \tag{18}$$

$$e = \Omega\frac{f(x)}{x}a(\theta), \tag{19}$$

$$\dot{\Omega} = -\eta\Omega(1 - \theta)f(x)s/b. \tag{20}$$

where $y(t) = (1 - \theta)\nu(t)Y(t)/L(t)$ is the productivity of effective labor $L(t)/\nu(t)$ in relative units, $e(t) = \nu(t)E(t)/L(t)$ is the pollution per unit of effective labor, and the average labor intensity in relative units x is determined by (12).

4.1 Balanced Growth Path

On the balanced growth path

$$Y(t) = Y_0\exp(\gamma t), M(t) = M_0\exp(\gamma t), J(t) = J_0\exp(\gamma t) \tag{21}$$

$x = x^* = const$ [1], so that from (13),(12),(16),(21)

$$\gamma = n + \varepsilon\alpha. \tag{22}$$

From (18) $y = const$ too. From (17) x^* is determined by the next equation

$$f(x^*) = \frac{b(\mu + n)}{s(1 - \varepsilon)(1 - \theta)}. \tag{23}$$

The aggregate consumption due to (3) $C(t) = (1 - \theta)Y(t) - bJ(t) = C_0 exp(\gamma t)$. Due to (22) the consumption per unit of effective labour $c = C(t)\nu(t)/L(t) = const$.

The pollution per unit of effective labor $e(t)$ (19) is decreasing along with a decrease in the value $\Omega(t)$ due to abatement technological progress (20). From (22) we can conclude that the growth rate of total emission along the balanced growth path

$$g_E = n + \varepsilon\alpha - g_A. \tag{24}$$

So that if $n \geq 0$ then a sustainable growth is guaranteed by the next conditions: $\varepsilon\alpha > 0$, and $g_A > n + \varepsilon\alpha$. The technological progress in goods production $\varepsilon\alpha$ garantees per capita income growth $C(t)/L(t)$. The technological progress in abatement g_A should be grater the growth of aggregate output $Y(t)$ to garantee the aggregate pollution $E(t)$ fall.

The growth rate γ of the balanced growth path (21) due to equation on the aggregate production capacity (16) and (23)

$$\gamma = \frac{(\mu + n)}{(1 - \varepsilon)} - \mu. \tag{25}$$

From here the balanced growth rate $\gamma > 0$ if $n > -\varepsilon\mu$.

From (22) $\alpha = (\gamma - n)/\varepsilon$. Substituting here γ from (25) we have

$$\alpha = \frac{(\mu + n)}{(1 - \varepsilon)}. \tag{26}$$

Different balanced growth paths, which are numbered different x^* (or s by virtue of (23)) are characterized by different effective per capita consumption $c(x^*)$.

$$c(x^*) = \frac{(1 - \theta)f(x^*) - b(\mu + n)/(1 - \varepsilon)}{x^*}. \tag{27}$$

It is natural to call the optimal balanced growth path the one that maximizes the value of $c(x^*)$. Since $c(+0) < 0$, than the maximum of effective per capita consumption is reached at a finite value $x^\#$. From the necessary conditions of maximum we find the equation defining $x^\#$:

$$f(x^\#) - x^\# f'(x^\#) = \frac{b(\mu + n)}{(1 - \theta)(1 - \varepsilon)}. \tag{28}$$

This relationship is called Solow rule. Here this rule is expressed in our variables x, it takes into account the parameters of environmental θ, technological b, μ, population growth rate n, and rate of scientific and technical progress ε.

4.2   The Endogenous Production Function

Paper [12] uses the standard exogenous Cobb-Douglas production function, in our notation $f_0(x) = x^\beta$, where β is a constant effective labor share, with $0 < \beta < 1$.

In contrast to [12], we will use here the endogenous production function $f(x)$ obtained in [1,6] according to the initial micro description of the dynamics of production capacities differentiated by the moments of creation for the balanced growth path of the model [1] with age limit on production capacities A:

$$f(x) = \frac{1}{\rho}\left\{1 - [1 - (\rho - \sigma)x]^{\rho/(\rho-\sigma)}\right\}, \tag{29}$$

where σ and ρ are defined by the next relations:

$$\sigma \stackrel{\text{def}}{=} \varepsilon + \frac{\mu}{\alpha} < \rho, \tag{30}$$

$$\rho \stackrel{\text{def}}{=} 1 + \frac{1}{\alpha A}W_0(-\alpha A e^{-\alpha A}). \tag{31}$$

Here $W_0(z)$ is the main branch of the Lambert W function [2,21], where $z \in (-1/e, 0)$. We have used the inequalities that are valid under normal economic conditions: the age limit of production capacities is much more than one year, $A \gg 1$, the share of new investments exceeds the positive rate of capacity degradation, $\alpha > \mu > 0$, the rate of technological progress in the production of goods is non-negative, $\varepsilon \geq 0$, so $\sigma > 0$, $\alpha A > 0$ and from (31) $\rho < 1$.

By definition of production capacity $f(x) \leq 1$, so from (29) we have

$$0 \leq x \leq \frac{1}{(\rho - \sigma)}\left[1 - (1 - \rho)^{1-\sigma/\rho}\right]. \tag{32}$$

In calculations, we should extend the function by condition $f(x) = 1$ for x above the right boundary of the interval (32).

Now we can find the derivative of the function $f(x)$ (29):

$$f'(x) = [1 - (1 - \rho)x]^{\sigma/(\rho-\sigma)}, \tag{33}$$

if x belongs to the interval (32), and $f'(x) = 0$ if $x \geq \left[1 - (1-\rho)^{1-\sigma/\rho}\right]/(\rho-\sigma)$.

From here we can find the second derivative of this production function:

$$f''(x) = -\frac{(1-\rho)\sigma}{(\rho-\sigma)}[1 - (1-\rho)x]^{(2\sigma-\rho)/(\rho-\sigma)}, \tag{34}$$

if x belongs to the interval (32), and $f''(x) = 0$ if $x \geq \left[1 - (1-\rho)^{1-\sigma/\rho}\right]/(\rho-\sigma)$.

This production function and it's derivatives are shown in the following Fig. 2.

The functions (29) and (33) are necessary to find the growth rate of emissions. From (2),(11),(16),(17),(20),(24) we have $g_A = \eta\alpha$ and

$$\frac{\dot{E}}{E} = g_E + \frac{f'(x)}{f(x)}\dot{x} = n - (\eta - \varepsilon)\alpha + \frac{f'(x)}{f(x)}\dot{x}. \tag{35}$$

Here on the Fig. 3 we use partly the same as on the Fig. 1 parameters of the Russian economy [1,14]: $\mu = 0.03155$, $\theta = 0.1$, $s = 0.19$, $n = 0.011$, $\varepsilon = 0.038115$, $\alpha = 0.11$, partly the new: $A = 25$, $\eta = 1.1$, $\rho = 0.993492$, and $\sigma = 0.633318$.

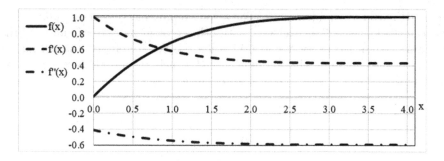

Fig. 2. Production function $f(x)$ and it's derivatives.

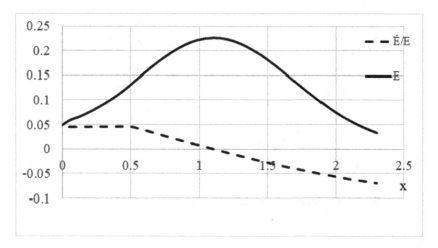

Fig. 3. Rate of change of aggregate emissions $\dot{E}/E$ and emissions E by effective labor on production capacity x.

5   Conclusions and Plans

This paper considers the simplest dynamic model of the economy, the Solow model, augmented by a description of the environmental consequences of economic growth, considering the costs of environmental remediation. This model allows describing the environment Kuznets curve.

The model of the ecological system has not been described in this paper. In order to build a model of a circular economy, it is necessary to consider an ecological system model, since sustainably growing ecological systems are circular, and their models look accordingly. Often models of ecological systems are described by considering the cycle of biogeochemical elements. For example, in [22] the carbon cycle is used to describe the impact of the economy on ecology. An example of the carbon cycle in a steppe ecosystem [23], as one of the simplest, is shown in Fig. 4. Here, the circles show the carbon stocks in the atmosphere A (closed to planet Earth as a whole), green phytomass of vegetation G, root system W, litter V, and humus H. The arrows represent the corresponding flows between stocks, as well as exchange with other ecological systems and with the economic system (recovery S and utilization F). Separately, it is worth noting the flux of the complex pollution indicator P, which has a detrimental effect on carbon fluxes that can lead to degradation of the ecosystem under consideration.

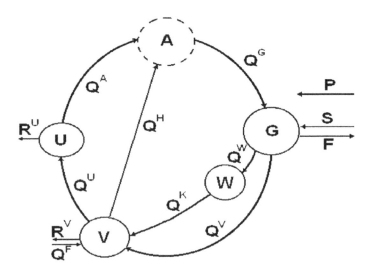

Fig. 4. Carbon cycle in the steppe ecosystem. Source: [23].

The carbon cycle in a forest ecosystem is described, for example, in [22]. As for the economic system, there is no closed cycle yet. The traditional linear scheme is still used: extraction of natural resources, production, distribution, consumption, waste disposal. Statistics state that 80% of consumer goods end their life in garbage bins within half a year. For the Russian economy the most polluting sphere is the extraction of natural resources, namely: extraction of fuel and energy minerals. They account for more than half of all garbage. At present the level of waste recycling in the USA is 35%, in Germany 50%, and in Russia only 5–7%. At the same time, more than 90% of garbage is sent to landfills and unauthorized dumps, so the amount of accumulated waste is growing every year. This situation is associated with the high material intensity and low resource efficiency of most production in the main sectors of the Russian economy, as well as with the raw material structure of the Russian economy. It is necessary to build a mathematical model of the closed-cycle economy for the theoretical study of environmental-economic interaction in the transition to such an economic scheme.

References

1. Olenev, N.N.: Identification of a production function with age limit for production capacities. Math. Models Comput. Simul. **12**(4), 482–491 (2020). https://doi.org/10.1134/S2070048220040134

2. Olenev, N.: Fluctuations of aggregated production capacity near balanced growth path. In: Olenev, N., Evtushenko, Yu., Jaćimović, M., Khachay, M., Malkova, V., Pospelov, I. (eds.) OPTIMA 2022, LNCS, vol. 13781, pp. 192–204. Springer, Heidelberg (2022). https://doi.org/10.1007/978-3-031-22543-7_14

3. Olenev, N.: Economy of Greece: an evaluation of real sector. Bull. Polit. Econ. **10**(1), 25–37 (2016)

4. Olenev, N.: Identification of an aggregate production function for Polish economy. Quant. Methods Econ. **19**(4), 430–439 (2018). https://doi.org/10.22630/MIBE.2018.19.4.41

5. Olenev, N.: Identification of an aggregate production function for the economy of Turkey. In: Proceedings of the Izmir International Congress on Economics and Administrative Sciences (IZCEAS 2018), pp. 1761–1770. DETAY YAYINCILIK, Izmir (2018)

6. Olenev, N.: Golden rule saving rate for an endogenous production function. In: Jaćimović, M., Khachay, M., Malkova, V., Posypkin, M. (eds.) OPTIMA 2019. CCIS, vol. 1145, pp. 267–279. Springer, Cham (2020). https://doi.org/10.1007/978-3-030-38603-0_20

7. Phelps, E.: The golden rule of accumulation: a fable for growth men. Am. Econ. Rev. **51**(4), 638–643 (1961)

8. Swan, T.W.: Economic growth and capital accumulation. Econ. Rec. **32**(2), 334–361 (1956)

9. Solow, R.M.: A contribution to the theory of economic growth. Q. J. Econ. **70**(1), 65–94 (1956)

10. Solow, R.M.: Technical change and the aggregate production function. Rev. Econ. Stat. **39**(3), 312–320 (1957)

11. Grossman, G., Krueger, A.: Economic growth and the environment. Q. J. Econ. **110**, 353–377 (1995)

12. Brock, W.A., Taylor, M.S.: The green Solow model. J. Econ. Growth **2010**(15), 127–153 (2010)

13. Copeland, B.R., Taylor, M.S.: North-South trade and the global environment. Quart. J. Econ. **109**, 755–787 (1994)

14. Olenev, N.N., Pechenkin, R.V., Chernetsov, A.M.: Parallel programming in MATLAB and its applications (In Russian). CC RAS, Moscow (2007)

15. State report on the state and protection of the environment of the Russian Federation in 2021. https://news.solidwaste.ru/2022/12/gosudarstvennyj-doklad-o-sostoyanii-i-ob-ohrane-okruzhayushhej-sredy-rf-v-2021-godu/. Accessed 1 Jul 2023

16. National Accounts – Analysis of Main Aggregates (AMA). http://unstats.un.org/unsd/snaama/dnllist.asp. Accessed 1 Jul 2023

17. Theil, H.: Economic Forecasts and Policy. North-Holland Publishing Company (1961)

18. Olenev, N.: Identification of a Russian economic model with two kinds of capital. In: Computer Algebra Systems in Teaching and Research VI, pp. 94–103. Siedlce University of Natural Sciences and Humanities, Siedlce (2017)

19. Johansen, L.: Production functions and the concept of capacity. Collect. Econ. Math. Econometr. **2**, 49–72 (1968)
20. Johansen, L.: Production functions: An integration of micro and macro, short run and long run aspects. North-Holland Publishing Company, Amsterdam (1972)
21. Corless, R.M., Gonnet, G.H., Hare, D.E.G., Jeffrey, D.J., Knuth, D.E.: On the lambert W function. Adv. Comput. Math. **5**, 329–359 (1996). https://doi.org/10.1007/BF02124750
22. Olenev, N.N., Petrov, A.A., Pospelov, I.G.: Ecological Consequences of Economic Growth. Math. Model. **10**(8), 17–32 (1998). (in Russian)
23. Demberel, S., Olenev, N.N., Pospelov, I.G.: An interaction model for livestock farming and ecosystem. Math. Comput. Simul. **67**(4–5), 335–342 (2004)

Two Balanced Growth Paths Based on an Endogenous Production Function

Nicholas Olenev[(✉)] [iD]

Federal Research Center "Computer Science and Control" of the Russian Academy of Sciences, Moscow, Russia
nolenev@mail.ru

Abstract. The endogenous production function presented in this paper allows for two solutions to the problem of finding balanced growth in a dynamic model of the economy. The endogenous production function is constructed on the basis of a microeconomic description of the dynamics of production capacities by technology. At the micro-level, it is assumed that the number of jobs is set at the time of creation of a production unit, and production capacity of this unit falls at a constant rate. In addition, an age limit of production capacity is set. Then it is possible to construct a production function and obtain its analytical expression under balanced growth path. This expression for the production function contains, among the parameters, the share of new capacities, the rate of decline of each capacity at the micro-level, the growth rate of scientific and technological progress, and the age limit of capacities. The rate of economic growth depends on these same parameters, which makes Solow problem of the golden growth rate for the endogenous production function more interesting.

Keywords: Endogenous production function · Balanced growth path · Production capacity · Age limit of capacities

1 Introduction

1.1 Houthakker-Johansen Model of Production

There is an extensive literature devoted to production functions represented in the form of capacity distribution by technology. For the first time, Hendrik S. Houthakker [1] noticed that there is no justification at any level of economic modeling (micro, meso, macro) to apply a neoclassical production function of the same given type with good mathematical properties, for example, the Cobb-Douglas production function. This work [1] showed that in order to obtain an aggregate Cobb-Douglas type production function at the macro-level it is necessary to have a Pareto distribution at the micro-level. Some time later this

The publication has been prepared with the support of the "Research and development program using the infrastructure of the Shared Center of FRC CSC RAS".

work was noticed by Robert Solow [2], and many researchers began to construct aggregate production functions and study their properties [3].

The next step in this direction was the use of the concept of production capacity instead of physical capital, which we owe to Leif Johansen [4,5]. Production capacity is the maximum possible amount of output, that is, this quantity is measured in the same units as output. As a result, the aggregate production function has a simple economic sense of the level of utilization of production capacity. Thanks to this approach, L. Johansen managed to construct aggregate production functions for some industrial sectors of the Swedish and Norwegian economies, in particular, for the tanker fleet, which can now be in demand in the construction of models of the blue economy (ocean development economy).

Igor Pospelov successfully used aggregated production functions to describe single-product economic models [6–8]. The papers [6,7] present a simulation model of an economy with this type of aggregated production function and give its analytical study.

The textbook [8] considers the construction of a production function and the simplest dynamic model of optimal economic growth. The aggregate production function of a sector of the economy emerges as a solution to the problem of optimal allocation of resources between enterprises of the sector. In the study of the optimal economic growth model, special attention is paid to the discussion of the meaning and implications of the turnpike property of the model.

The paper [9] of Alexander Petrov and Igor Pospelov presents a description of the first part of the research area "System Analysis of an Developing Economy", which describes and investigates the theory of production functions.

Alexander Shananin built a mathematical theory of such production functions, proved the equivalence of the concept of production function and profit function [10–12]. In [10], estimates of variations in the production function and profit function, based on locally summarized distributions of capacities by technology, are given through variations in the distribution of these capacities. In [11] the initial representativeness of profit functions is investigated. The paper [12] considers new problems of integral geometry related to the Houthakker-Johansen production models, taking into account the substitution of production factors.

In [13], a new class of aggregate production functions of the economic sector was obtained for condition that each production capacity decreases with a constant rate with a constant number of workers employed at it. This type of production functions contains among the parameters the growth rate of the sector of the economy.

If we fix the age limit of production capacity in this model, we obtain a discussed in this paper new class of endogenous production functions for the sector of the economy.

In a dynamic model of the economy, an equilibrium growth path is a steady state of growth in which "level variables", such as in our model total output and total production capacity, grow at a constant rate, and the relationships between key model variables are stable. Paul Romer [14] called this concept of a steady state of growth "the balanced growth path".

Depending on the initial conditions and the economic policy pursued, the economic system may fall into one of the possible characteristic growth regimes. As shown in [15], most of these modes are fluctuating, which corresponds to complex values of the growth rate and the corresponding complex branches of the Lambert W function. These fluctuating modes can be damped, oscillatory, or transitioning to other modes, such as a balanced growth path. All this depends on what initial conditions add up.

The present paper for the first time finds that there are two possible balanced growth paths, one with a positive growth rate (increasing path) and one with a negative growth rate (decreasing path). These two regimes correspond to the two real branches of the Lambert W function (W_0, W_{-1}).

For each of these balanced growth paths, analytical expressions for the production function are found and presented here for the first time, and its simplest properties are discussed.

1.2   Production Function with Limited Age of Capacities

The papers [15–21] present an endogenous production function of a branch of the economy or the national economy as a whole, built on the basis of the initial microeconomic description of the dynamics of production capacities with a given age limit, and an identification of its parameters. It is assumed that the capacity of the production unit decreases from the moment of its creation at a given constant rate, and the number of jobs on it remains unchanged until it is liquidated at the moment of reaching the age limit.

The paper [16] shows how the parameters of the macroeconomic model of the real sector can be estimated from the Greek statistics on the use of GDP based on the initial microeconomic description of production by comparing the macroeconomic indicators calculated by the model and their statistical counterparts. Papers [17–19] solve similar problems according to data from Russia, Poland and Turkey, respectively. The paper [20] presents an analytical expression for the age-limited aggregate production function, the results of its identification according to Russian data, and their economic interpretation. In the paper [21] for this production function, which contains among the parameters the growth rate and the age limit of capacities, the golden rule of savings is presented. The paper [15] shows that in addition to the highways of balanced growth, there are many oscillatory modes in the model under study.

Production capacity in accordance with [4,5] is determined by the maximum possible output of products, that is, unlike capital, it is determined in the same units as output.

The main hypothesis on which the microeconomic description of the dynamics of production capacities m is based states [17,20] that at the moment τ of creating a production unit, its technological characteristics and the number of jobs are set. At the same time, this number of jobs in a production unit does not change until the age limit is reached, but its capacity decreases at a constant rate $\mu > 0$. Then, the labor intensity λ of a production unit grows at the same rate μ to maintain the number of jobs $m\lambda$.

When constructing an endogenous macroeconomic production function based on a description of the microeconomic dynamics of production capacities, additional assumptions must be made about the structure of the distribution of new capacities and about how the least labor intensity ν available at time t (the best technology) changes. Here we will assume, as in [15–21], that all new capacities $J(t)$ in monent t use the best technology, that is, they have the least known labor input $\nu(t)$.

This work will explicitly consider the two balanced volume turnpikes allowed by the model, one with balanced growth and the other with balanced decline (two balanced growth paths).

2   Problem Statement

If t is the current moment in time, and τ is the moment of capacity creation, then $t - \tau$ is the age of this production capacity. In these Lagrange variables, the decline in production capacity with age is described by the relation $m(t, \tau) = J(\tau)exp(-\mu(t-\tau))$, and the growth of labor intensity is described by the equation $\lambda(t, \tau) = \nu(\tau)exp(\mu(t-\tau))$, while the number of jobs in production capacity does not change with age, $\lambda(t, \tau)m(t, \tau) = \nu(\tau)J(\tau)$.

The total production capacity $M(t)$ is determined by the age limit of capacities $A(t)$.

$$M(t) = \int_{t-A(t)}^{t} J(\tau)e^{-\mu(t-\tau)}d\tau. \tag{1}$$

Differentiating (1) with respect to time and dividing by $M(t)$, we find the growth rate of the total production capacity

$$\frac{1}{M(t)}\frac{dM}{dt} = \alpha(t) - \mu - \alpha(t-A)\left(1 - \frac{dA}{dt}\right)\frac{M(t-A)}{M(t)}e^{-\mu A(t)}, \tag{2}$$

where $\alpha(t) = J(t)/M(t)$ is a share of new production capacities in the total production capacity, $A = A(t)$ in expressions $\alpha(t - A)$ and $M(t - A)$.

We will assume, as in [15–21], that the rate of decrease in the best labor input $\nu(t)$ is proportional to the share of new capacities.

$$\frac{1}{\nu(t)}\frac{d\nu}{dt} = -\varepsilon\alpha(t), \tag{3}$$

where $\varepsilon > 0$ chracterized the growth rate of scientific and technological progress.

The production function expresses the dependence of output $Y(t)$ on production factors: total labor $L(t)$ and total production capacity $M(t)$. Here, this function is determined parametrically through the optimal loading of production capacities by the labor resources used.

$$Y(t) = \int_{t-a(L)}^{t} J(\tau)e^{-\mu(t-\tau)}d\tau, \quad L(t) = \int_{t-a(L)}^{t} \nu(\tau)J(\tau)d\tau, \tag{4}$$

where $a(L) = a(L(t))$ means the age of the oldest capacity loaded by the used labor resources $L(t)$. Here $a(L) \leq A(t)$, so that $Y(t) \leq M(t)$.

Let $x(t)$ be the ratio of the average labor input of capacities $L(t)/M(t)$ to the least labor input $\nu(t)$, that is, the average labor input in relative units,

$$x(t) = \frac{L(t)}{\nu(t)M(t)}. \tag{5}$$

Then the parametric expression for the production function (4) in the new notation $Y(t) = M(t)f(x(t))$ will be written as

$$f(x(t)) = \int_{t-a(x)}^{t} \alpha(\tau)\frac{M(\tau)}{M(t)}e^{-\mu(t-\tau)}d\tau, \tag{6}$$

$$x(t) = \int_{t-a(x)}^{t} \alpha(\tau)\frac{M(\tau)\nu(\tau)}{M(t)\nu(t)}d\tau. \tag{7}$$

The function $f(x(t))$ has the meaning of loading the total capacity $M(t)$, $0 \leq f(x(t)) \leq 1$. In the following text we will call it the production function.

In a special case, with a constant growth rate γ of the total capacity $M(t)$, a fixed age limit $A \gg 1$ and a constant share of new capacities $0 < \alpha < 1$ from the equations (2–3), (6–7) you can get an analytical expression for the specific production function (meaning the total capacity load) [21]:

$$f(x(t)) = \frac{\alpha}{\varphi}\left\{1 - \left[1 - \frac{\varphi - \mu - \varepsilon\alpha}{\alpha}x(t)\right]^{\varphi/(\varphi-\mu-\varepsilon\alpha)}\right\}, \tag{8}$$

where $\varphi = \gamma+\mu$ is a real root $\varphi = \varphi(\alpha, \mu, A)$ of the next transcendental equation

$$\frac{\varphi}{\alpha} = 1 - e^{-\varphi A}, \tag{9}$$

obtained from (1) when $\gamma = const$, $\alpha = const$, and $A = const$. Recall that in [15] we considered complex roots of the equation equivalent to (9) which was written in different dimensionless variables than here. These complex roots describe oscillatory trajectories.

In the next section, we will show that there are two real roots of the equation (9), which provide two different balanced growth paths, and then we will give an interpretation of this situation.

3   Two Balanced Growth Paths

If we rewrite the transcendental equation on φ (9) in the form of

$$\alpha A(\varphi/\alpha - 1)e^{\alpha A(\varphi/\alpha-1)} = -\alpha Ae^{-\alpha A}, \tag{10}$$

then by definition of Lambert W function [15,22] from (10) we have

$$\alpha A(\varphi/\alpha - 1) = W(-\alpha Ae^{-\alpha A}). \tag{11}$$

The argument of the Lambert W function in (11) is negative, because in economic terms the share of new capacities $0 < \alpha < 1$, and the age limit $A \gg 1$. For a negative real argument $z = a = -\alpha A e^{-\alpha A}$ on the interval $(-1/e, 0)$ there are two real branches of the Lambert W function (see Fig. 1).

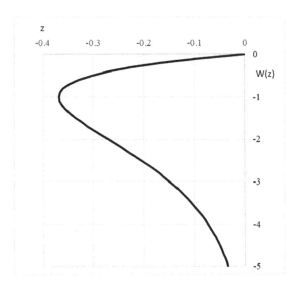

Fig. 1. The real branches of the Lambert W function with negative argument z.

Theorem 1. *There are two modes of balanced growth path for the production function (8).*

In the first mode, the production function (8) is

$$f(x(t)) = \frac{1}{(1-\beta)} \left\{ 1 - \left[1 - \left(1 - \varepsilon - \beta - \frac{\mu}{\alpha}\right) x(t) \right]^{\frac{(1-\beta)}{(1-\varepsilon-\beta-\mu/\alpha)}} \right\}, \qquad (12)$$

where $\beta = \beta(\alpha A) > 0$.

In the second mode, the production function does not depend on the parameter A (age limit), or rather on the dimensionless parameter αA, the second production function actually depends only on two dimensionless parameters: the growth rate of scientific and technological progress ε and the ratio of the degradation rate to the share of new production capacities in the total production capacity μ/α:

$$f(x(t)) = \frac{1}{(\varepsilon + \mu/\alpha)} \ln \left[1 + \left(\varepsilon + \frac{\mu}{\alpha}\right) x(t) \right]. \qquad (13)$$

Proof. In the first case, when $a \geq -1/e$ and the value of the function $w \geq -1$, the single-valued function $W_0(a)$ will be defined as the main branch of the function $W(z)$. In this case, due to (11)

$$\varphi = \alpha + W_0(-\alpha A e^{-\alpha A})/A. \qquad (14)$$

The argument of the main branch in (14) is negative, the value of the Lambert W function on this part $(-1/e, 0)$ of the main branch is also negative, so $\varphi < \alpha$. If we introduce a dimensionless positive parameter

$$\beta = \beta(\alpha A) = -\frac{1}{\alpha A} W_0(-\alpha A e^{-\alpha A}) > 0, \tag{15}$$

then according (14)–(15) $\varphi = \alpha(1 - \beta)$, and the production function (8) can be written as (12).

The condition $f(x) \leq 1$ (total output does not exceed total production capacity) imposed on output leads to the condition $x \leq 1/(1 - \varepsilon - \mu/\alpha)$ and the need to complete the definition of the production function (8) by the next condition $f(x) = 1$ for $x \geq 1/(1 - \varepsilon - \mu/\alpha)$.

Note that for $\beta \to 0$ $(A \to \infty)$ the function (12) transforms, up to different notation $(x = L/M$ in [13], $x = L/(\nu M)$ as defined by (5)), into the known production function [13]:

$$f(x(t)) = 1 - \left[1 - \left(1 - \varepsilon - \frac{\mu}{\alpha}\right) x(t)\right]^{1/(1-\varepsilon-\mu/\alpha)}. \tag{16}$$

In the second case, when $a \geq -1/e$, and the value of the function $w \leq -1$, a single-valued function $W_{-1}(a)$ will be defined as an additional branch of the function $W(z)$. Moreover, due to [22] $W_{-1}(-\alpha A e^{-\alpha A}) = -\alpha A$ and therefore due to (11) $\varphi = 0$, and the production function (8) (total capacity load) does not depend on the age limit A and increases logarithmically with the growth of average labor input as (13).

The condition $f(x) \leq 1$ imposed on the total output leads to the condition $x \leq \left(e^{\varepsilon+\mu/\alpha} - 1\right)/(\varepsilon + \mu/\alpha)$ and the need to complete the definition of the production function (13) by the next condition $f(x) = 1$ for $x \geq \left(e^{\varepsilon+\mu/\alpha} - 1\right)/(\varepsilon + \mu/\alpha)$.
Q.E.D.

The economic meaning of the equality $\varphi = 0$ in the second growth regime (according to [21] $\varphi = \gamma + \mu$) is the negative growth rate γ of the total production capacities (1) equal to $-\mu$, at which the volume of new production capacity is equal to the volume of capacities dismantled due to exceeding the age limit. A regime of economic decline similar to the path (13) was observed in the Russian economy in the 90 s of the last century.

4   Properties of the Endogenous Production Function

If in the time interval under study there may be a change in various characteristic modes for the economic model, this production function and its parameters can be estimated numerically based on the initial microeconomic description of the model, as it is done in [16–20]. Papers [17,20] estimate the average capacity age limit for the national economy of Russia on the one-product model. It gives $A = 25$. This means that in 2017, the production capacities of the 1980 s and the first two years of 90 s (when investments in new capacities were still significant) on average cease to influence cost inflation (1992+25=2017). From now on, inflation depends only on the monetary measures taken, and the real sector of the economy affects inflation less and less [17,20].

4.1 Production Function (12)

Let us introduce the notation

$$B = 1 - \beta, \tag{17}$$

$$D = \varepsilon + \frac{\mu}{\alpha}, \tag{18}$$

where $0 < D < B < 1$, and by virtue of (15)

$$B = 1 - \beta(\alpha A) = 1 + \frac{1}{\alpha A} W_0(-\alpha A e^{-\alpha A}). \tag{19}$$

In addition, by (14), and by the equality $\varphi = \gamma + \mu$ we have a connection of B with the growth rate γ.

$$B = \frac{\gamma + \mu}{\alpha}. \tag{20}$$

The expression for our production function (12) looks more concise due to these notations (17),(18):

$$f(x) = \frac{1}{B} \left\{ 1 - [1 - (B - D) x]^{B/(B-D)} \right\}, \quad 0 \le x \le x_{max}, \tag{21}$$

where $x = x(t)$ due to (5) and $f(x) = 1$, if $x \ge x_{max}$. Here

$$x_{max} = \frac{1}{(B - D)} \left\{ 1 - [1 - B]^{(B-D)/B} \right\} < \frac{1}{(B - D)}. \tag{22}$$

First derivative of the production function (21)

$$f'(x) = \left\{ 1 - [1 - (B - D) x]^{D/(B-D)} \right\}, \quad 0 \le x \le x_{max}. \tag{23}$$

Second derivative of the production function (21)

$$f''(x) = -D \left\{ 1 - [1 - (B - D) x]^{(2D-B)/(B-D)} \right\}, \quad 0 \le x \le x_{max}. \tag{24}$$

The production function (21) and its derivatives (23),(24) are shown in the Fig. 2. The calculation on this figure used the values of the production function's parameters identified according to the characteristic data of the Russian economy 1970–2019 [20]: average age limit of production capacities $A = 25$ years, share of new production capacities in the total production capacity $\alpha = 0.11$, depreciation rate of production capacities $\mu = 0.03155$, the growth rate of scientific and technological progress $\varepsilon = 0.3465$. Then we have $z = -\alpha A e^{-\alpha A} = -0.1758$, $W_0(z) = -0.2188$, $\beta = 0.07956$, $\varphi = 0.10124$, $B = 0.9204$, $D = 0.6333$, $x_{max} = 1.9015$.

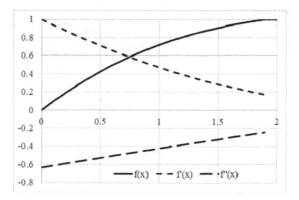

Fig. 2. The production function $f(x)$ (12) and its' derivatives $f'(x)$ and $f''(x)$.

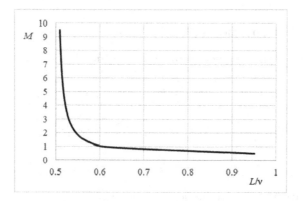

Fig. 3. The isoquant $Mf((L/\nu)/M)) = Y_c$ for production function (12).

On the line segment $0 \leq x \leq x_{max}$, the technical rate of substitution $\rho_{L/\nu,M}$ or the marginal rate of technical substitution of the effective labor L/ν for production capacity M is calculated by the next formula.

$$\rho_{L/\nu,M} = -\frac{dM}{d(L/\nu)} = \frac{f(x)}{f'(x)} - x. \tag{25}$$

Here $f(x)$ can be taken from the expression (21) and $f'(x)$ from (23).

Figure 3 presents an isoquant for production function (12) drawn with the above values of the parameters and $Y_c = 0.5$. At any point on the isoquant, the marginal rate of technical substitution $\rho_{L/\nu,M}$ is the absolute value of the slope of the isoquant at that point. The equation for the isoquant was found from (21) and the relation $Y_c = Mf((L/\nu)/M))$:

$$\frac{L}{\nu} = \frac{M}{(B-D)}\left\{1 - \left[1 - \frac{Y_cB}{M}\right]^{B/(B-D)}\right\}. \tag{26}$$

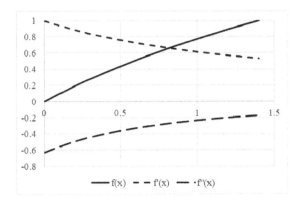

Fig. 4. The production function $f(x)$ (13) and its' derivatives $f'(x)$ and $f''(x)$.

4.2   Production Function (13)

The expression for the production function (13) due to notation (18) looks as:

$$f(x) = \frac{1}{D} ln\,[1 + Dx]\,, \quad 0 \le x \le x_{max}, \tag{27}$$

where $x = x(t)$ due to (5), and $f(x) = 1$, if $x \ge x_{max}$. Here

$$x_{max} = \frac{1}{D} \left[e^D - 1\right]. \tag{28}$$

First derivative of the production function (27)

$$f'(x) = \frac{1}{1 + Dx} > 0, \quad 0 \le x \le x_{max}. \tag{29}$$

Second derivative of the production function (27)

$$f''(x) = -\frac{D}{(1 + Dx)^2} < 0, \quad 0 \le x \le x_{max}. \tag{30}$$

The Fig. 4 presents the production function (27) and its first derivatives (29),(30) on the parameters of the Russian economy [20]: $\alpha = 0.11$, $\mu = 0.03155$, $\varepsilon = 0.3465$. In this case $D = 0.6333$, $x_{max} = 1.3956$.

The first and second derivatives of this production function (27) are higher than the corresponding values of that production function (21), so this production function (27) reaches its maximum value equal to one earlier than the first production function (21).

The marginal rate of technical substitution of the effective labor L/ν for production capacity M, $\rho_{L/\nu,M}$, is also calculated by (25). But now $f(x)$ can be taken from the expression (27) and $f'(x)$ from (29), $0 \le x \le x_{max}$, where x_{max} is taken from (28).

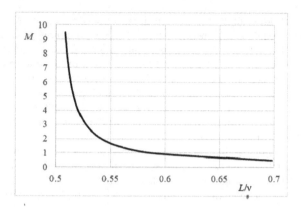

Fig. 5. The isoquant $Mf((L/\nu)/M)) = Y_c$ for production function (13).

An isoquant for production function (13) is presented on Fig. 5, $Y_c = 0.5$. The equation for the isoquant for (27):

$$\frac{L}{\nu} = \frac{M}{D}\left[e^{DY_c/M} - 1\right]. \tag{31}$$

The isoquant $Y = Y_c$ for this production function (13) lies above the isoquant for that production function (12). This means that more production resources are required for the same output in the second mode.

5 Conclusions

This paper shows that for an endogenous production function built on the hypothesis that production capacity decreases at a constant rate, keeping the number of jobs until it reaches a fixed age limit, there are two balanced growth paths: one with a positive growth rate and the other with a negative growth rate. For these two cases, the production function is represented by different analytical expressions, some properties of which are given in the paper.

References

1. Houthakker, H.S.: The pareto distribution and the Cobb-Douglas production function in activity analysis. Rev. Econ. Stud. **23**(1), 27–32 (1955–1956)
2. Solow, R.M.: Some Recent Developments in the Theory of Production. In: Brown, M. (ed.): The Theory and Empirical Analysis of Production, pp. 25–53. Columbia University Press, New York (1967)
3. Levhari, D.: A note on Houthakker's aggregate production function in a Multifirm industry. Econometrica **36**(1), 27–32 (1968)
4. Johansen, L.: Production Functions and the Concept of Capacity. Recherches Recentes sur la Fonction de Production, Collection. Economie Mathematique et Econometrie **2**, 49–72 (1968)

5. Johansen, L.: Production Functions: An Integration of Micro and Macro, Short Run and Long Run Aspects. North-Holland Publishing Company, Amsterdam (1972)
6. Petrov, A.A., Pospelov, I.G.: Mathematical modelling of socio-economic system. In: Marchuk, G.I. (eds.) Modelling and Optimization of Complex System. LNCIS, vol. 18. Springer, Berlin, Heidelberg (1979). https://doi.org/10.1007/BFb0004170
7. Pospelov, I.G.: On the system-analysis of a developing economy - an analytic investigation of an imitational model. Eng. Cybern. **18**(2), 20–30 (1979)
8. Pospelov, I.G.: One-Product Description of the Reproduction of the Economy (in Russian). MIPT, Moscow (2015)
9. Petrov, A.A., Pospelov, I.G.: Systems analysis of developing economics - on the theory of production functions. 1. Eng. Cybern. **17**(2), 10–18 (1979)
10. Shananin, A.A.: Investigation of a class of production functions arising in a Macrodescription of economic systems. USSR Comput. Math. Math. Phys. **24**(6), 127–134 (1984)
11. Shananin, A.A.: Study of a class of profit functions arising in a macro description of economic systems. USSR Comput. Math. Math. Phys. **25**(1), 34–42 (1985)
12. Shananin, A.A.: Inverse problems in economic measurements. Comput. Math. Math. Phys. **58**(2), 170–179 (2018). https://doi.org/10.1134/S0965542518020161
13. Olenev, N.N., Petrov, A.A., Pospelov, I.G.: Model of change processes of production capacity and production function of industry. In: Samarsky, A.A., Moiseev, N.N., Petrov, A.A. (Eds.): Mathematical Modelling: Processes in Complex Economic and Ecologic Systems (in Russian), pp. 46–60. Nauka, Moscow (1986). https://doi.org/10.13140/RG.2.1.3938.8880
14. Romer, P.: Increasing returns and long-run growth. J. Polit. Econ. **94**(5), 1002–1037 (1986)
15. Olenev, N.: Fluctuations of Aggregated Production Capacity Near Balanced Growth Path. In: Olenev, N., Evtushenko, Yu., Jaćimović, M., Khachay, M., Malkova, V., Pospelov, I. (eds.) OPTIMA 2022, LNCS, vol. 13781, pp. 192–204. Springer, Heidelberg (2022). https://doi.org/10.10007/978-3-031-22543-7_14
16. Olenev, N.: Economy of Greece: an evaluation of real sector. Bull. Polit. Econ. **10**(1), 25–37 (2016)
17. Olenev, N.N.: Parameter identification of an endogenous production function. CEUR-WS **1987**, 428–435 (2017)
18. Olenev, N.: Identification of an aggregate production function for polish economy. Quant. Methods Econ. **XIX**(4), 430–439 (2018)
19. Olenev, N.: Identification of an aggregate production function for the economy of Turkey. In: Proceedings of the Izmir International Congress on Economics and Administrative Sciences (IZCEAS 2018) on New Trends in Economics and Administrative Sciences, pp. 1761–1770. DETAY YAYINCILIK, Izmir (2018)
20. Olenev, N.N.: Identification of a production function with age limit for production capacities. Math. Models Comput. Simul. **12**(4), 482–491 (2020). https://doi.org/10.1134/S2070048220040134
21. Olenev, N.: Golden rule saving rate for an endogenous production function. In: Jaćimović, M., Khachay, M., Malkova, V., Posypkin, M. (eds.) OPTIMA 2019. CCIS, vol. 1145, pp. 267–279. Springer, Cham (2020). https://doi.org/10.1007/978-3-030-38603-0_20
22. Corless, R.M., Gonnet, G.H., Hare, D.E.G., Jeffrey, D.J., Knuth, D.E.: On the lambert w function. Adv. Comput. Math. **5**, 329–359 (1996). https://doi.org/10.1007/BF02124750

Analysis of Import Substitution Processes Taking into Account Industry Specifics in the Regional Economy

Anastasiia Rassokha[1,2(✉)] and Natalia Obrosova[2,3]

[1] Moscow Institute of Physics and Technology, Dolgoprudny, Russia
[2] Federal Research Center "Computer Science and Control" of RAS, Moscow, Russia
`arta13@list.ru`
[3] Center of Fundamental and Applied Mathematics, Lomonosov MSU, Moscow, Russia

Abstract. Analysis of import substitution processes is one of the urgent problems for many countries in the context of deglobalization of the world economy. Mathematical methods of convex analysis make it possible to construct a model of input-output balance with non-linear production functions. It can be identified from the official statistics of input-output tables. This model allows us to describe the substitution of production factors. The elasticity of substitution can be estimated using the solution of the Yang dual problem. With the help of the developed model for industries with a significant share of imports, the possibility of replacing imported production factors with domestic ones is studied on the example of the Russian economy.

Keywords: Factor substitution · Input-output model · Input-output tables · Dual problem

Introduction

Currently, the global economy is experiencing deglobalization trends. The expansion and complication of mutual economic ties between different countries of the world, observed in recent decades, is being replaced by a reverse process: the economies are becoming more closed. This process was especially pronounced during the pandemic. The Russian economy, in addition to global trends, is currently subject to sanctions. The sanctions are mainly aimed at reducing the volume of exports and making the import of goods and services less accessible. This undoubtedly has an impact on the prospects for Russia's economic development, and therefore needs to be analyzed.

With all this, both the role of imported production factors and the decrease in their availability are not the same for different sectors of the economy. Of course, first of all, the current situation affects industries for which the cost of imported production factors constitutes a significant share of all production costs. However, in addition to the share of imports, it is necessary to take into account its causes. There are several reasons for importing factors of production. Among the main ones, one can list the profitability of importing compared to domestic production or the fundamental difficulty of domestic production caused by the unavailability of raw materials or technologies. In addition, for some types of goods for which it is possible to establish domestic production, this process can take a long time.

The purpose of this work is to analyze the reasons for imports and the possibility of subsequent import substitution for various industries at the level of a mathematical model.

Traditionally, to date, for the analysis of complex changes occurring with various economic sectors, the classical linear model of Leontief's input-output balance is used ([1]). However, for our purposes, it is not applicable, since the linear model does not allow the substitution of production factors or changes in production technologies. Several attempts have been made to generalize the Leontiev model and construct a model of a nonlinear input-output balance. Thus, a series of papers was published in which the network model of the US economy was built and studied: [2–7]. In these works, Cobb-Douglas functions were used instead of the Leontief functions as production functions of industries, which corresponds to the assumption that the norms of financial costs per unit of output are constant. The substitution of production factors in such a model is already allowed, but the elasticity of substitution is fixed. At the same time, in work [8], it was shown on the example of various countries of the world that the Leontief model gives good results for describing the economies of the developing countries of the world, for countries with highly developed economies, the model with Cobb-Douglas production functions shows good results, and for intermediate countries, to which includes Russia, both models give serious errors. This means that intermediate cases need to be considered.

A more general model, which includes both the case of Cobb-Douglas functions and the case of Leontief functions as limiting cases, is the model with production functions with constant elasticity of substitution (CES). Early work with production functions of this kind suggested the possibility of their use, but the model was applied to rather artificial examples. In addition, they did not develop a mathematical apparatus for working with such models. In 2017, in [9], a model with production functions of the CES type was applied to estimate the elasticity of substitution of production factors in the sectors of the Japanese economy. However, the approach of the authors of the work required, in addition to input-output tables, detailed information on price indices for the products of industries for the period under review. For the Russian economy, this approach is inapplicable, since the nomenclature of industries for input-output tables is critically different from the nomenclature in which statistical services calculate product price indices.

A comprehensive description of the intersectoral balance model, using a minimum of statistical information for identification and allowing the missing information to be restored, was made in works [10, 11]. In addition, in the same works, mathematical results were formulated and proved that substantiate the correctness of such models. This approach was continued and developed in works [12–14]. The technology developed in them and substantiated mathematical results will be used below in this work.

1   Description of Substitution of Production Factors

The possibility of substituting imported production factors with domestic ones should be taken into account when modeling the production of industries, that is, it should be described by the type of production functions. The source of data for such modeling is the input-output tables data published by Rosstat. The most up-to-date data available for analysis is the 2017–2020 supply and use tables [15]. We will assume that during the period under consideration, the elasticity of import substitution was constant for each of the industries identified by the statistics. Then the production functions of industries will be functions with constant elasticity of substitution (CES). Since the type of production functions should make it possible to assess the possibility of substituting imports with domestic products, we will single out two arguments in them: imports and domestic products. Thus, the production functions of industries will have the form

$$F_j(D_j, Im_j) = \left(\alpha_j D_j^{-\rho_j} + \beta_j Im_j^{-\rho_j} \right)^{-\frac{1}{\rho_j}}, \ j = 1, \ldots, m, \qquad (1)$$

where j is the industry number, D_j is the volume of consumption of domestic products in base year prices, Im_j is consumption of imports in base year prices, $\alpha_j \geq 0$ and $\beta_j \geq 0$ are coefficients describing the ratio between the volume of consumption of imported and domestic production factors and simultaneously performing the function of calibration, ρ_j is the exponent in the corresponding CES-function, the value of which describes the possibility of substitution of production factors. The value ρ_j is chosen from the set $[-1, 0) \cup (0, +\infty)$. Negative values of ρ_j correspond to the substitution of production factors (the case $\rho_j = -1$ corresponds to complete substitution), positive values correspond to the complementarity of production factors (the limiting case $\rho_j = +\infty$ corresponds to the Leontief production function with complete absence of substitution).

We will try to choose the parameters of the production functions of industries in such a way that the pattern they describe best approximates the statistical data.

2   Dual Description of Input-Output Balance

For the most complete account of the available statistical information, we need a dual description of production, detailed in [10,14]. The Yang dual functions of production functions are the functions

$$q_j(p_D, p_{Im}) = \inf \left\{ \frac{p_D D_j + p_{Im} Im_j}{F_j(D_j, Im_j)} \ \middle| \ D_j \geq 0, Im_j \geq 0, F_j(D_j, Im_j) > 0 \right\},$$

$j = 1, \ldots, m$, which are price indices for the products of the respective industries. Here p_D and p_{Im} are price indices for domestic and imported products, respectively. The Yang dual of production functions of the form (1) will, in turn, have the form

$$q_j(p_D, p_{Im}) = \left(c_{1j}(p_D)^{\frac{\rho_j}{1+\rho_j}} + c_{2j}(p_{Im})^{\frac{\rho_j}{1+\rho_j}} \right)^{\frac{1+\rho_j}{\rho_j}}, \quad j = 1, \ldots, m,$$

where $c_1 = \alpha_j^{\frac{1}{1+\rho_j}}$, $c_2 = \beta_j^{\frac{1}{1+\rho_j}}$.

For price indices, the following relations must hold (see, for example, [13,14]):

$$\frac{\partial q_j}{\partial p_D} \cdot \frac{1}{q_j} = \frac{D_j}{Y_j}, \quad \frac{\partial q_j}{\partial p_{Im}} \cdot \frac{1}{q_j} = \frac{Im_j}{Y_j}, \quad j = 1, \ldots, m, \tag{2}$$

where Y_j is output of j-th industry in base year prices.

In addition, the necessary equality is obvious

$$Y_j = F_j(D_j, Im_j), \quad j = 1, \ldots, m. \tag{3}$$

Note that both in the production function and in its dual function of the price index, all parameters and arguments, except for p_D and p_{Im}, have an index corresponding only to the number of the current industry under consideration. Therefore, in the future, to simplify the notation, we will omit the industry number, since all such functions will be considered for each of the industries separately.

We will select the parameters α, β and ρ in such a way that the statistical data best satisfy the (3) and (2) ratios over the entire period under consideration (2017–2020). Then we can formulate an optimization problem for each industry, from the solution of which we will find the optimal values of the parameters α, β and ρ:

$$\Phi(\alpha, \beta, \rho) \to \min,$$

there

$$\Phi(\alpha, \beta, \rho) = \sum_{t=2017}^{2020} \left[\left(1 - \frac{F(D(t), Im(t))}{Y(t)} \right)^2 + \right.$$

$$+ \left(\frac{D(t)}{Y(t)} - \frac{\partial q(p_D, p_{Im})}{\partial p_D} \cdot \frac{1}{q(p_D(t), p_{Im}(t))} \right)^2 + \tag{4}$$

$$\left. + \left(\frac{Im(t)}{Y(t)} - \frac{\partial q(p_D, p_{Im})}{\partial p_{Im}} \cdot \frac{1}{q(p_D(t), p_{Im}(t))} \right)^2 \right].$$

Here t is year number, $Y(t)$, $D(t)$, $Im(t)$, $p_D(t)$, $p_{Im}(t)$ are output, consumption of domestic and imported products, price indices for domestic and imported products, respectively, in the year t. These values should be taken from the statistical data or reconstructed from them. The values α, β and ρ, by which we optimize the functional, are included in the functions $F(D(t), Im(t))$, $q(p_D(t), p_{Im}(t))$, $\frac{\partial q(p_D, p_{Im})}{\partial p_D}$, $\frac{\partial q(p_D, p_{Im})}{\partial p_{Im}}$.

However, the difficulty of directly finding the minimum of the functional by the parameters of the production function lies in the fact that the arguments of the production function are the consumption of domestic and imported products in physical terms, or, equivalently for this type of problem, in the prices of any one time period, which is considered the base. At the same time, statistical data (supply and use table data) are given at prices of the current period. That is, to bring all values to one point in time, we need to know the price indices for imports and domestic products for various industries. The volume of import consumption in current prices is a statistical value, as the import price index we will consider the price index of the dual-currency basket. The output of industries at current prices is also known. However, the industry nomenclature that provides commodity price index data is critically at odds with the industry nomenclature used to compile the supply and use tables. Therefore, price indices for the products of industries need to be calculated.

3 Finding Price Indices for Industries' Products

To calculate price indices for products of industries, we use the scheme described in [8, 12]. In order to use the described technique, it is necessary to have symmetrical input-output tables for the period under consideration (2017–2020). We will construct them according to the tables of resources and use, taking into account the hypothesis of the constancy of technology in the industry [16]. This hypothesis can be applied, since in modern Russian statistics the share of by-products produced by industries is very small.

Let us describe a model for the optimal distribution of resources based on a non-linear input-output balance. We assume that m of pure industries are connected by mutual deliveries of products as factors of production. Let us denote by X_i^j the output of the i-th industry, which is used as a production factor in the production process in the j-th industry, and by $X^j = (X_1^j, \ldots, X_m^j)$ - costs of the j-th branch of the PF produced by the entire set of branches. We will also assume that in the process of production industries use primary resources (n types) as production factors. Denote by $l^j = (l_1^j, \ldots, l_n^j)$ the vector of primary resource costs by the j-th industry, and by $G_j(X^j, l^j)$ is the production function of the j-th industry, i.e. dependence of the output of the j-th industry on the costs of production factors. We will assume that the production functions of industries have neoclassical properties, i.e., they are concave, monotonically nondecreasing, continuous functions on R_+^{m+n}, vanishing at zero. In addition, we assume that $F_j(X^j, l^j)$ are functions that are positively homogeneous of the first degree. We will say that such functions belong to the class Φ_{m+n}.

Let us denote by $X^0 = (X_1^0, \ldots, X_m^0)$ the volumes of deliveries of products produced by the branches under consideration to external consumers. We assume that the demand of external consumers is described using the utility function $G_0(X^0)$. Suppose the function $G_0(X^0) \in \Phi_m$. Let us also assume that the supply of primary resources to the considered group of industries is limited by the volumes $l = (l_1, \ldots, l_n) \geq 0$, and consider the problem of the optimal distribution of these resources between industries in order to maximize the utility function of external consumers for balance constraints on primary resources and products manufactured by industries:

$$G_0(X^0) \to \max, \tag{5}$$

$$G_j(X^j, l^j) \geq \sum_{i=0}^{m} X_j^i, \ j = 1, \ldots, m, \tag{6}$$

$$\sum_{j=1}^{m} l^j \leq l, \tag{7}$$

$$X^0 \geq 0, \ X^1 \geq 0, \ldots, X^m \geq 0, \ l^1 \geq 0, \ldots, l^m \geq 0. \tag{8}$$

We will assume that the "internal" production functions $G_j(X^j, l^j)$ and the utility function of final consumers $G_0(X^0)$ are Cobb-Douglas functions:

$$G_j(X^j, l^j) = K_j \prod_{i=1}^{m} (X_i^j)^{a_i^j} \prod_{i=1}^{n} (l_i^j)^{b_i^j}, \ j = 1, \ldots, m,$$

$$G_0(X^0) = K_0 \prod_{i=1}^{m} (X_i^0)^{a_i^0}.$$

This type of functions corresponds to the assumption of the constancy of the norms of the financial costs of industries, which is quite consistent with the system of estimates currently adopted in Russia.

We will identify the described model by the restored symmetrical tables of the input-output balance for 2017–2020. Such tables consist of three quadrants and describe financial flows between m pure industries, n primary resources and k final consumers. The value Z_i^j, $i = 1, \ldots, m$, $j = 1, \ldots, m$, located in the first quadrant at the intersection of the ith row and jth column, equals the amount of money that the j-th industry paid for the products of the i-th industry. The value Z_{m+i}^j, $i = 1, \ldots, n$, $j = 1, \ldots, m$, which is in the second quadrant at the intersection of the $m + i$th row of the table and the jth column is equal to the amount of money the jth industry paid for the ith primary resource. Finally, the value Z_i^{m+j}, $i = 1, \ldots, m$, $j = 1, \ldots, k$, which is in the third quadrant at the intersection of the ith row and $m + j$-th column of the table is equal to the amount of money that the j-th end consumer paid for the products of the i-th industry. For the statistical data to be correct, the sum of the elements of the j-th row must obviously be equal to the sum of the elements of the j-th column for any $j = 1, \ldots, m$:

$$\sum_{i=1}^{m+n} Z_i^j = \sum_{i=1}^{m+k} Z_j^i, j = 1, \ldots, m.$$

To identify the model, we will aggregate the third quadrant into a common final consumption column. In addition, since we consider only the balance of domestic products, that is, we exclude the import line from the tables, it is necessary to make appropriate adjustments to the values of final consumption. We will allocate two primary resources. In one of the primary resources, we will allocate wages, in the second we will combine the profit of the industry, the use of fixed capital and some types of taxes. In this sum, for most industries, the main role is played by profit.

It is shown in [12] that when identifying a model for any one year with number t, the parameters of production functions should be determined as follows. Let

$$Z^0(t) = \left(Z_1^0(t), \dots, Z_m^0(t)\right), Z_i^0(t) = \sum_{j=m+1}^{m+k} Z_i^j(t)$$

is final consumption vector,

$$A_0(t) = \sum_{i=1}^{m} \sum_{j=m+1}^{m+k} Z_i^j(t)$$

is total final consumption,

$$A_j(t) = \sum_{i=1}^{m+n} Z_i^j(t) = \sum_{i=1}^{m+k} Z_j^i(t), \ j = 1, \dots, m$$

is the total costs of the j-th industry or the total funds received by it. Then if we define

$$a_i^j(t) = \frac{Z_i^j(t)}{A_j(t)}, \ i = 1, \dots, m, \ j = 0, \dots, m, \ t = 2017, \dots, 2020,$$

$$b_i^j(t) = \frac{Z_{m+i}^j(t)}{A_j(t)}, \ i = 1, \dots, n, \ j = 1, \dots, m, \ t = 2017, \dots, 2020,$$

$$K_j(t) = \frac{A_j(t)}{\prod_{i=1}^{m}(Z_i^j(t))^{a_i^j(t)} \prod_{i=1}^{n}(Z_{m+i}^j(t))^{b_i^j(t)}}, \ j = 0, \dots, m, \ t = 2017, \dots, 2020,$$

solving the (5)–(8) problem will reproduce the input-output table data for the corresponding year, i.e., the inverse problem will be solved in this way. However, since we have access to statistical data for 4 years, for a more accurate description of the real situation and to exclude the influence of random fluctuations, we will use all 4 available input-output tables for identification. In this case, it is impossible to reproduce exactly all 4 tables. For the best approximation of the data of all available tables at the same time (in the sense of a square deviation), the parameters of production functions should be set as follows:

$$a_i^j = \frac{1}{4} \sum_{t=2017}^{2020} a_i^j(t), \ i = 1, \dots, m, \ j = 0, \dots, m,$$

$$b_i^j = \frac{1}{4} \sum_{t=2017}^{2020} b_i^j(t), \ i = 1, \dots, n, \ j = 1, \dots, m,$$

$$K_j = \left(\prod_{t=2017}^{2020} K_j(t) \right)^{1/4}, \ j = 0, \dots, m.$$

Yang's dual functions to production functions and utility functions will be the functions of the cost of production of industries

$$h_j(p, s) = \frac{1}{G_j\left(a^j, b^j\right)} p_1^{a_1^j} \dots p_m^{a_m^j} s_1^{b_1^j} \dots s_n^{b_n^j}, \ j = 1, \dots, m,$$

and consumer price index

$$h_0(p) = \frac{1}{G_0\left(a^0\right)} p_1^{a_1^0} \dots p_m^{a_m^0}.$$

Here, the arguments of the dual functions are the prices for the products of industries $p = (p_1, \dots, p_m)$ and the prices for primary resources $s = (s_1, \dots, s_n)$. Then, if we define $A = \left\| a_i^j \right\|_{j=1,\dots,m}^{i=1,\dots,m}$, $B = \left\| b_k^j \right\|_{k=1,\dots,n}^{j=1,\dots,m}$ and E is the identity matrix $m \times m$, the following proposition is true:

Proposition 1 ([12]). *Let the matrix A be productive. Then the task*

$$h_0(p) \to \max,$$

$$h_j(p, s) \geq p_j \geq 0, \ j = 1, \dots, m,$$

has a solution of the form $p_j = s_1^{c_1^j} \dots s_n^{c_n^j}, \ j = 1, \dots, m,$ where the matrix $C = \left\| c_j^k \right\|_{j=1,\dots,m}^{k=1,\dots,n} = (E - A^*)^{-1} B^*.$

Thus, using the wage index and the GDP index as prices for primary resources, we can find the vector of prices for industry products $p(t)$ for each time period. We will use them to bring the output of industries to the prices of the initial period of time. Calculating $X_i^j(t) = \frac{Z_i^j(t)}{p_i(t)}, \ i = 1, \dots, m, \ j = 1, \dots, m,$ $l_i^j(t) = \frac{Z_{m+i}^j(t)}{s_i(t)}, \ i = 1, \dots, n, \ j = 1, \dots, m,$ as $D_j(t)$ we will take the value of the "internal" production function for the corresponding industry:

$$D_j(t) = G_j\left(X^j(t), l^j(t)\right).$$

4   Substitution Elasticity Estimate

Consider industries in which the share of imports among production factors is significant (at least 5% of the total costs of the industry). For industries with a smaller share of imports, its substitution is not so critical, and, in addition, due to the high relative error of calculations, it may be difficult to interpret

Table 1. Characteristics of the substitution of imported and domestic inputs. Calculations and statistics.

Industry name	No	Import share	ρ	$\left(\frac{\beta}{\alpha}\right)^{\frac{1}{1+\rho}}$
Forestry and logging	1	0.048	−0.75	0.024
Fisheries and fish farming	2	0.103	−0.63	0.091
Food, beverage, tobacco	3	0.093	−0.81	0.067
Manufacture of textiles, clothing, leather and leather goods	4	0.210	0.68	0.234
Woodworking and manufacture of articles of wood and cork, except furniture, manufacture of articles of straw and materials for plaiting	5	0.067	−0.85	0.034
Manufacture of paper and paper products	6	0.119	−0.8	0.108
Manufacture of chemicals and chemical products; Manufacture of rubber and plastic products	7	0.147	−0.76	0.141
Manufacture of medicines and materials used for medical purposes	8	0.217	−0.65	0.260
Manufacture of other non-metallic mineral products	9	0.077	−0.82	0.047
Metallurgical production	10	0.075	−0.9	0.045
Manufacture of fabricated metal products, except machinery and equipment	11	0.106	−0.78	0.087
Manufacture of computers, electronic and optical products	12	0.202	0.22	0.218
Manufacture of electrical equipment	13	0.196	−0.68	0.219
Manufacture of machinery and equipment not included in other groups	14	0.156	0.24	0.152
Manufacture of motor vehicles, trailers and semi-trailers	15	0.311	−0.53	0.433
Manufacture of other vehicles and equipment	16	0.129	−0.8	0.121
Manufacture of furniture, other finished products	17	0.154	0.33	0.159
Repair and installation of machinery and equipment	18	0.123	−0.13	0.107
Waste water collection and treatment; collection, processing and disposal of waste; processing of secondary raw materials; provision of pollution control and other services related to waste disposal	19	0.069	−0.82	0.050
Construction	20	0.073	−0.69	0.052
Wholesale and retail trade and repair of motor vehicles and motorcycles	21	0.099	−0.71	0.078
Water transport activities	22	0.112	−0.88	0.127
Air and space transport activities	23	0.209	−0.73	0.478
Hotel and catering activities	24	0.059	−0.83	0.028
Publishing activities; Printing and copying of information media	25	0.110	−0.77	0.089
Production of motion pictures, video films and television programs, publication of sound recordings and music; activities in the field of television and radio broadcasting	26	0.092	−0.77	0.072
Telecommunication activities	27	0.080	−0.9	0.057
Development of computer software, consulting services in this area and other related services; information technology activities	28	0.071	−0.86	0.046
Activities in the field of architecture and engineering design; technical testing, research and analysis	29	0.069	−0.89	0.048
Research and development	30	0.096	−0.84	0.076
Advertising and market research	31	0.061	−0.8	0.046
Activities of travel agencies and other organizations providing tourism services	32	0.067	−0.05	0.193
Health care activities	33	0.082	−0.86	0.061
Repair of computers, personal and household items	34	0.094	−0.57	0.093

the results. For these industries, we will solve the problem of minimizing the functional (4), the results are given in Table 1.

Here the value $\left(\frac{\beta}{\alpha}\right)^{\frac{1}{1+\rho}}$ means the ratio of the use of imports to the volume of use of domestic production factors.

For clarity, we depict the results obtained on the graph 1. For industries whose parameter ratios differ noticeably from the majority, we will sign the corresponding points with the numbers of the corresponding industry.

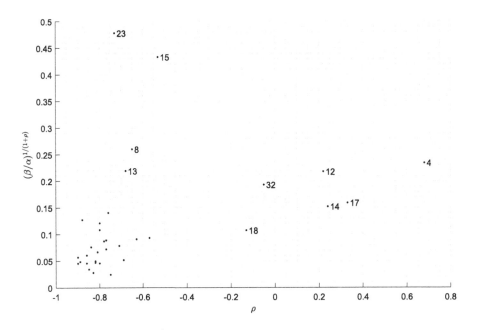

Fig. 1. The ratio of parameters of production functions for various industries.

It can be seen that for most industries both the value of ρ and the ratio of the use of imports to the use of domestic products are quite small. At the model level, we can conclude that the import substitution procedure is quite affordable for them.

In addition to the general group of industries, which correspond to the points concentrated in the lower left part of the graph, two groups of industries that differ from them can be distinguished.

The first of these, which includes sectors 13 (production of electrical equipment), 8 (production of medicines and materials used for medical purposes), 15 (production of motor vehicles, trailers and semi-trailers) and 23 (activities of air and space transport), differs by a rather high share of imports among production factors, however, the values of ρ for them do not exceed -0.5. Since the values of the parameter ρ close enough to -1 correspond to good substitutability of production factors, this group of industries can be interpreted as industries for which the substitution of imports by domestic products is quite possible, but unprofitable.

The second distinguished group of industries differs, first of all, in rather large values of ρ, which indicates poor import substitution. This group includes industries 18 (repair and installation of machinery and equipment), 32 (activities of travel agencies and other organizations providing services in the field of tourism), 12 (manufacture of computers, electronic and optical products), 14 (manufacture of machinery and equipment, not included in other groups), 17 (manufacture of furniture, other finished products), 4 (manufacture of textiles, clothing, leather and leather products). At the same time, we note that the level of imports is quite high in all the listed industries. For further research, it is this group of industries that requires a more detailed clarification of the reasons for the poor substitution of imports: is such substitution fundamentally impossible (as, for example, for the tourism sector), or is there not enough internal high-tech or high-quality production of some types of production factors used for such substitution.

5   Conclusion

The non-linear intersectoral balance model was applied to analyze the possibility of substituting imported and domestic production factors for various sectors of the Russian economy. For industries with a significant share of imports, the elasticity of substitution of imported production factors by domestic ones was calculated. Depending on the ratio of the parameters of production functions, these industries were divided into 3 groups, for which the prospects for import substitution are different. The results of the model calculations as a whole correspond to the ideas of economists about the import dependence of industries, which indicates the adequacy of the applied model. Therefore, the model can be used to analyze the problems of transition to the technological sovereignty of the Russian economy.

Acknowledgements. This work was supported by the Russian Science Foundation grant no. 23-21-00429.

References

1. Leontief, W.: Quantitative input and output relations in the economic systems of the United States. Rev. Econ. Stat. **18**(3), 105–125 (1936)
2. Acemoglu, D., Ufuk, A., William, K.: Networks and the macroeconomy: an empirical exploration. In: National Bureau of Economic Research Macroeconomics Annual (Martin Eichenbaum and Jonathan Parker, eds.), vol. 30, pp. 276–335 (2016)
3. Acemoglu, D, Ozdaglar, A., Tahbaz-Salehi, A.: Cascades in networks and aggregate volatility. NBER Working Papers 16516, National Bureau of Economic Research Inc, (2010)

4. Acemoglu, D., Asuman, O., Tahbaz-Salehi, A.: Microeconomic origins of macroeconomic tail risks. Am. Econ. Rev. **107**(1), 54–108 (2017)
5. Acemoglu, D., Vasco, M.C., Asuman, O., Tahbaz-Salehi, A.: The network origins of aggregate fluctuations. Econometrica **80**(5), 1977–2016 (2012)
6. Acemoglu, D., Asuman, O., Tahbaz-Salehi, A.: Systemic risk and stability in financial networks. Am. Econ. Rev. **105**(2), 564–608 (2015)
7. Acemoglu, D., Azar, P. D.: Endogenous production networks. Working paper (2018)
8. Obrosova, N., Shananin, A., Spiridonov, A.: On the comparison of two approaches to intersectoral balance analysis. J. Phys.: Conf. Ser. **2131**(2) (2021)
9. Kim, J., Nakano, S., Nishimura, K.: Multifactor CES general equilibrium: models and applications. Econ. Model. **63**, 115–127 (2017)
10. Shananin, A.A.: Young duality and aggregation of balances. Dokl. Math. **102**(1), 330–333 (2020). https://doi.org/10.1134/S1064562420040171
11. Shananin, A.A.: Problem of aggregating of an input-output model and duality. Comput. Math. Math. Phys. **61**(1), 153–166 (2021)
12. Rassokha, A.V., Shananin, A.A.: Inverse problems of the analysis of input-output balances. Math. Models Comput. Simul. **13**(6), 943–954 (2021). https://doi.org/10.1134/S2070048221060193
13. Shananin, A., Rassokha, A.: Inverse problems in analysis of input-output model in the class of CES functions. J. Inverse Ill-Posed Probl. **29**(2), 305–316 (2021)
14. Obrosova, N., Shananin, A., Spiridonov, A.: Nonlinear input-output model with nested CES technologies for the analysis of macroeconomic effects of a foreign trade shock. Lobachevskii J. Math. **4**(1), 401–417 (2023)
15. Federal State Statistics Service of Russia. https://rosstat.gov.ru/statistics/accounts. Accessed 12 May 2023
16. Miller, R.E., Blair, P.D.: Input-Output Analysis: Foundations and Extensions. Prentice Hall, Englewood Cliffs, N.J. (1985)

Features of Optimal Drilling in Gas Fields

Alexander K. Skiba$^{(\boxtimes)}$

Federal Research Center "Computer Science and Control",
Russian Academy of Sciences, Vavilov str. 44, block 2, 119333 Moscow, Russia
a.k.skiba@mail.ru

Abstract. In this paper, we consider a continuous aggregated dynamic model for the development of a group of gas fields. The model is based on the assumption of proportionality between the change in the borehole flow rate and the current natural gas production. We believe that the drilling company carries out consistent drilling of fields. The drilling company does not return to work on previously drilled fields. We formulate the problem of maximizing the cumulative accumulated production of natural gas over a fixed period of time under the current constraints on capital investment. It is shown that the direct solution of the original problem is reduced to numerous computational procedures, which strongly depend on the number of fields in the group. To solve the original problem, we make a number of simplifying assumptions that allow us to significantly expand the set of admissible trajectories and reduce the original problem to a fixed-time optimal control problem with a free right end. It is shown that the optimal solution exists. To search for it, the Pontryagin's maximum principle is used. A theorem and two corollaries are formulated and proved. The study shows that some optimal solutions to the extended problem are among the feasible trajectories of the original problem. Two numerical algorithms for finding the optimal solution are proposed. Generalizing conclusions are drawn.

Keywords: Aggregated dynamic model of gas field development · Maximization of cumulative production · Sequence of putting fields into development · Applied problem of optimal control · Optimal control · Pontryagin's maximum principle

1 Introduction

The presence of minerals in Russia is an important factor in its economic well-being. It is one of the largest raw material exporters, providing resources not only for its territories but for the whole world. In many industries, our country occupies a leading position, having a huge share of the extraction of resources among other countries. The role of Russian natural resources is very important for the whole world. This is confirmed by the fact that, against the backdrop of a difficult geopolitical situation that has been established in recent years, new packages of sanctions imposed by the EU, the US, and other foreign countries

© The Author(s), under exclusive license to Springer Nature Switzerland AG 2023
N. Olenev et al. (Eds.): OPTIMA 2023, LNCS 14395, pp. 287–300, 2023.
https://doi.org/10.1007/978-3-031-47859-8_21

have led to serious inflationary processes in the global commodity market. This sanctions pressure has done much more harm to the initiators of such a policy than to Russia itself. This indicates the ineffectiveness of the sanctions and their revision.

One of the important minerals is natural gas. Natural gas is an environmentally friendly mineral that is a mixture of gaseous hydrocarbons of natural origin and consists mainly of methane with the addition of other alkanes [1]. Sometimes it may contain a certain amount of carbon dioxide, nitrogen, hydrogen sulfide, and helium. As a result, natural gas is an extremely valuable raw material from which individual components or simpler impurities can be isolated.

The main consumers of natural gas in the Russian market are the electric power industry (39%), the residential sector (population) (11.5%), and the domestic sector (8.5%). This is almost 60% of the consumed gas in total. The next most important consumer is industry (more than 30% of consumption). The "other" section (approximately 10.5%) includes gas losses during transportation through gas pipelines in the case of incidents on gas pipelines and equipment. The source of losses is the technological equipment used for its transportation, which, due to wear or the device's operation, has gas leaks.

Heating is the most important application for gas in the residential sector. In second place in importance is the heating of water. In private homes, heating systems based on gas-fired air heaters are usually used. Gas is also used in boiler plants. Such installations produce steam or hot water for circulation in radiators or heating pipes. In warm areas, room gas heaters are also used. The improvement of household gas stoves for cooking is carried out through the use of closed burners and multifunctional ovens.

The consumption of energy resources by the country's agriculture is increasing every year. The main consumers of gas in agricultural production are animal husbandry and crop production, auxiliary and communal services, and warehouses of agricultural products. Liquid gas is indispensable for heating cowsheds, stables, and grain dryers. Gas helps in the storage and transportation of agro-industrial products. Large consumers of natural gas in agricultural production are greenhouses that need energy for heating, heating irrigation water, and producing steam.

For the transportation of gas and gas condensate, rail, water, road, and pipeline modes of transport are used. Associated (petroleum) gas separated from oil enters the gas processing plant through pipelines, where propane and butane are separated from it. Then it is sent to the domestic or industrial gas supply systems of cities and towns. Since the 1950 s, the method of sea transportation of liquefied natural gas (methane) in special tankers (methane carriers) has become widespread. Methane makes up the bulk of natural gas. If methane at atmospheric pressure is cooled to $-162°$ C, it becomes liquid. Pipeline transport is the main type of intra-continental natural gas transport. Gas in its gaseous state is transported through pipelines (gas pipelines) after compression (compression) by compressors.

Since natural gas (very often also called blue fuel) lies at a sufficiently deep depth underground (1–6 km), its extraction requires the implementation of a number of engineering and technical measures. In the bowels of the earth, the gas is located in voids connected by cracks. Moreover, it is located there under very high pressure, significantly exceeding the pressure of the atmosphere on the earth's surface.

Boreholes are used to extract natural gas. As a result of drilling a borehole (to equalize the pressure and increase the flow, several evenly spaced boreholes are usually drilled on the territory of the field), then laying it out by casing pipes filled with cement from the outside, a natural draft arises. Thus, the most valuable natural fuel goes outside, where it is cleaned and further supplied to consumers.

For many years, the Department of Mathematical Methods of the Regional Programming of the Federal Research Center "Computer Science and Control" of RAS has been working on the construction and improvement of models for the development of gas fields; various optimization problems have been solved [2–4]. Interesting problems using pipeline throughput constraints include maximizing accumulated production and maximizing shelf lengths [5]. The Pontryagin maximum principle in the classical formulation [6, 7] was used as the mathematical apparatus for the problems to be solved. Existence theorems for optimization problems were also applied [8].

However, until now, when improving models, restrictions on drilling and the construction of new boreholes have not been taken into account. Although the main costs fell on these works, this article corrects the existing shortcoming.

2 Construction of a Model for a Group of Gas Fields, Posing and Solving Optimisation Problems

Before proceeding with the construction of the model, we introduce the following notation, which takes real values:

We introduce the following notation:

- by T, we denote the planning horizon;
- by t, we denote the current time ($0 \leq t \leq T$);
- by $Q_i(t)$, we denote the current gas production at the i-th field;
- by $q_i(t)$, we denote the flow rate of production borehole at the i-th field;
- by q_i^0, we denote the initial flow rate of production borehole at the i-th field;
- by $n_i(t)$, we denote the number of boreholes commissioned per unit of time at the i-th field;
- by $\bar{n}$, we denote the maximum opportunity to commission new boreholes at the i-th field;
- by $N_i(t)$, we denote the operating stock of production boreholes at the i-th field;
- by N_i^0, we denote the initial stock of production boreholes at the i-th field;
- by $V_i(t)$, we denote the recoverable gas reserve remaining in the deposit at the i-th field;

- by V_i^0, we denote the initial recoverable gas reserve at the i-th field;
- by c_i, we denote the cost of construction of one borehole at the i-th field;
- by h_i, we denote the average depth of the i-th deposit;
- by γ, we denote the specific capital investment per unit of borehole length;
- by $v_i(t)$, we denote the mechanical speed of drilling boreholes by one enterprise at the i-th field;
- by $\bar{v}$, we denote the maximum mechanical speed of drilling boreholes by one enterprise;
- by K, we denote the capital investment allocated for the construction of new boreholes per unit of time.

The entered notation uses the index i, which takes integer values in the range from 1 to m. We make the following approximate assumptions:

- The group consists of m deposits.
- Each field corresponds to one deposit.
- At any moment, the gas field is covered by a uniform grid of producing boreholes.
- The control of the dynamic process of field development is carried out by choosing the drilling speed of the borehole $v_i(t)$, which satisfies the constraint described below (6).
- Drilling a borehole, its development, and putting it into the development of the field occur at the same time.
- The entire recoverable gas reserve can be produced using any number of boreholes.
- The flow rates of all boreholes in the field are the same.
- The reserve of boreholes is not created;
- The drilling of deposits by the enterprise is carried out consistently, and in the future, it will not return to work at these deposits.

The cost of construction of one borehole and the increase in their number at the time of t at the i-th field are determined by the formulas:

$$c_i = h_i \gamma; \tag{1}$$

$$n_i(t) = \frac{v_i(t)}{h_i}. \tag{2}$$

We describe a development model for a group of gas fields with mutually influencing boreholes [5–7]. Dependencies are established between the variables, presented as a system of ordinary differential equations:

$$\dot{N}_i = n_i(t) = \frac{v_i(t)}{h_i}, \tag{3}$$

$$\dot{q}_i = -\frac{q_i^0}{V_i^0} q_i(t) N_i(t) = -\alpha_i^0 q_i(t) N_i(t), \tag{4}$$

$$\dot{V}_i(t) = -Q_i(t) = -q_i(t)N_i(t) \tag{5}$$

when restricted

$$0 \leq n_i(t) \leq \bar{n}_i \text{ or}$$

$$0 \leq v_i(t) \leq \bar{v} \tag{6}$$

with initial conditions

$$V_i^0 > 0, \tag{7}$$

$$q_i^0 > 0, \tag{8}$$

$$N_i^0 = 0. \tag{9}$$

In the description of the differential Eq. (4), we have introduced another additional notation:

$$\alpha_i^0 = \frac{q_i^0}{V_i^0}. \tag{10}$$

The rationale for the dynamics of the flow rate of production borehole (4) is given in [4]. Capital investments in the development of fields are mastered by one drilling company, and they cannot exceed the maximum capabilities of drilling rigs. Therefore

$$K = \sum_{i=1}^{m} c_i n_i(t) = \gamma \sum_{i=1}^{m} v_i(t) = \gamma \bar{v}.$$

If, in reality, the investment is less than $\gamma \bar{v}$, then we reduce the maximum drilling rig speed $\bar{v}$ by the corresponding amount. Therefore, we consider the fulfilment of the last relationship to be fair. From here, we get

$$\sum_{i=1}^{m} v_i(t) = \bar{v}. \tag{11}$$

From (3) and (4), taking into account (2), (8), and (9), we arrive at the following formulas:

$$N_i(t) = \int_0^t n_i(\theta)d\theta = \int_0^t \frac{v_i(\theta)}{h_i}d\theta, \tag{12}$$

$$q_i(t) = q_i^0 \exp[-\alpha_i^0 \int_0^t (t-\theta)n_i(\theta)d\theta] = q_i^0 \exp[-\alpha_i^0 \int_0^t (t-\theta)\frac{v_i(\theta)}{h_i}d\theta]. \tag{13}$$

It follows from (12) and (13) that $N_i(t) \geq 0$ and $q_i(t) > 0$.

Obviously, the maximum accumulated gas production for a fixed period $[0, \tau]$ of drilling the i-th field is achieved when $v_i(t) = \bar{v}$. In this case, the equalities (12) and (13), taking into account (8) and (9), are represented as follows:

$$N_i(t) = \frac{\bar{v}}{h_i}t; \tag{14}$$

$$q_i(t) = q_i^0 \exp(-\alpha_i^0 \frac{\bar{v}}{2h_i}t^2). \tag{15}$$

The current and accumulated gas production are described by the formulas:

$$Q_i(t) = q_i^0 \frac{\bar{v}}{h_i} t \exp[-\alpha_i^0 \frac{\bar{v}}{2h_i} t^2]; \tag{16}$$

$$\int_0^t Q_i(\theta)d\theta = \int_0^t q_i(\theta)N_i(\theta)d\theta = \frac{q_i^0 - q_i(t)}{\alpha_i^0} = \frac{q_i^0}{\alpha_i^0}[1 - \exp(-\alpha_i^0 \frac{\bar{v}}{2h_i} t^2)]. \tag{17}$$

Problem 1. *We fix the sequence of drilling m fields. By $\tau_0 = 0$, we denote the drilling start time in the first field. The end time is τ_1. The time duration of the borehole drilling of the first deposit in order is equal to $\tau_1 - \tau_0$. If $\tau_1 - \tau_0 = 0$, then this field is not drilled and, accordingly, is not developed. A similar relationship is established for other deposits. The start time of drilling the last m-th field is τ_{m-1}, and the time of completion is $\tau_m = T$. In this case, the following relations:*

$$\sum_{i=1}^m (\tau_i - \tau_{i-1}) = T \text{ and } \tau_i \geq \tau_{i-1}, \quad i = 1, 2, \ldots, m. \tag{18}$$

are fulfilled. Production at each developed field begins the moment the boreholes are drilled. The borehole drilling rate at the i_1-th field during the entire time period $\tau_{i_1} - \tau_{i_1-1}$ is constant and equals bar$v$. Upon completion of borehole drilling, production at the field continues in depletion mode until the end of the planned period of T. For the remaining $i \neq i_1$ fields, the borehole drilling rate $v_i(t)$ in the interval from τ_{i_1-1} to τ_{i_1} is equal to zero. It is required to find the maximum cumulative production for a group of gas fields on the set of all sequences of drilling fields:

$$\int_0^T \sum_{i=1}^m q_i(t)N_i(t)dt \to \max. \tag{19}$$

For a fixed sequence of drilling fields, we apply the Weierstrass theorem, according to which a continuous function reaches its maximum value on a closed, bounded set. The set of all enumerations of drilling sequences is finite. Therefore, a solution to Problem 1 exists.

Let's count the number of possible situations that may arise when solving this problem. Suppose that when solving the problem for the maximum accumulated production, all m deposits are drilled sequentially. Such orders (permutations) are equal to $m!$. Suppose that only $m - 1$ of m fields are being drilled, then the number of such sequences is equal to

$$(m - 1)!m = m! = \Gamma(m + 1).$$

Here the gamma function having the property $\Gamma(m + 1) = m\Gamma(m)$ is given by the improper integral

$$\Gamma(m) = \Gamma(m, s = 0) = \int_{s=0}^{\infty} t^{m-1}e^{-t}dt.$$

Assume that k fields are being drilled ($k = 1, 2, ..., m$). In this case, we calculate the number of sequences using the formula:

$$C_m^k k! = \frac{m!}{(m-k)!} = \frac{\Gamma(m+1)}{\Gamma(m+1-k)}.$$

The total number of all orders for deposit commissioning is given by the formula:

$$P(m) = \sum_{k=1}^{m} \frac{m!}{(m-k)!} = e\Gamma(m+1,1) - 1 = e \int_1^\infty t^m e^{-t} dt - 1. \qquad (20)$$

Let us calculate the number of possible sequences of drilling fields depending on the specific number of deposits. To do this, we use the formula (20):

$$P(1) = 1; \quad P(5) = 325; \quad P(10) = 9864100; \quad P(15) = 3.554627472075 * 10^{12}.$$

We observe a sharp increase in the number of drilling sequences. Consequently, a direct approach to solving Problem 1 leads to huge calculations. Indeed, the optimization problem is solved for each sequence of field input data. Next, we compare the results of solving optimization problems on the set of all input sequences and choose the maximum value. For a relatively small set of fields equal to 15 units, we will have to sort through more than three and a half trillion options and not only sort through them but also solve the problem of maximizing the total cumulative production for each option.

The simplest algorithms for finding the optimal solution are as follows: To find the global maximum, you first need to scan the area under consideration with some steps, calculate all local values, and then choose the largest one from them. Another option would be a simple scan with the calculation of the values of the function, which allows us to select a subdomain of the largest values of the function from it and search for the global extremum, already in its vicinity.

The numerical solution of this problem by the method of enumeration of options is practically not possible. Therefore, it is necessary to carry out a complete analytical study of Problem 1 and find a realistic algorithm for the numerical solution of the problem. It is proposed to reduce Problem 1 to the optimal control problem. To do this, it is necessary to weaken some of the above assumptions for the formulation of Problem 1. A drilling enterprise can simultaneously work on several fields. A temporary break in the drilling of deposits is also possible. During this break, the company works in other fields.

Let us formulate an optimal control problem.

Problem 2. *For a system of differential equations (3) and (4) with initial conditions (7)-(9), it is required to find the m dimensional vector function $\tilde{\mathbf{v}}(t)$, satisfying the constraints (11) and*

$$v_i(t) \geq 0, \qquad i = 1, 2, \ldots, m, \qquad (21)$$

and the corresponding trajectory $(\tilde{\mathbf{q}}(t), \ \tilde{\mathbf{N}}(t))$, delivering the maximum value to the functional (19). The right end of the optimal trajectory is considered free.

The optimal control in Problem 2 exists. This follows from the theorem given in [8, § 4.2].

For further research, the following notation will be useful to us:

$$a_i = \ln(\frac{q_i^0}{h_i}); \ \kappa = \int_0^T (T - t)\bar{v}dt = \frac{1}{2}\bar{v}T^2; \ \mu_i = \frac{\alpha_i^0}{h_i} \int_0^T (T - t)\tilde{v}_i(t)dt. \quad (22)$$

We will assume that the deposits are numbered in descending order of a_i values. We assume that all a_i are different.

Let us take into consideration the set of all deposits $\Omega \equiv \{1, 2, \ldots, m\}$.

Theorem 1. *The solution to Problem 2 is not unique in the general case. The set Ω is divided into two non-intersecting subsets, one of which may be empty. Such a partition exists, and it is determined by the only natural number $\tilde{l}$ ($1 \leq \tilde{l} \leq m$):*

a) for i, $1 \leq i \leq \tilde{l}$, the values of $\tilde{v}_i(t)$ are not defined; we know only their integral relations

$$\mu_i = \left(\sum_{k=1}^{\tilde{l}} \frac{h_k}{\alpha_k^0}\right)^{-1} \left[\kappa + \sum_{k=1}^{\tilde{l}} \frac{h_k}{\alpha_k^0}(a_i - a_k)\right] \geq 0; \quad (23)$$

b) for j, $\tilde{l} < j \leq m$, $\tilde{v}_j(t) \equiv 0$ and

$$\kappa + \sum_{k=1}^{\tilde{l}} \frac{h_k}{\alpha_k^0}(a_j - a_k) < 0. \quad (24)$$

Proof. Let us explain the formulation of Theorem 1 in more detail. The group of gas fields is divided into two non-overlapping subsets. The first subset contains deposits under development. It is not empty, since non-zero capital investments are allocated to the entire group at each moment. The second subset contains deposits excluded from development. It may be empty. For example, a group consisting of one field contains only one non-empty subset. In any case, despite the possible negative technological and economic characteristics, a single deposit from the first subset will be developed.

1) To solve Problem 2, we use the Pontryagin maximum principle [6,7]. We write out the Hamiltonian, the adjoint system, and the transversality conditions:

$$H = \sum_{k=1}^m (-\psi_i \alpha_i^0 N_i q_i + \chi_i \frac{v_i}{h_i} + N_i q_i);$$

$$\dot{\psi}_i = (\psi_i \alpha_i^0 - 1)N_i; \quad (25)$$

$$\dot{\chi}_i = (\psi_i \alpha_i^0 - 1)q_i; \quad (26)$$

$$\psi_i(T) = \chi_i(T) = 0. \quad (27)$$

After simple transformations (25), taking into account (4), and (27), we get

$$(\psi_i \alpha_i^0 - 1)q_i = const = -q_i(T).$$

From this and from (26), (27) follows:

$$\chi_i(t) = q_i(T)(T - t). \tag{28}$$

Taking into account (28), we obtain the maximum condition for the Hamiltonian with respect to the control at $\mathbf{v} = \tilde{\mathbf{v}}$ in the form:

$$\sum_{k=1}^{m} \frac{\tilde{q}_i(T)}{h_i} \tilde{v}_i(t) = \max_{\mathbf{v}} \sum_{k=1}^{m} \frac{\tilde{q}_i(T)}{h_i} v_i(t). \tag{29}$$

Thus, the necessary optimality condition in the problem under consideration takes the form (29) under the restriction (11), (21).

2) Let us study the solution (29) to Problem 2. Suppose that for some i_1, the strict inequality

$$\frac{q_{i_1}(T)}{h_{i_1}} > \max_{i \neq i_1} \frac{q_i(T)}{h_i}. \tag{30}$$

holds.

In this case, it follows from (29) that the control $\tilde{\mathbf{v}}$ maximizing the Hamiltonian has only one nonzero component $\tilde{v}_{i_1}(t) = \bar{v}$, the remaining components $\tilde{v}_i(t) \equiv 0$. Economically, this means that drilling takes place only at the i_1-th field. It is easy to show that for a sufficiently large planning period $[0, T]$, such management is not optimal.

Indeed, let T be such that for some $i_2 \neq i_1$, the inequality

$$a_{i_1} - a_{i_2} = \ln(\frac{q_{i_1}^0 h_{i_2}}{q_{i_2}^0 h_{i_1}}) < \frac{\alpha_{i_1}^0 \bar{v} T^2}{h_{i_1} 2} \tag{31}$$

holds.

In this case, from (4), (12), (31), and

$$\bar{v}\frac{T^2}{2} = \int_0^T (T - t)\bar{v}dt = \kappa,$$

it follows that

$$\frac{q_{i_1}(T)}{h_{i_1}} = \frac{q_{i_1}^0}{h_{i_1}} \exp[-\alpha_{i_1}^0 \int_0^T (T - t)\frac{\bar{v}}{h_{i_1}}dt] = \frac{q_{i_1}^0}{h_{i_1}} \exp[-\alpha_{i_1}^0 \frac{\bar{v}T^2}{2h_{i_1}}] < \frac{q_{i_2}^0}{h_{i_2}} = \frac{q_{i_2}(T)}{h_{i_2}},$$

which contradicts assumption (30).

Consider the set $\Omega \equiv \{1, 2, \ldots, m\}$ and its subsets. Suppose that for some subset of ω, the condition

$$\frac{\tilde{q}_i(T)}{h_i} = const_{i \in \omega}(i) \equiv \exp(\lambda^\omega) > \max_{j \in \Omega/\omega} \frac{\tilde{q}_j(T)}{h_j}. \tag{32}$$

is fair.

From (11), (21), (29), and (32), it follows that $\widetilde{v}_j(t) \equiv 0$ for $j \in \Omega/\omega$, i.e. $\mu_j = 0$. At the same time, $\widetilde{v}_i(t)$ for $i \in \omega$ is not uniquely defined. It is only clear that $\mu_i \geq 0$. Taking this into account, we can integrate only the equality (11) and derive the condition

$$\sum_{i \in \omega} \frac{h_i}{\alpha_i^0} \mu_i = \kappa. \tag{33}$$

From (13), we get

$$\widetilde{q}_i(T) = q_i^0 \exp(-\mu_i) = h_i \exp(a_i - \mu_i), \ i \in \omega. \tag{34}$$

where

$$\mu_i = a_i - \ln(\frac{\widetilde{q}_i(T)}{h_i}) = a_i - \lambda^\omega. \tag{35}$$

Finally, from (33) and (35), it follows:

$$\lambda^\omega = (\sum_{i \in \omega} \frac{h_i}{\alpha_i^0})^{-1}(\sum_{i \in \omega} \frac{h_i}{\alpha_i^0} a_i - \kappa). \tag{36}$$

Substituting (36) into (35), we obtain a formula for determining

$$\mu_i = (\sum_{k \in \omega} \frac{h_k}{\alpha_k^0})^{-1}[\sum_{k \in \omega} \frac{h_k}{\alpha_k^0}(a_i - a_k) + \kappa]. \tag{37}$$

Note that the conditions (37) can be satisfied by different controls $\widetilde{v}_i(t)$, $i \in \omega$.

3) Let us find out what properties $\omega = \widetilde{\omega}$ has that provide the optimal solution. Let us take into consideration the quantities

$$\nu_i^\omega \equiv a_i - \lambda^\omega, \tag{38}$$

which we will formally calculate for all $i \in \Omega$. Note that $\nu_i^\omega = \mu_i$ for $i \in \omega$. From the definition(38), it follows that

$$\nu_s^\omega - \nu_\eta^\omega = a_s - a_\eta,$$

and this difference does not depend on ω, i.e., the numbers ν_i^ω form a decreasing sequence for any ω.

Let the non-strict inequality $\nu_{j_1}^{\widetilde{\omega}} \geq 0$ hold for some $j_1 \in \Omega/\widetilde{\omega}$.

From (34), (35), and (38) follows the inequality $\frac{\widetilde{q}_{j_1}(T)}{h_{j_1}} = \frac{q_{j_1}^0}{h_{j_1}} = \exp(a_{j_1}) \geq \exp(\lambda^{\widetilde{\omega}})$, which contradicts the assumption (32). Therefore,

$$\nu_j^{\widetilde{\omega}} < 0, \ j \in \Omega/\widetilde{\omega}. \tag{39}$$

Let $\nu_{i_1}^{\widetilde{\omega}} < 0$ be satisfied for some $i_1 \in \widetilde{\omega}$. This assumption contradicts the condition (21), and therefore

$$\nu_i^{\widetilde{\omega}} \geq 0, \ i \in \widetilde{\omega}. \tag{40}$$

The properties (39) and (40) mean that $\tilde{\omega}$ consists of some number $\tilde{l}$ (still unknown) of the first numbers from Ω and allow us to write the relations (37), (39), and (40) in the form (23), (24). Further in this section, we will consider only subsets of ω of this kind.

4) Let us analyze the changes in the values of ν_i^ω, which occur when passing from $\omega' \equiv \{1, 2, \ldots, (l-1)\}$ to $\omega'' \equiv \{1, 2, \ldots, l)\}$. From (36) and (38), we get

$$\nu_i^{\omega''} = (\sum_{k=1}^{l} \frac{h_k}{\alpha_k^0})^{-1}[(\sum_{k=1}^{l-1} \frac{h_k}{\alpha_k^0})\nu_i^{\omega'} + (a_i - a_l)\frac{h_l}{\alpha_l^0}]. \qquad (41)$$

Whence it follows that for $i \in \omega''$, the inequality $\nu_i^{\omega'} \geq 0$ entails $\nu_i^{\omega''} \geq 0$, and from the inequality $\nu_i^{\omega''} < 0$ flows $\nu_i^{\omega'} < 0$. Putting $i = l$ in the formula (41), we get that $\nu_l^{\omega'}$ and $\nu_l^{\omega''}$ are negative or non-negative for any value of l.

5) Denote by $d(\omega)$ the number of non-negative numbers among the quantities ν_i^ω. It follows from (41) that $d(\omega)$ is a non-decreasing function of $|\omega|$ (that is, the number of elements in the subset ω). The above properties (39) and (40) can be written as

$$\tilde{l} = |\tilde{\omega}| = d(\omega) \equiv \tilde{d}.$$

Let us prove that $\tilde{\omega}$ exists. Indeed, otherwise there would be ω' and ω'' such that

$$d(\omega') > |\omega'| = l = |\omega''| - 1 > d(\omega'') - 1.$$

Mutually exclusive consequences follow from the left and right inequalities: $\nu_l^{\omega'} \geq 0$ and $\nu_l^{\omega''} < 0$. Let $\tilde{\omega}$ be found. It is obvious that $\nu_{\tilde{l}}^{\tilde{\omega}} \geq 0$ and $\nu_{\tilde{l}+1}^{\tilde{\omega}} < 0$. For the subset $\hat{\omega} \equiv \tilde{\omega}/\tilde{l}$, we obtain, as proved in item 4, that $\nu_{\tilde{l}}^{\hat{\omega}} \geq 0$, i.e., $|\hat{\omega}| < d(\hat{\omega})$, and therefore $\hat{\omega}$ cannot be optimal. Similarly, considering subsets with even smaller $|\omega|$, it can be shown that if $|\omega| < \tilde{l}$, then $|\omega| < d(\omega)$. For the subset $\hat{\omega} \equiv \tilde{\omega} \bigcup (\tilde{l} + 1)$, we get $\nu_{\tilde{l}+1}^{\hat{\omega}} < 0$, i.e., $|\hat{\omega}| > d(\hat{\omega})$, and therefore, $\hat{\omega}$ cannot be optimal. Similarly, it is shown that if $|\omega| > \tilde{l}$, then $|\omega| > d(\omega)$. So, we have come to the conclusion that the optimal subset of $\tilde{\omega}$, and hence the number $\tilde{l}$, is uniquely determined.

6) Note, finally, that those controls $\tilde{\mathbf{v}}(t)$, which we mentioned above as "optimal", satisfy only the necessary optimality conditions.

In order to prove that they are indeed optimal, we replace the Lagrange problem with the equivalent Mayer problem. Really, from (4), we get

$$\int_0^T \sum_{i=1}^{m} Q_i(t)dt = \sum_{i=1}^{m} \frac{V_i^0}{q_i^0} \int_0^T [-\dot{q}_i(t)]dt = \sum_{i=1}^{m} V_i^0[1 - \frac{q_i(T)}{q_i^0}]. \qquad (42)$$

As follows from (13) and from the assumption (32): the quantities $q_i(T)$ for $i \in \omega$ have the same values for all possible types of controls satisfying conditions (23). For $j \in \Omega/\omega$, it follows that $\tilde{v}_j(t) \equiv 0$. Hence, $\tilde{q}_j(T)$ are also uniquely determined.

Since there exists an optimal control in the cumulative production maximization problem and all controls satisfy the necessary optimality conditions, deliver the same value to the functional (19), we can conclude that these and only these controls are optimal. The theorem has been proven.   □

7) Consider the question of finding the subset $\tilde{\omega}$. If $d(\Omega) = m$ or $d(\{1\}) = 1$, then the question of finding $\tilde{\omega}$ is solved: $\tilde{\omega}$ is Ω or $\{1\}$, respectively. If neither of the two equalities is satisfied, then the following iterative procedure can be proposed.

Find ω_1 such that $|\omega_1| = d(\Omega)$. If $d(\omega_1) < d(\Omega)$, then we find such ω_2 that $|\omega_2| = d(\omega_1)$, and so on until we come to $\tilde{\omega} = \omega_n$, the encouraging result of $d(\omega_n) = d(\omega_{n-1})$. We can't "skip" $\tilde{\omega}$, because if there were such ω_s and ω_{s+1} that

$$d(\omega_{s+1}) > |\omega_{s+1}| = l = d(\omega_s) < |\omega_s|,$$

then the left and right inequalities would have mutually exclusive consequences: $\nu_{l+1}^{\omega_s} \geq 0$ and $\nu_{l+1}^{\omega_s} < 0$. You can start not with Ω but with $\{1\}$, then move on to ω_1, so that $|\omega_1| = d(\{1\})$, and so on until eventually we move on to the same $\tilde{\omega}$.

The described procedure is, in some way, an analogue of Newton's method. Another procedure similar to the half-division method can be proposed. Let $d(\{1\}) > 1$, $d(\Omega) < m$. Let's put $|\omega_1| = [\frac{m+1}{2}]$. If $d(\omega_1) < [\frac{m+1}{2}]$, then we can conclude that $1 < \tilde{l} < [\frac{m+1}{2}]$. If $d(\omega_1) > [\frac{m+1}{2}]$, then $[\frac{m+1}{2}] < \tilde{l} < m$, $|\omega_2| = [\frac{|\omega_1|+1}{2}]$ etc.

8) It should be noted that the conditions of Theorem 1 are satisfied under different controls. For example, it is possible to develop $\tilde{l}$ of the first deposits at once with drilling rig speeds of $\tilde{v}_i(t) = \frac{\mu_i h_i}{\alpha_i^0 \kappa}$ or vice versa. All drilling rigs initially work on the same field of $(\tilde{v}_{i_1}(t) = \bar{v}, \tilde{v}_j(t) = 0, 1 \leq i_1 \leq \tilde{l}, j \neq i_1)$, then on the other, on the third, etc., and the order is arbitrary. The optimal control $\tilde{\mathbf{v}}(t)$ can obviously have an infinite number of breaks, i.e., not be a piecewise-continuous function (such a field development policy is, of course, completely unrealistic, which indicates the limits of the model's applicability).

9) It is of interest to investigate some limiting cases of the problem under consideration, which is done in the following corollaries.

Corollary 1. *There exists T^{**} such that $\tilde{\omega} = \Omega$ for $T > T^{**}$, i.e., with a sufficiently long planning period, all deposits are developed.*

This assertion follows from the continuity of $\nu_i^{\Omega}(T)$ as a function of T and the fact that

$$\lim_{T \to \infty} \nu_i^{\Omega}(T) = \infty.$$

Corollary 2. *There exists T^* such that $\tilde{\omega} = \{1\}$ for $T < T^*$, i.e., with a sufficiently small planning period, only one first deposit is developed.*

This assertion follows, again, from the continuity of $\nu_i^{\{1\}}(T)$, as a function T, and the fact that

$$\lim_{T \to 0} \nu_i^{\{1\}}(T) = \ln \frac{q_i^0}{h_i} - \ln \frac{q_1^0}{h_1} \equiv \nu_i^{\{1\}}(0) < 0 \ \text{ for } i > 1.$$

We emphasize that in this case, the solution becomes unambiguous:

$$\widetilde{v}_1(t) = \bar{v}, \ \widetilde{v}_i(t) \equiv 0, \ i \neq 1.$$

10) The set of acceptable trajectories for Problem 1 is a subset of the acceptable trajectories for Problem 2. This subset includes some optimal trajectories related to Problem 2. So they are also optimal trajectories for Problem 1.

The procedure for finding optimal solutions is as follows: In accordance with Theorem 1, all optimal trajectories for problem 2 are searched, their main parameters are calculated, and two subgroups of deposits are uniquely determined: a subgroup of drilled deposits and a subgroup of undeveloped deposits. Among the optimal elements found, all acceptable trajectories related to problem 1 are selected. They are optimal for both problems 1 and 2. These include any procedures for drilling deposits. The maximum value of the functional does not depend on the sequence of drilling deposits. It is the same for all deposits.

3   Conclusion

The development of the field requires large capital investments, which are mainly for the construction of boreholes. Therefore, it is advisable for a drilling company, in accordance with the development project, to focus its attention on performing all work on one field. Upon completion of them, transfer legal and other rights to the operating enterprise and proceed to drilling the next field in order.

The question arises about the sequence of the choice of deposits. It turns out that to view an insignificant number of deposits, a large number of iterations will be required. In the work, a formula was derived for calculating the number of iterations. For example, to view all 15 deposits, you will need to sort through more than 3.5 trillion options. Also, for each option, it is necessary to solve the problem of maximizing the cumulative accumulated production at a fixed time interval. Sorting through such a large number of options presents a certain complexity for modern computing machines. The difficulty that has arisen can be overcome by solving an optimization problem, and based on the analytical results obtained, a computational process for finding the maximum solution can be built.

The initial problem is to find the maximum accumulated production in a finite time period for a group of gas fields with a restriction on capital investment. The drilling company carries out sequential drilling of deposits. The company does not return to work on previously drilled fields. It is proposed to solve this problem by searching for a solution to another extended problem. It is assumed that the company can drill any number of deposits at the same time.

It is also possible to return to work on previously drilled fields. Any valid option for drilling deposits in the original problem is included in the acceptable version of the extended problem. The converse is not true.

The extended problem is an optimal control problem. Using the Pontryagin maximum principle, we solve it. A subgroup of developed deposits is clearly identified. Optimal controls are not explicitly defined. Only their integral relationships are determined. Any control satisfying integral relations is optimal, and it reaches the same value of accumulated production. Two procedures for searching for a single subgroup of developed deposits are proposed. It is shown that with small planning periods, only one field is being developed, even with poor technical and economic characteristics. With large planning periods, the entire group of gas fields is drilled.

The results of the extended task study are transferred to the original task. Therefore, any sequence of drilling of deposits satisfying integral ratios leads to the same value of cumulative accumulated production. The resulting ambiguity of the optimization problem solution can be used to select a specific option based on another, not necessarily formalized, criterion.

References

1. Vyakhirev, R., Korotaev, Yu., Kabanov, N.: Theory and Experience of Gas Recovery. Nedra, Moscow (1998)
2. Khachaturov, V., Solomatin, A., Zlotov, A.: Planning and Design of Development of Oil and Gas Producing Regions and Deposits: Mathematical Models, Methods, Application. LENAND, Moscow (2015)
3. Margulov, R., Khachaturov, V., Fedoseev, A.: System Analysis in Long-term Planning of Gas Production. Nedra, Moscow (1992)
4. Skiba, A.: Dynamic model analysis of gas deposit developments. In: Eleventh International Conference on Management of large-scale system development (MLSD 2018), Moscow, Russia, October 1–3, 2018, pp. 619–622, IEEE Xplore Digital Library (2019)
5. Skiba, A.K.: Maximization of the accumulated extraction in a gas fields model. In: Evtushenko, Y., Jaćimović, M., Khachay, M., Kochetov, Y., Malkova, V., Posypkin, M. (eds.) OPTIMA 2018. CCIS, vol. 974, pp. 453–469. Springer, Cham (2019). https://doi.org/10.1007/978-3-030-10934-9_32
6. Pontryagin, L., Boltyanskii, V., Gamkrelidze, R., Mishechenko, E.: The Mathematical Theory of Optimal Processes. Interscience Publishers, Wiley, New York (1962)
7. Moiseev, N.: Elements of the theory of optimal systems. Science, Moscow (1975)
8. Lee, E., Markus, L.: Foundations of Optimal Control Theory. Wiley, New York (1967)

Applications

On Solving the Robust Transfer Line Balancing Problem with Parallel Tasks and Interval Processing Times

Pavel Borisovsky[✉][iD]

Sobolev Institute of Mathematics SB RAS, Novosibirsk, Russia
pborisovsky@ofim.oscsbras.ru

Abstract. The paper considers a problem of designing a transfer line, in which machines can execute blocks of parallel tasks. It is assumed that some tasks may have uncertain processing times and it is required to find a line configuration to maximize a stability radius, which is a widely used metric for evaluating a robustness of a system. The earlier studies on this problem did not impose any limits on the possible increase of the task processing times, which could lead to solutions that correspond to rather improbable or impossible situations from the practical point of view. In this paper, a more realistic assumption that the processing time may vary within a certain interval is made. A way to calculate the stability radius under this assumption is proposed and a metaheuristic optimization approach is developed. In the experimental study, the solutions under the old and the new settings are compared and discussed.

Keywords: Transfer line · Uncertainty · Stability radius · Metaheuristic

1 Introduction

Automatic machining lines take a substantial place in the modern industry. Generally, a line consists of a sequence of machines (workstations) that perform a certain set of mechanical operations (tasks). Each workpiece (machined part) passes through the line by means of a conveyor belt so that all the required tasks are performed. The line balancing optimization problems consist in finding the assignment of the given set of tasks to workstations and scheduling them in order to achieve the best performance of the line. Comprehensive surveys of such problems and their solution methods can be found in [2,3].

In this paper, a special case of lines with a complex parallel-serial structure is of interest. Its particular feature is that the tasks can be united in blocks where they are performed in parallel and each machine executes a set of blocks consequently. The usual tasks precedence and takt time constraints are taken

This research is supported by Russian Science Foundation grant 22–71–10015.

into account. This problem was introduced in [7] and it is know as a Transfer Line Balancing Problem (TLBP).

The robust version of the problem assumes that some tasks have uncertain processing times. A widely-used criterion for such problems is a maximization of the *stability radius* introduced in [13]. For TLBP it was considered in [12], where a Mixed Integer Programming (MIP) model and a greedy heuristic were developed. This study was extended in [5], where special conditions concerning inseparable or conflicting groups of tasks (*inclusion/exclusion* constraint) were taken into account. In both papers it is assumed that the extra processing times can be arbitrary large. As it is discussed in [12], this may lead to solutions that correspond to improbable or impossible situations, in which only one task has a large time excess and for the others the excess is equal to zero. To overcome this, in this paper an upper limit on the extra time is introduced (in other words, the processing times may vary within some given time intervals). The uncertain formulations with interval processing times are quite known for other line balancing problems. For example, a branch-and-bound algorithm for the robust assembly line balancing problems was developed in [10]. More references can be found in the review paper [11]. There seem to be no any such research on the considered problem with blocks of parallel tasks. Note that there is also an alternative approach [9], in which a different concept of *stability factor* is introduced and investigated for the simple assembly line. Application of this concept to the TLBP and its comparison to the interval approach could be of interest but requires a separate study.

In this paper, it is shown that computing of the stability radius can be generalized for the new formulation, although it is expressed in a more complex way, which is not convenient for MIP modeling. Instead, a parallel metaheuristic is developed and implemented for executing on a Graphics Processing Unit (GPU). This approach is evaluated on the test instances from the literature and it demonstrated a competitive performance compared to the previously developed algorithms. The analysis of results showed that the solutions under the limited and unlimited settings may significantly differ, which confirms the importance of the new formulation.

The paper is structured as follows. Section 2 gives a general problem description, reproduces the expression for computing the stability radius without limits on the processing times, and formulates the problem with interval times. In Sect. 3, a way to compute the stability radius under the new assumption is obtained and validated. Section 4 describes the metaheuristic approach and Sect. 5 presents its experimental evaluation and comparison of solutions with limited and unlimited processing times. Section 6 concludes the study.

2   Problem Description

The structure of the considered transfer line can be described as follows. The processing of each machined part requires to perform n *tasks*. The nominal *processing times* are denoted by $t = (t_1, ..., t_n)$. The tasks can be united in *blocks*, where they are executed in parallel, so the block time is defined by the maximal processing time of its tasks. A *machine (workstation)* is a tool that performs a set of blocks in a sequence, and its workload time is equal to the sum of the block times. For each machine, the workload time is limited above by a given constant T, which represents the line *cycle time (takt time)*. It is an important parameter that defines the throughput of the line. Parameters r_{max} and b_{max} are the limits on the number of tasks in one block and the number of blocks on one machine accordingly. The maximal number of machines is denoted by m.

The precedence relations are represented by a partial order on the set of tasks. For each task i, the set of precedent tasks is given. It consists of tasks that must be finished before the start of i, i.e. they must be allocated either to the precedent machines or to the precedent blocks of the same machine.

For simplicity, the widely known inclusion/exclusion constraints (see, e.g., [5]) are not considered in this study, although they can be incorporated if necessary. An illustration of a transfer line is given in Fig. 1, where tasks are shown as horizontal bars and their lengths reflect the processing times.

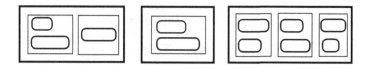

Fig. 1. Transfer line with 11 tasks, six blocks, and three machines.

The problem considered here asks to construct a set of blocks and assign them to machines to design a required structure of the transfer line. Each task must be executed exactly once respecting the aforementioned requirements. In the deterministic formulation, the optimization criterion is usually consists in either minimization of the number of machines given the fixed cycle time or vice versa. The criterion in the uncertain scenario is discussed below.

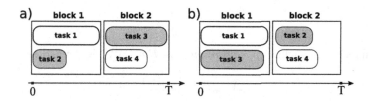

Fig. 2. Example of solutions with poor robustness (a) and good robustness (b).

The robust formulation assumes that some tasks are considered as *uncertain*, which means that their execution times may exceed their nominal values. Let $V = \{1, \ldots, n\}$ be the set of all tasks and $\widetilde{V} \subseteq V$ be the set of uncertain tasks. Informally, the problem asks to find a solution that remains feasible when the execution times are increased. This is illustrated in Fig. 2, which shows two different solutions for a line that has a single machine with two blocks each one executing two tasks ($m = 1, r_{\max} = 2, b_{\max} = 2$). The filled bars correspond to the uncertain tasks. The x-axis represent the time scale. In the solution (a), even a small increase of t_3 will violate the cycle time constraint, so this solution cannot be considered as robust. In the solution (b), there is a certain time reserve and both t_2 and t_3 can be slightly increased without violating the feasibility.

For a formal definition of robustness, a concept of *stability radius* can be used. It is defined as follows: first introduce a ball in a space of possible time deviations of the uncertain tasks:

$$B(\varepsilon) \stackrel{def}{=} \{\boldsymbol{\xi} \in \mathbb{R}^n \mid \xi_j \geq 0 \text{ for } j \in \widetilde{V}, \ \xi_j = 0 \text{ for } j \notin \widetilde{V},$$

$$\text{and } \|\boldsymbol{\xi}\| \leq \varepsilon\}, \quad \varepsilon \geq 0. \tag{1}$$

Then for a given solution S and a vector of processing times $\boldsymbol{t}$ the stability radius is given by

$$\rho(S, \boldsymbol{t}) \stackrel{def}{=} \max\{\varepsilon \geq 0 \mid \forall \boldsymbol{\xi} \in B(\varepsilon) \text{ solution } S$$

$$\text{remains feasible if } \boldsymbol{t} \text{ is replaced by } \boldsymbol{t} + \boldsymbol{\xi}\}. \tag{2}$$

In this study, we consider the case of l_1 norm. Since only non-negative time deviations are of interest, i.e. $\boldsymbol{\xi} \geq 0$ the l_1 norm is defined as $\|\boldsymbol{\xi}\|_1 = \sum_j \xi_j$ (in what follows, for any vector $\boldsymbol{\xi}$, the notation $\boldsymbol{\xi} \geq 0$ means that $\xi_j \geq 0$ for each j). For a given feasible solution of the TLBP, an expression to evaluate the stability radius is obtained in [12] and it is as follows. Let us call a block or a machine uncertain if it contains at least one uncertain task. Denote by

V_k the set of all tasks in block k;
$\widetilde{V}_k$ the set of all uncertain tasks in block k;
$\tau_k = \max_{j \in V_k} t_j$ the nominal block working time of block k.
For any uncertain block k on machine p define its *save time* $\Delta_k^{(p)}$ as

$$\Delta_k^{(p)} := \tau_k - \max_{j \in \widetilde{V}_k} t_j.$$

For any uncertain machine p a *minimal save time* $\Delta_{\min}^{(p)}$ is defined as

$$\Delta_{\min}^{(p)} := \min_{k \in \widetilde{U}_p} \Delta_k^{(p)},$$

where $\widetilde{U}_p$ is the set of uncertain blocks on machine p. Then the stability radius ρ corresponding to l_1-norm for a given feasible solution S can be calculated as follows

$$\rho(S, \boldsymbol{t}) = \min_{p \in \widetilde{W}} \left(T - \sum_{k \in U_p} \tau_k + \Delta_{\min}^{(p)} \right), \tag{3}$$

where $\widetilde{W}$ is the set of uncertain machines in the line and U_p is the set of all blocks on machine p.

The optimization problem asks to find the solution S that is feasible w.r.t. the given nominal times t and has the maximal $\rho(S, t)$. In [12] expression (3) was used to develop a MIP model of the robust TLBP, which was further used in [5] in the MIP-based matheuristic optimization algorithm.

In (3) it is assumed that extra processing times of uncertain tasks are not limited. As it is discussed in [12], this approach has a drawback from the practical point of view because the stability radius is reached on vector $\boldsymbol{\xi}$ (see (2)), in which there is only one large non-zero component: $\boldsymbol{\xi} = (0, .., \rho, ..0)$. This could be very improbable or even impossible in the real-life environment. To overcome this, in this paper we make a more realistic assumption that each component of $\boldsymbol{\xi}$ is limited by a given constant L, i.e. the definition of $B(\varepsilon)$ is modified as follows:

$$B(\varepsilon) \overset{def}{=} \{\boldsymbol{\xi} \in \mathbb{R}^n \mid 0 \le \xi_j \le L \text{ for } j \in \widetilde{V}, \; \xi_j = 0 \text{ for } j \notin \widetilde{V},$$

$$\text{and } \|\boldsymbol{\xi}\| \le \varepsilon\}, \quad \varepsilon \ge 0. \quad (4)$$

In other words, we assume that the actual execution time of some uncertain task j may vary within interval $[t_j, t_j + L]$. In this sense, this is similar to the formulation [10], but for the different type of machining line.

Since formula (3) cannot be used with this assumption, the goal of this paper is to elaborate a way to calculate the stability radius, develop an optimization approach, and compare the solutions under the old and new settings. The new formulation is a generalization of the old one, because when $L = \infty$ (or even $L = T$), they are equivalent, so the problem remains strongly NP-hard. A more general case, in which each task has its own time limit, could be also considered, but in this study it is omitted for the sake of simplicity.

Note that in the optimization problem, the feasibility space is defined by the nominal times t. The extra times are used only for evaluation of the objective function. This means that introducing the limit L does not influence on the feasibility of solutions but implies only a different optimization criterion. As it is shown further, the new formulation may lead to solutions with much larger stability radii.

3   Computing the Stability Radius for the Interval Processing Times

First, suppose that the line consists of a single machine. Let us fix some feasible solution S and let $b_1, ..., b_K$ be the set of blocks of the given machine and $\Delta_1, ..., \Delta_K$ be their save times. Let $\tilde{n}$ be the number of uncertain tasks processed on this machine. Note that there is a trivial upper bound on the stability radius given by $L\tilde{n}$, and if it is reached on some solution, it is clearly optimal.

Computing ρ is based on the ideas of [12]. It is illustrated in Fig. 3. Starting from the block with the minimal save time let us extend its longest uncertain task time until it either reaches L or violates the cycle time constraint (i.e. the

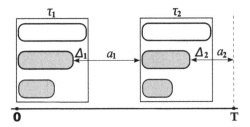

Fig. 3. Computing the stability radius.

total machine worktime exceeds T). In the picture, the extra task times are depicted with "⟷", a_1 and a_2 denote the resulting additional block times. Let $R = T - \sum_{k=1}^{K} \tau_k$ be the time reserve of the machine. From the construction, we may see that $\Delta_1 + a_1 = L$ and $a_1 + a_2 = R$. Recall that we use the l_1 norm so the stability radius is obtained as $\rho = ||\mathbf{\Delta} + \mathbf{a}|| = (\Delta_1 + a_1) + (\Delta_2 + a_2)$. In case the threshold T can not be reached, then the solution always remains feasible and $\rho = L\tilde{n}$.

For the formal description, let us assume w.l.o.g. that the set of blocks is sorted by increasing of their save times: $\Delta_1 \leq ... \leq \Delta_K$.

Algorithm 1. Computing the stability radius for one machine

1: Let $\rho := 0$, $R := T - \sum_{k=1}^{K} \tau_k$.
2: For $k = 1$ to K do:
    2.1. Let $a_k := L - \Delta_k$.
    2.2. If $a_k < 0$ then
           stop and return $L\tilde{n}$.
    2.3. If $R < a_k$ then
           set $a_k := R$, $\rho := \rho + \Delta_k + a_k$, stop and return ρ.
    Else
           $R := R - a_k$, $\rho := \rho + \Delta_k + a_k$.
3: Return $L\tilde{n}$.

Note that in case of unlimited processing times, the algorithm stops after the first iteration and returns the same result as of (3).

Theorem 1. *Algorithm 1 correctly computes the stability radius of one machine.*

Proof. We have to prove two statements: a) for any vector $\boldsymbol{\xi} \geq 0$ such that $||\boldsymbol{\xi}|| \leq \rho$ the solution remains feasible for processing times $\mathbf{t} + \boldsymbol{\xi}$; and b) for any scalar $e > 0$ there exists a vector $\boldsymbol{\xi} \geq 0$ such that $||\boldsymbol{\xi}|| = \rho + e$ and the solution is infeasible for processing times $\mathbf{t} + \boldsymbol{\xi}$ (excluding the special case $\rho = L\tilde{n}$).

a) It is easy to see that in cases when the algorithm returns $L\tilde{n}$ the solution is feasible for any $\boldsymbol{\xi}$ s.t. all $\xi_j \leq L$. Suppose the algorithm returns ρ at step 2.3. Denote by F the last iteration number, at which the algorithm stops. According to the algorithm,

$$\rho = \sum_{k=1}^{F}(\Delta_k + a_k) \text{ and } \sum_{k=1}^{F} a_k = R.$$

Also note that since $a_k = L - \Delta_k$ and Δ_k are sorted by ascending, the sequence Δ_k consists of the smallest possible values and so the sequence a_k consists of the largest possible values (except maybe the last one). For the convenience, let us call this statement a "property (S)".

Choose some $\xi > 0$ such that $\|\xi\| \leq \rho$. We may assume that in each block k there is only one uncertain task $i^{(k)}$ with positive value of corresponding component of ξ which will be denoted by $\xi^{(k)}$. Also, let $t^{(k)}$ be the processing time of task $i^{(k)}$, $\delta^{(k)} := \tau_k - t^{(k)}$, and $a^{(k)} := t^{(k)} + \xi^{(k)} - \tau_k$. Let us consider only the blocks with positive $a^{(k)}$ and denote their number by G. Then

$$\sum_{k=1}^{G} \xi^{(k)} = \sum_{k=1}^{G} \delta^{(k)} + \sum_{k=1}^{G} a^{(k)}.$$

Assume to the contrary that the solution is infeasible, i.e. $\sum_{k=1}^{G} a^{(k)} > R$. Due to the property (S) this means that the number of $a^{(k)}$ is not less than the number of a_k (i.e., $G \geq F$). Then using again the property (S)

$$\sum_{k=1}^{G} \delta^{(k)} \geq \sum_{k=1}^{F} \Delta_k.$$

Combining this together, we obtain

$$\|\xi\| \geq \sum_{k=1}^{G} \delta^{(k)} + \sum_{k=1}^{G} a^{(k)} > \sum_{k=1}^{G} \delta^{(k)} + R \geq \sum_{k=1}^{F} \Delta_k + R =$$

$$\sum_{k=1}^{F} \Delta_k + \sum_{k=1}^{F} a_k = \rho,$$

which is a contradiction.

b) Again it is enough to consider only the case when the algorithm returns at step 2.3. In this case $\|a + \Delta\| = \rho$ and $\|\tau + a\| = T$ so increasing the last value a_k by any small value will violate the feasibility. Q.E.D.

If a line consists of many machines it is easy to see that ρ can be computed as follows.

Algorithm 2. Computing the stability radius for many machines

---
1: For each machine j compute ρ_j by Algorithm 1.
2: If $\rho_j = L\tilde{n}_j$ for all j then
      return $L\tilde{n}$.
   Else
      return $\min_j \rho_j$.

Although the computation can be done in polynomial time and it is rather fast in practice, it is quite complex to be used within the MIP approach. The modeling of sorting the set of block and checking that the total extra time reaches T can be done in principle by means of introducing appropriate binary variables, but this will lead to a cumbersome model that will be hardly useful. Note that a relatively simple MIP model for the unlimited case [12] is rather hard for solving with Gurobi even for small-sized problem instances. To cope with this, a metaheuristic optimization algorithm is proposed.

4   Optimization Approach

The optimization problem considered in this section asks to find a solution of the transfer line balancing problem that satisfies the technical constraints described in Sect. 2 and maximizes the stability radius. The goal consists in development of a generic metaheuristic approach suitable for solving the problem in both variants: with limited or unlimited uncertain processing times.

To this end, a GPU-accelerated hybrid evolutionary algorithm is implemented. Note that GPU computing has become an interesting direction in optimization because it provides an order of magnitude of computing power comparing to the traditional CPUs, if the computations can be highly parallelized. A great number of successful GPU implementations of different metaheuristic algorithms are known, see, e.g., [6,8].

The algorithm developed in this study combines the ideas of a random local search and the "Go with the Winners" (GWW) replacement scheme proposed in [1]. This choice is explained by the simplicity of this scheme and its good searching ability. This algorithm can be easily parallelized and implemented on GPU and shows a fast computing speed. In [4] this approach was proposed and implemented for solving the classic Flow-Shop problem and one industrial scheduling problem. In the experiments it showed a superior performance comparing to the canonical genetic algorithm and several evolutionary algorithms from the literature. In this paper, an adaptation of this approach to the considered problem is presented.

As in many evolutionary algorithms, a solution encoding and a mutation operator are introduced. Any solution is represented by three vectors: π is a permutation of tasks, u is a binary vector such that $u_i = 1$ means that i-th task is the last one in the block, and similarly v is such that $v_i = 1$ if i-th task is the last one on the machine (i.e., the value 1 marks the end of a current block or machine). Consider, for example, the following values:

$\pi = (3, 5, 7, 1, 9, 4, 2, 8, 6)$

$u = (0, 1, 0, 1, 0, 0, 1, 0, 1)$

$v = (0, 0, 0, 1, 0, 0, 0, 0, 1)$

It corresponds to the line configuration with two machines, the first machine contains two blocks of tasks {3,5} and {7,1}, the second machine contains blocks {9,4,2} and {8,6}. Clearly, the last values, u_n and v_n, may not be taken into account, because they always correspond to the last task in a block or a station. Moreover, $v_i = 1$ always implies $u_i = 1$.

Initial solutions are built at random. Permutation π is generated according to the uniform distribution. Vector u is filled initially with zeros, then $n/r_{\max}$ random positions are filled with ones (this corresponds to creation of $n/r_{\max}$ blocks). Vector u is built similarly creating m machines.

Since the solution encoding is rather complex, the mutation operator consists of several minor mutations that modify different parts of the given solution. The widely known *swap* and *insertion* mutations (denoted by M_{swap} and M_{ins}) are applied to π. In M_{swap}, two positions i and j are selected at random and the values π_i and π_j are swapped. In M_{ins}, a randomly chosen element π_i is inserted to a new random position j. In addition, mutations for modification of u and v are introduced. In operator M_{mov}^u, a position i is chosen at random and if $u_i = 1$, it is swapped with either u_{i-1} or u_{i+1} (the choice is made with probability $1/2$). This action corresponds to reduction or extension of some block. If $u_i = 0$ then this operator does not do anything. Similarly, operator M_{mov}^v does the same to vector v. In operator M_{flip}^u, a randomly chosen element u_i is flipped, i.e., $u_i := 1 - u_i$. A similar mutation M_{flip}^v could be also considered, but it is not used in this study, because in the test instances it is assumed that the number of machines is exactly m. In case this is not true, it is reasonable to solve the problem several times with different values of m. The resulting entire mutation combines the described procedures calling them with some probabilities that must be adjusted experimentally.

To cope with the precedence constraints, a correction procedure is used. In this procedure, the most-left element without any predecessors is found in π and is moved to a new (initially empty) permutation π'. This is repeated until all the elements are moved. As a result, π' that is consistent with the partial order is built and it is used further instead of π. Unfortunately, this approach can not prevent the precedent tasks to appear in the same block, so in this case the penalty is applied.

To evaluate the objective function, the encoding vectors are decoded to a line structure and penalties are applied for each violation of the $r_{\max}$, $b_{\max}$, cycle time constraint, and precedent tasks in the same block. For each such occurrence, a large negative value $-P$ is added. Then the stability radius is computed according to formula (3) or Algorithm 2, depending on the problem settings. The resulting value combining the penalty and the stability radius is considered as an objective to be maximized.

The informal idea of the used hybrid algorithm is as follows. Several initial solutions are generated and a parallel random local search is applied to each solution independently. It consists in mutating each solution many times and choosing the best offspring for the next generation. After several iterations, the GWW step is made, in which a subset of the worst solutions are eliminated and replaced by the duplicates of the best solutions. This process is repeated until a certain stopping criterion is met. A formal scheme is given below.

Each iteration of the algorithm performs NH mutations and objective evaluations. They can be done independently in parallel, which suits especially well for a GPU. The tunable parameters N, H, and Q should be adjusted manually

---

Algorithm 3. Hybrid metaheuristic algorithm

1: Randomly generate initial solutions (population) $x_1, ..., x_N$.
2: Repeat within a given iterations or time limit:
    2.1. Apply random local search: repeat K times:
        2.1.1. For each solution x_k produce H offspring by means of mutation
               and choose the best one (denoted by x'_k).
        2.1.2. If x'_k is better than x_k than replace x_k by x'_k.
    2.2. Apply "Go With the Winners" step:
        2.2.1. Sort the set of solutions $x_1, ..., x_N$ by decreasing
               of their objective values.
        2.2.2. Replace the Q worst solutions $x_{N-Q+1}, ..., x_N$ by the copies
               of the Q best solutions $x_1, ..., x_Q$.
3: Return the best found solution.

---

depending on the problem scale and GPU characteristics. Note that this scheme provides a good scalability, i.e. the more computing cores are available in a GPU device, the better speed and accuracy can be expected if the parameters are properly adjusted.

5   Experimental Evaluation

The optimization algorithm was coded in C++ and CUDA and was run on a system with Tesla V100 GPU. During the preliminary testing, the following values of the tunable parameters were set: the population size is $N = 512$, the number of offspring of each solution is $H = 64$, the size of the population part for replacement in the GWW is equal to ten percent of the population size: $Q = 51$. For the termination criterion the iterations number was used. Each run of the random local search was given $K = 10000$ iterations; the main loop at step 2 was limited by ten iterations. Tuning the mutation was the most tricky part. At first, one of the procedures M_{swap} and M_{ins} is chosen with probability 0.4 and none of them is used with probability 0.2. Then both M^u_{mov} and M^v_{mov} are applied, and finally M^u_{flip} is called with probability $1/n$. These setting shows quite an appropriate performance on the considered data.

    The same benchmark set as in [12] is used[1]. Series S1 and S2 contain small sized instances with $n = 25, m = 5$, series S3 contain larger instances with $n = 50, m = 10$. In each instance, a fixed random permutation of tasks is given to define the set of uncertain tasks. Namely, given a parameter $ratio \in [0, 1]$, the first $\lceil ratio * n \rceil$ tasks in this permutation are considered as uncertain.

    The results for $r_{max} = 2$ and $ratio = 0.5$ are given in Table 1. The first ten instances of series S1 and S3 are presented, which is quite enough to demonstrate the observed effects. The problems were solved in two modes. First, the algorithm was run for maximization of objective function computed according to (3),

---

[1] The test instances are available at http://pagesperso.ls2n.fr/~gurevsky-e/data/R-TLBP.zip.

which is further denoted by ρ_U and corresponds to the previous (unlimited) problem formulation. The results are collected in the parts of the tables entitled by "optimize unlimited". On each instance, ten independent runs were made and the maximal value of ρ_U was obtained and presented in the column ρ_U^{best}. In addition, for each solution, the stability radius in the new formulation (denoted by ρ_L) was computed according to Algorithm 2. If there were several solutions with the same maximal ρ_U, the corresponding values of ρ_L were taken on the average and reported in column ρ_L^{avg}. Then the algorithm was run vice versa optimizing ρ_L and reporting average ρ_U of the solutions with the best ρ_L. The upper limit L was set to 20 for S1 and 10 for S3. These results are given in the part entitled by "optimize limited".

Table 1. Comparison of solutions for unlimited and limited cases for $r_{max} = 2$

Series S1 $n = 25, m = 5$	optimize unlimited		optimize limited		Series S3 $n = 50, m = 10$	optimize unlimited		optimize limited	
$L = 20$	ρ_U^{best}	ρ_L^{avg}	ρ_U^{avg}	ρ_L^{best}	$L = 10$	ρ_U^{best}	ρ_L^{avg}	ρ_U^{avg}	ρ_L^{best}
S1.0	30.4	30.4	20.59	260	S3.0	15.5	15.6	14.4	16.2
S1.1	27.3	27.3	16.52	38.6	S3.1	17.6	17.6	17.4	17.6
S1.2	29.0	29.0	21.9	30.5	S3.2	17.9	17.9	9.0	18.0
S1.3	34.5	34.5	20.99	260	S3.3	11.6	11.6	10.2	11.6
S1.4	28.5	28.5	14.26	36.7	S3.4	14.0	14.0	13.4	14.0
S1.5	32.3	32.3	21.54	260	S3.5	14.1	14.1	13.8	14.1
S1.6	28.5	28.5	20.17	35.4	S3.6	20.7	22.9	7.9	250
S1.7	38.5	38.5	20.62	260	S3.7	20.7	22.1	16.7	22.1
S1.8	32.4	32.4	21.7	260	S3.8	24.2	24.8	15.2	250
S1.9	24.4	24.4	23.87	24.4	S3.9	4.4	4.4	4.4	4.4

We can see that in the unlimited case the results are very close to the ones of the MIP approach [12] (they can be found in the supplementary materials to [12], Tables B.1 and B.3). Note that one run of the metaheuristic algorithm takes about one second on the instances of series S1 and about 50 s on the instances of S3, which is rather fast comparing to the MIP solver. The values ρ_L are mostly the same as ρ_U except for several instances of S3. In the other case when ρ_L is optimized, the situation is different. First, we note that ρ_L^{best} is mostly much larger than ρ_U^{best}, which means that when upper limit is taking into account, much better solutions can be obtained. For quite many solutions the value $L\tilde{n}$ was reached (it equals $20 \cdot 13 = 260$ for S1 and $10 \cdot 25 = 250$ for S3). Second, for the optimal (or best found) solutions w.r.t ρ_L criterion, the corresponding ρ_U are rather small and far below ρ_U^{best}. This means that the two problem formulations are indeed very different and the unlimited case cannot serve as an approximation of the limited case.

Experiments with $r_{max} = 3$ were also done and the results are given in Table 2. As before, the maximal found values of ρ_U are close to the ones of the previous study (see Tables B.4 and B.6 of the supplementary materials to [12]) and the maximal values of ρ_L are better than ρ_U. In series S1, mostly the optimal solutions with $\rho_L = L\tilde{n}$ were found.

The experiments with other *ratio* settings were also done, in general they show similar effects and for brevity they are not reported here. For *ratio* = 0.25, there are more cases with $\rho_L = L\tilde{n}$. For *ratio* = 0.75, on the contrary, ρ_U and ρ_L are quite close to each other. Clearly, in the edge case when *ratio* = 1 all the tasks are uncertain and all $\Delta_k = 0$, so it is easy to see that $\rho_U = \rho_L$ unless L is small enough and $\rho_L = L\tilde{n}$.

Table 2. Comparison of solutions for unlimited and limited cases for $r_{\max} = 3$

Series S1 $n = 25, m = 5$	optimize unlimited		optimize limited		Series S3 $n = 50, m = 10$	optimize unlimited		optimize limited	
$L = 20$	ρ_U^{best}	ρ_L^{avg}	ρ_U^{avg}	ρ_L^{best}	$L = 10$	ρ_U^{best}	ρ_L^{avg}	ρ_U^{avg}	ρ_L^{best}
S1.0	34.7	58.4	25.2	260	S3.0	15.5	15.7	14.2	16.4
S1.1	38.6	38.6	26.2	260	S3.1	18.4	18.4	17.8	18.4
S1.2	36.9	36.9	23.5	260	S3.2	17.9	17.9	6.3	18.1
S1.3	41.1	260	31.8	260	S3.3	12.0	12.0	4.45	15.4
S1.4	36.4	37.0	25	260	S3.4	14.0	14.0	14.0	14.0
S1.5	40.2	260	32.1	260	S3.5	17.2	17.3	15.8	17.3
S1.6	37.8	37.8	20.5	260	S3.6	20.7	22.9	10.9	250
S1.7	41.0	260	32.	260	S3.7	22.1	22.0	16.3	23.9
S1.8	39.1	39.1	21.4	260	S3.8	27.8	29.6	16.6	250
S1.9	24.4	24.4	23.6	24.4	S3.9	4.4	4.4	4.4	4.4

6   Conclusions

This paper provides a generalization of the earlier study on evaluating and maximizing the stability radius of the transfer line balancing problems with uncertain processing times. The introduced assumption that the extra processing times are bounded from above allows to obtain more adequate solutions from the practical point of view. A metaheuristic GPU-accelerated optimization algorithm is developed. In the experiments, it demonstrated good approximation quality and computing speed in comparison to the previously published results. It is shown that considering limited processing times in many cases leads to much better values of the stability radius, which could not be obtained under the old settings. In the future research, it would be worthwhile to consider a more general case with different intervals of processing times and to compare this approach to the stability factor optimization. A research toward the MIP modeling could be also of interest, although the form of the objective function may lead to a very complex model with many binary variables that will be hardly applicable for production scale instances. Instead, the branch-and-bound method and other metaheuristic approaches seem to be more promising directions.

Acknowledgements. The computing cluster Tesla of Sobolev Institute of Mathematics, Omsk Department was used in the experiments.

References

1. Aldous, D., Vazirani, U.: "Go with the winners" algorithms. In: Proceedings of 35th Annual Symposium on Foundations of Computer Science, Santa Fe, NM, USA, pp. 492–501. IEEE (1994). https://doi.org/10.1109/SFCS.1994.365742

2. Battaïa, O., Dolgui, A.: A taxonomy of line balancing problems and their solution approaches. Int. J. Prod. Econ. **142**, 259–277 (2013)

3. Battaïa, O., Dolgui, A.: Hybridizations in line balancing problems: a comprehensive review on new trends and formulations. Int. J. Prod. Econ. **250**(5), 108673 (2022)

4. Borisovsky, P.: A parallel "Go with the winners" algorithm for some scheduling problems. J. Appl. Ind. Math. (in press)

5. Borisovsky, P., Battaïa, O.: MIP-based heuristics for a robust transfer lines balancing problem. In: Olenev, N.N., Evtushenko, Y.G., Jaćimović, M., Khachay, M., Malkova, V. (eds.) OPTIMA 2021. LNCS, vol. 13078, pp. 123–135. Springer, Cham (2021). https://doi.org/10.1007/978-3-030-91059-4_9

6. Cheng, J.R., Gen, M.: Accelerating genetic algorithms with GPU computing: a selective overview. Comput. Ind. Eng. **128**, 514–525 (2019)

7. Dolgui, A., Guschinsky, N., Levin, G.: On problem of optimal design of transfer lines with parallel and sequential operations. In: 7th IEEE International Conference on Emerging Technologies and Factory Automation, Barcelona, Spain, vol. 1, pp. 329–334. IEEE (1999). https://doi.org/10.1109/ETFA.1999.815373

8. Essaid, M., Idoumghar, L., Lepagnot, J., Brévilliers, M.: GPU parallelization strategies for metaheuristics: a survey. Int. J. Parallel Emergent Distrib. Syst. **34**(5), 497–522 (2019)

9. Gurevsky, E., Rasamimanana, A., Pirogov, A., Dolgui, A., Rossi, A.: Stability factor for robust balancing of simple assembly lines under uncertainty. Discret. Appl. Math. **318**, 113–132 (2022)

10. Gurevsky, E., Hazir, Ö., Battaïa, O., Dolgui, A.: Robust balancing of straight assembly lines with interval task times. J. Oper. Res. Soc. **64**, 1607–1613 (2013)

11. Hazir, Ö., Dolgui, A.: A review on robust assembly line balancing approaches. IFAC-PapersOnLine **52**(13), 987–991 (2019)

12. Pirogov, A., Gurevsky, E., Rossi, A., Dolgui, A.: Robust balancing of transfer lines with blocks of uncertain parallel tasks under fixed cycle time and space restrictions. Eur. J. Oper. Res. **290**(3), 946–955 (2021)

13. Sotskov, N.: Stability of an optimal schedule. Eur. J. Oper. Res. **55**, 91–102 (1991)

On Cluster Editing Problem with Clusters of Small Sizes

Alexander Kononov[1(✉)] and Victor Il'ev[1,2]

[1] Sobolev Institute of Mathematics SB RAS, Novosibirsk, Russia
alvenko@math.nsc.ru, iljev@mail.ru
[2] Dostoevsky Omsk State University, Omsk, Russia

Abstract. In the cluster editing problem, one has to partition the set of vertices of a graph into disjoint subsets (called clusters) minimizing the number of edges between clusters and the number of missing edges within clusters. We consider a version of the problem in which cluster sizes are bounded from above by a positive integer s. This problem is NP-hard for any fixed $s \geqslant 3$. We propose polynomial-time approximation algorithms for this version of the problem. Their performance guarantees are equal to $5/3$ and $5/2$ for the cases $s = 3$ and $s = 4$, respectively. We also show that the cluster editing problem is APX-complete for the case $s = 3$ even if the maximum degree of the graphs is bounded by 4.

Keywords: Cluster editing · Approximation algorithm · Performance guarantee

1 Introduction

Clustering is the problem of grouping an arbitrary set of objects so that objects in each group are more similar to each other than to those in other clusters. In other words, it is required to partition a given set of objects into some pairwise nonintersecting subsets (clusters) so that the sum of the number of similarities between the clusters and the number of missing similarities inside the clusters would be minimum.

One of the most visual formalizations of clustering is the graph clustering [20]. In the 60–70s of the last century, a number of papers addressed the graph approximation problem [10,11,22,24]. In this problem, the similarity relation between objects is given by an undirected graph whose vertices are in one-to-one correspondence to the objects and whose edges connect similar objects. The goal is to partition the set of vertices of a graph into disjoint subsets (called clusters) minimizing the number of edges between clusters and the number of missing edges within clusters. The number of clusters may be given, bounded, or undefined. Various formulations and interpretations of the graph approximation problem can be found in [1,10,11,22,24].

Later, the graph approximation problem was repeatedly and independently rediscovered and studied under various names (Correlation Clustering, Cluster

N. Olenev et al. (Eds.): OPTIMA 2023, LNCS 14395, pp. 316–328, 2023.
https://doi.org/10.1007/978-3-031-47859-8_23

Editing, etc. [3, 4, 21]). Recently, the unweighted version of the problem was given the name of the cluster editing problem, whereas the correlation clustering problem refers to problem statements with arbitrary edge weights [19, 21, 23]. This is due to the fact that the problem without weights has the following natural interpretation. Given a graph G. It is required to determine the smallest number of edge insertions and deletions that must be carried out in order to transform G into a disjoint union of cliques.

For the convenience of the following presentation, we will introduce some definitions below.

We consider *ordinary* graphs, i.e., undirected graphs without loops and multiple edges. A graph G is called a *cluster graph* if every connected component of G is a complete graph [21]. Let $\mathcal{M}(V)$ be the family of all cluster graphs on the set of vertices V, let $\mathcal{M}_k(V)$ be the family of all cluster graphs on the vertex set V having exactly k connected components, and let $\mathcal{M}_{\leqslant k}(V)$ be the family of all cluster graphs on V having at most k connected components, $2 \leqslant k \leqslant |V|$.

If $G_1 = (V, E_1)$ and $G_2 = (V, E_2)$ are graphs on the vertex set V, then the distance $d(G_1, G_2)$ between them is defined as

$$d(G_1, G_2) = |E_1 \Delta E_2| = |E_1 \setminus E_2| + |E_2 \setminus E_1|,$$

i.e., $d(G_1, G_2)$ is the number of distinct edges in G_1 and G_2.

In the 1960s-1980s the following three graph approximation problems were studied. They can be considered as various formalizations of the cluster editing problem [10, 11, 14, 22, 24]:

Problem CE. Given an ordinary graph $G = (V, E)$, find a graph $M^* \in \mathcal{M}(V)$ such that

$$d(G, M^*) = \min_{M \in \mathcal{M}(V)} d(G, M).$$

Problem CE$_\mathbf{k}$. Given an ordinary graph $G = (V, E)$ and an integer k, $2 \leqslant k \leqslant |V|$, find a graph $M^* \in \mathcal{M}_k(V)$ such that

$$d(G, M^*) = \min_{M \in \mathcal{M}_k(V)} d(G, M).$$

Problem CE$_{\leqslant \mathbf{k}}$. Given an ordinary graph $G = (V, E)$ and an integer k, $2 \leqslant k \leqslant |V|$, find a graph $M^* \in \mathcal{M}_{\leqslant k}(V)$ such that

$$d(G, M^*) = \min_{M \in \mathcal{M}_{\leqslant k}(V)} d(G, M).$$

The first theoretical results related to the graph approximation problem were obtained in the middle of the last century. In 1964, Zahn [24] studied Problem **CE** for the graphs of some special form. In 1971, Fridman [10] determined the first polynomially solvable case of Problem **CE**. He showed that Problem **CE** for a graph without triangles can be reduced to constructing maximum matching in this graph. In 1986, Křivánek and Morávek [18] showed that problem **CE** is NP-hard, but their article remained unnoticed.

In 2004, Bansal, Blum, and Chawla [3] and independently Shamir, Sharan, and Tsur [21] proved again that Problem **CE** is NP-hard. In [21] it was also proved that Problem $\mathbf{CE_k}$ is NP-hard for every fixed $k \geqslant 2$. In 2006, a more simple proof of this result was published by Giotis and Guruswamy [12]. In the same year, Ageev, Il'ev, Kononov, and Talevnin [1] independently proved that Problems $\mathbf{CE_2}$ and $\mathbf{CE_{\leqslant 2}}$ are NP-hard even for cubic graphs, and derived from this that both Problems $\mathbf{CE_k}$ and $\mathbf{CE_{\leqslant k}}$ are NP-hard for every fixed $k \geqslant 2$.

In [3], a simple 3-approximation algorithm for Problem $\mathbf{CE_{\leqslant 2}}$ was proposed. A randomized polynomial-time approximation scheme for Problem $\mathbf{CE_{\leqslant 2}}$ was presented in [1]. Giotis and Guruswamy [12] proposed a randomized polynomial-time approximation scheme for Problem $\mathbf{CE_{\leqslant k}}$ (for every fixed $k \geqslant 2$). In 2008, pointing out that the complexity of the polynomial time approximation scheme of [12] deprives it the prospects of practical use, Coleman, Saunderson and Wirth [8] proposed a 2-approximation algorithm for Problem $\mathbf{CE_{\leqslant 2}}$, applying a local search procedure to the feasible solutions obtained by the 3-approximation algorithm from [3]. For Problem $\mathbf{CE_2}$, a 2-approximation algorithm was proposed by Il'ev, Il'eva, and Morshinin [17]. As regards to Problem $\mathbf{CE}$, it was shown in 2005 [5] that Problem $\mathbf{CE}$ is APX-hard, and a 4-approximation algorithm was developed. In 2008, a 2.5-approximation algorithm for Problem $\mathbf{CE}$ was presented by Ailon, Charikar, and Newman [2]. Finally, in 2015, a $(2.06 - \varepsilon)$-approximation algorithm for Problem $\mathbf{CE}$ was proposed [7].

The cluster editing problem with bounded cluster sizes was independently introduced by Il'ev and Navrotskaya [15] and Puleo and Milenkovic [19]. Puleo and Milenkovic [19] provided a 6-approximation LP rounding algorithm and a randomized 7-approximation algorithm. The first algorithm uses a region growing technique [5,9] to obtain a solution from the relaxed solution of the corresponding integer linear program. The running time of the algorithm is incurred by the LP solver. The randomized algorithm is based on the cc-pivot algorithm [2]. While it has an approximation ratio worse than the ratio of 6, it does not require solving a linear program and has a low complexity.

The cluster editing problem with clusters no larger than 2 is equivalent to the well-known maximal matching problem and can be solved in polynomial time. In our paper, we present polynomial-time algorithms with performance guarantee better than 6 for the problems in which the cluster sizes are bounded from above by 3 and 4. We note that our algorithms use a technique different from the techniques used in [19]. First, we introduce two simple data reduction rules that transform a given instance to a simplified one. Second, for the simplified instance we use a greedy heuristic that finds a maximal under inclusion collection of vertex-disjoint cliques of a given size. Additionally, we prove that the cluster editing problem in which the cluster sizes are bounded from above by 3 is APX-hard. Since this problem is also in APX, then it is APX-complete.

Note that the same greedy heuristic was applied by Chataigner et al. [6] for approximate solving the $\mathcal{K}_r$-packing problem. In this problem one has to find in a given graph G a family of vertex-disjoint cliques of sizes $t \leq r$ that maximizes the number of covered edges of G. Our technique of studying worst-case behaviour of the greedy heuristic is adaptation and extension of the technique used by Chataigner et al.

Note also that in [6] authors refer to a more complicated heuristic of Hurkens and Schrijver [13] that allows to find more accurate solutions to the problem of finding the maximum family of vertex-disjoint k-cliques in a graph G. However, the heuristic of Hurkens and Schrijver does not improve the worst-case performance of our algorithms.

In Sect. 2, we give the statement of the clustering problem with clusters of bounded sizes and study some properties of optimal solutions. In Sect. 3, we present a $\frac{5}{3}$-approximation algorithm for the cluster editing problem in which the cluster sizes are bounded from above by 3 and prove that this problem is APX-hard. In Sect. 4, we propose a $\frac{5}{2}$-approximation algorithm for the problem in which the cluster sizes are bounded from above by 4. Conclusion summarizes the results of the work.

2 Preliminaries

In contrast to Problems $\mathbf{CE_k}$ and $\mathbf{CE_{\leqslant k}}$, where the restrictions on the number of clusters are imposed, we now discuss the problem of graph clustering with clusters of bounded sizes.

Let $\mathcal{M}^{\leqslant s}(V)$ be the family of all cluster graphs on V such that the size of each connected component is at most some integer s, $2 \leqslant s \leqslant |V|$.

Problem $\mathbf{CE^{\leqslant s}}$. Given an ordinary n-vertex graph $G = (V, E)$ and an integer s, $2 \leqslant s \leqslant n$, find $M^* \in \mathcal{M}^{\leqslant s}(V)$ such that

$$d(G, M^*) = \min_{M \in \mathcal{M}^{\leqslant s}(V)} d(G, M).$$

In [3], in the proof of NP-hardness of Problem $\mathbf{CE}$ without any constraints on the number and sizes of clusters, it is actually shown that Problem $\mathbf{CE^{\leqslant 3}}$ is NP-hard. Later, in [15], it was shown that Problem $\mathbf{CE^{\leqslant s}}$ is NP-hard for every fixed $s \geqslant 3$. In [16], it was shown that the problem is polynomially solvable on graphs without triangles.

We present some properties of optimal solutions. In the formulation of our properties, we use the following notation. Given a graph $G = (V, E)$ and a nonempty set of vertices $V' \subseteq V$, let $\delta(V')$ denote the cut defined by V' (i.e., the set of edges between V' and $V \setminus V'$). We denote by K_r a complete graph on r vertices and let Q_4 be the 4-vertex graph with 5 edges. A complete graph on 3 vertices is called *triangle*.

Remark 1. Given an instance G of $\mathbf{CE^{\leqslant s}}$ with $s \geq 3$. Let $H \subseteq G$, where $H = K_s$ and $|\delta(H)| \leq s - 1$. Then there is an optimal solution that contains H as a component.

Indeed, let $M \in \mathcal{M}^{\leqslant 3}(V)$ and $x < s$ be the maximal number of vertices of H that belong to the same component of M. Then, at least $x(s - x)$ edges between vertices of H made a contribution to the objective function. Remove all vertices of H from all the components of the cluster graph M where they lie, create a new component H and add it to M. The value of objective function decreases at least $x(s - x) - (s - 1) \geq 0$ for $x > 0$ and $s \geq x + 1$.

Remark 2. Given an instance G of $\mathbf{CE}^{\leqslant s}$ with $s \geq 4$. Let $H \subseteq G$, where $H = Q_4$ and $|\delta(H)| \leq 1$. Then there is an optimal solution that contains a clique on the set of vertices of H as a component.

Indeed, let vertices of H belong to different components of an optimal solution M^*. Then at least two edges between vertices of H make a contribution to the objective function. Remove all vertices of H from all components of M^* where they lie, create the new 4-clique on vertices of H and add it to M^*. The new component contributes one edge insertion and at most one edge deletion to the objective function. Thus, the value of the objective function does not increase.

Remark 3. Given an instance G of $\mathbf{CE}^{\leqslant s}$ with $s \geq 3$. Let $H \subseteq G$, where $H = K_2$ and $|\delta(H)| \leq 1$. Then there is an optimal solution that contains H as a component.

Remark 4. Given an instance G of $\mathbf{CE}^{\leqslant s}$ with $s \geq 4$. Let $H \subseteq G$, where $H = K_3$ and $|\delta(H)| \leq 1$. Then there is an optimal solution that contains H as a component.

The proofs of these statements are similar to the previous one.

Thus, we can preprocess the given graph $G = (V, E)$ by sequentially finding a subgraph satisfying one of the conditions in the above remarks, adding the found subgraph to the solution and removing it from G with all edges incident to its vertices.

Property 1. Given an instance of $\mathbf{CE}^{\leqslant 3}$. Let G be the graph obtained from a given graph after preprocessing.

a) If $H \subseteq G$ and $H = K_2$, then $|\delta(H)| \geq 2$.
b) If $H \subseteq G$ and $H = K_3$, then $|\delta(H)| \geq 3$.

Property 2. Given an instance of $\mathbf{CE}^{\leqslant 4}$. Let G be the graph obtained from a given graph after preprocessing.

a) If $H \subseteq G$ and $H = K_2$, then $|\delta(H)| \geq 2$.
b) If $H \subseteq G$ and $H = K_3$, then $|\delta(H)| \geq 2$.
c) If $H \subseteq G$ and $H = Q_4$, then $|\delta(H)| \geq 2$.
d) If $H \subseteq G$ and $H = K_4$, then $|\delta(H)| \geq 4$.

3   The Cluster Editing Problem with Clusters of Size 3

In this section, we present a $\frac{5}{3}$-approximation algorithm for Problem $\mathbf{CE}^{\leqslant 3}$ and show that this problem is APX-hard.

Let $\mathcal{F}$ be a fixed family of graphs. A set of pairwise vertex disjoint subgraphs of G, each isomorphic to an element of $\mathcal{F}$ is called an $\mathcal{F}$-*packing* in G. An $\mathcal{F}$-packing that is maximal under inclusion is called a *maximal $\mathcal{F}$-packing* [6,13].

We note that a maximal $\{K_r\}$-packing can clearly be computed in polynomial time for any fixed r. We greedily pick an arbitrary clique of size r and remove all vertices of this clique with all incident edges. We note that this "folklore" heuristic coincides with the heuristic $HS(r,1)$ proposed by Hurkens and Schrijver [13].

Remark 5. The cardinality of any maximal $\{K_r\}$-packing in a graph is greater or equal to $\frac{1}{r}$ of the cardinality of a maximum $\{K_r\}$-packing.

Indeed, let $\mathcal{K}_r$ be a maximal $\{K_r\}$-packing and $\mathcal{E}_r$ be a largest set of pairwise disjoint cliques of size r in G. At least one vertex of each clique from $\mathcal{E}_r$ must belong to some clique in $\mathcal{K}_r$ and we get $|\mathcal{E}_r| \leq r|\mathcal{K}_r.|$

To avoid redundant notation, we use $\mathcal{K}_r$ to denote both a set of cliques of size r and the subgraph of G, which is the union of these cliques.

Consider an approximation algorithm for Problem $\mathbf{CE}^{\leqslant 3}$. Let G be obtained from a given graph after preprocessing, i.e. Property 1 holds.

Algorithm 1: Approximation algorithm for Problem $\mathbf{CE}^{\leqslant 3}$

Input: A graph $G = (V, E)$ satisfying conditions of Property 1.
Output: A cluster graph $M \in \mathcal{M}^{\leqslant 3}(V)$.
1 Find a maximal $\{K_3\}$-packing of G, say $\mathcal{K}_3$.
2 $G \leftarrow G - \mathcal{K}_3$.
   /* Delete all vertices of $\mathcal{K}_3$ with all incident edges.   */
3 Find a maximum matching in G, say $\mathcal{K}_2$.
4 **return** $\mathcal{K}_2 \cup \mathcal{K}_3$

Note that the family of cliques $\mathcal{A} = \mathcal{K}_2 \cup \mathcal{K}_3$ returned by Algorithm 1 induces the cluster graph $M \in \mathcal{M}^{\leqslant 3}(V)$ – an approximate solution to Problem $\mathbf{CE}^{\leqslant 3}$ on the graph $G = (V, E)$.

Theorem 1. *Let $G = (V, E)$ be an arbitrary graph. Then*

$$\frac{d(G, M)}{d(G, M^*)} \leqslant \frac{5}{3}, \tag{1}$$

where M^* is an optimal solution to Problem $\mathbf{CE}^{\leqslant 3}$ on the graph G, M is the cluster graph constructed by Algorithm 1.

Proof. Note that if bound (1) will be proved for the graph obtained from a given graph G after preprocessing, then (1) will be true for G too. So, further suppose w.l.o.g. that G is the graph obtained from a given graph after preprocessing.

We note that there exists an optimal solution M^* that contains the only K_2- and K_3-subgraphs of the graph G. Indeed, let three vertices belong to the same component of M^* and induce a 2-edge chain in G. We can split this component into two components. The first component consists of two vertices connected by an edge and the second is an isolated vertex. The value of the objective function will not change in this case.

Let $\mathcal{O}$ be a family of cliques inducing an optimal solution M^* to Problem **CE$^{\leqslant 3}$**.

We denote by $\mathcal{E}_i$ the collection of all i-cliques in $\mathcal{O}$, $i = 2, 3$. Let $N(\mathcal{O})$ and $N(\mathcal{A})$ be the number of edges in $\mathcal{E}_2 \cup \mathcal{E}_3$ and $\mathcal{K}_2 \cup \mathcal{K}_3$, respectively. We have $N(\mathcal{O}) = 3|\mathcal{E}_3| + |\mathcal{E}_2|$. Moreover, Property 1 implies that $d(G, M^*) \geq \frac{3|\mathcal{E}_3|+2|\mathcal{E}_2|}{2}$. Chataigner et al. [6] proved that

$$N(\mathcal{A}) \geq 2|\mathcal{E}_3| + |\mathcal{E}_2|. \tag{2}$$

For the convenience of readers, we present our own proof of this claim. We will associate to each clique $C \in \mathcal{E}_2$ at least one edge from $\mathcal{K}_2 \cup \mathcal{K}_3$, and to each clique $C \in \mathcal{E}_3$ at least two edges from $\mathcal{K}_2 \cup \mathcal{K}_3$.

Consider a clique $K \in \mathcal{K}_3$ and vertices u, v, w belong to K.

(a) Let $C \in \mathcal{E}_3$ and $C = K$. Then all three edges of K are associated with C.

(b) Let $C \in \mathcal{E}_3 \cup \mathcal{E}_2$ and vertices u, v belong to C, but $w \notin C$. We associate the edges uv and uw to C.

(c) Let $C \in \mathcal{E}_3 \cup \mathcal{E}_2$ and $\{w\} = K \cap C$. We associate the edge vw to C.

As a result, if a clique $C \in \mathcal{E}_3 \cup \mathcal{E}_2$ has i common vertices with K, then i edges of K are associated to C. Moreover, each such edge is associated to only one clique $C \in \mathcal{E}_3 \cup \mathcal{E}_2$.

Further, consider the graph $G' = G - \mathcal{K}_3$. There are no triangles in G'. Let E^* be the set of edges in $G' \cap (\mathcal{E}_3 \cup \mathcal{E}_2)$. Then E^* forms a matching in G'.

(d) Let $\mathcal{C} = \{C \in \mathcal{E}_3 \cup \mathcal{E}_2 \mid C \cap E^* \neq \emptyset\}$. Since $\mathcal{K}_2$ is a maximum matching in G', then $|E^*| \leq |\mathcal{K}_2|$. Hence, we can associate to each component $C \in \mathcal{C}$ one edge from $\mathcal{K}_2$; if $C \in \mathcal{E}_3$, then this is the second edge associated to C.

Thus, at least two edges from $\mathcal{K}_2 \cup \mathcal{K}_3$ are associated to each clique $C \in \mathcal{E}_3$ and at least one edge from $\mathcal{K}_2 \cup \mathcal{K}_3$ is associated to each clique $C \in \mathcal{E}_2$. Hence, we get (2).

From (2) we have, $d(G, M) \leq d(G, M^*) + |\mathcal{E}_3|$, and

$$\frac{d(G, M)}{d(G, M^*)} \leq 1 + \frac{|\mathcal{E}_3|}{d(G, M^*)} \leq 1 + \frac{2|\mathcal{E}_3|}{3|\mathcal{E}_3| + 2|\mathcal{E}_2|} \leq \frac{5}{3}.$$

Theorem 1 is proved.

Theorem 2. *Problem* **CE**$^{\leqslant 3}$ *is APX-hard.*

Proof. We prove that Problem **CE**$^{\leqslant 3}$ is APX-hard on graphs with maximum degree 4. Since we have shown above that this problem has a constant approximation algorithm, we can conclude that it lies in APX.

Chataigner et al. [6] show that the $\{K_2, K_3\}$-packing problem is APX-hard on graphs with maximum degree 4 via a L-reduction from the restricted version of the MAX SAT problem. An $\{K_2, K_3\}$-packing in a graph G is a set of pairwise vertex-disjoint subgraphs of G, each isomorphic to either K_3 or K_2. The $\{K_2, K_3\}$-packing problem is the problem of finding the maximum number of edges that can be covered by an $\{K_2, K_3\}$-packing.

Let G be an instance for the $\{K_2, K_3\}$-packing problem and Problem **CE**$^{\leqslant 3}$. We note that any feasible solution of the $\{K_2, K_3\}$-packing problem is also a feasible solution to Problem **CE**$^{\leqslant 3}$. Let $f(Sol)$ and $g(Sol)$ be the values of a feasible solution Sol of the $\{K_2, K_3\}$-packing problem and Problem **CE**$^{\leqslant 3}$, respectively. We have $f(Sol) = m - g(Sol)$, where m is the number of edges in G. Hence, if OPT is an optimal solution to the $\{K_2, K_3\}$-packing problem, then OPT is an optimal solution to Problem **CE**$^{\leqslant 3}$, and we have $f(OPT) = m - g(OPT)$. Moreover, we have $f(OPT) - f(Sol) = m - g(OPT) - m + g(Sol) = g(Sol) - g(OPT)$.

Let G be a graph with maximum degree 4, then it is easy to see that $f(OPT) \geq m/7$. Indeed, let z be a number of edges in any maximal matching M in G. Each edge of M has at most 6 edges that are incident to it. Since each edge in the graph G either belongs to OPT or is incident to one of the edges from OPT, we have $m \leq 7z \leq 7f(OPT)$.

Suppose we have an α-approximation algorithm for Problem **CE**$^{\leqslant 3}$. Let Sol be a solution produced by this algorithm. Hence, Sol is a feasible solution of the $\{K_2, K_3\}$-packing problem. We have

$$f(Sol) = f(OPT) + g(OPT) - g(Sol) \geq m - \alpha g(OPT) = m - \alpha(m - f(OPT))$$

$$= \alpha f(OPT) - (\alpha - 1)m \geq \alpha f(OPT) - 7(\alpha - 1)f(OPT) = (7 - 6\alpha)f(OPT).$$

We obtain that if there exists an α-approximation algorithm for Problem **CE**$^{\leqslant 3}$, then there exists a $(7 - 6\alpha)$-approximation algorithm for the $\{K_2, K_3\}$-packing problem. In particular, if for any $\varepsilon > 0$ there exists a $(1 + \varepsilon)$-approximation algorithm for Problem **CE**$^{\leqslant 3}$, then there exists a $(1 - 6\varepsilon)$-approximation algorithm for the $\{K_2, K_3\}$-packing problem. Since the $\{K_2, K_3\}$-packing problem is APX-hard on graphs with maximum degree 4, we obtain that Problem **CE**$^{\leqslant 3}$ is APX-hard as well.

Theorem 2 is proved.

4   The Cluster Editing Problem with Clusters of Size 4

In this section, we present an approximation algorithm for Problem $\mathbf{CE}^{\leqslant 4}$. Let G be the graph obtained from a given graph after preprocessing, i.e. Property 2 holds.

Algorithm 2: Approximation algorithm for Problem $\mathbf{CE}^{\leqslant 4}$

Input: A graph $G = (V, E)$ satisfying conditions of Property 2.
Output: A cluster graph $M \in \mathcal{M}^{\leqslant 4}(V)$.
1 Find a maximal $\{K_4\}$-packing of G, say $\mathcal{K}_4$.
2 $G \leftarrow G - \mathcal{K}_4$.
   /* Delete all vertices of $\mathcal{K}_4$ with all incident edges.   */
3 Find a maximal $\{K_3\}$-packing of G, say $\mathcal{K}_3$.
4 $G \leftarrow G - \mathcal{K}_3$.
   /* Delete all vertices of $\mathcal{K}_3$ with all incident edges.   */
5 Find a maximum matching in G, say $\mathcal{K}_2$.
6 **return** $\mathcal{K}_2 \cup \mathcal{K}_3 \cup \mathcal{K}_4$

Note that the family of cliques $\mathcal{A} = \mathcal{K}_2 \cup \mathcal{K}_3 \cup \mathcal{K}_4$ returned by Algorithm 2 induces the cluster graph $M \in \mathcal{M}^{\leqslant 4}(V)$ – an approximate solution to Problem $\mathbf{CE}^{\leqslant 4}$ on the graph $G = (V, E)$.

Theorem 3. *Let $G = (V, E)$ be an arbitrary graph. Then*

$$\frac{d(G, M)}{d(G, M^*)} \leqslant \frac{5}{2}, \tag{3}$$

where M^* is an optimal solution to problem $\mathbf{CE}^{\leqslant 4}$ on the graph G, M is the cluster graph constructed by Algorithm 2.

Proof. In our proof, we adapt and extend the method used by Chataigner et al. [6]. Note that if bound (3) will be proved for the graph obtained from a given graph G after preprocessing, then (3) will be true for G too. So, further suppose w.l.o.g. that G is the graph obtained from a given graph after preprocessing.

Let $\mathcal{O}$ be a family of connected subgraphs in $M^* \cap G$. Note that there exists an optimal solution M^* to Problem $\mathbf{CE}^{\leqslant 4}$ on G such that only K_2-, K_3-, K_4 and Q_4-subgraphs of the graph G are in $\mathcal{O}$. Indeed, evidently that if a clique $C = K_2 \in \mathcal{O}$, then $C \subseteq G$. Suppose that a clique $C = K_3 \in \mathcal{O}$ is obtained from some 3-vertex subgraph H of G by adding one edge e. Then instead of adding e we can delete from H another edge and obtain another cluster graph $M' \in \mathcal{M}^{\leqslant 4}(V)$ with the same value of the objective function: $d(G, M') = d(G, M^*)$. Finally, suppose that a clique $C = K_4 \in \mathcal{O}$ is obtained from some 4-vertex subgraph H of G by adding at least 2 edges. Then instead of their adding we can delete from H at most 2 edges and obtain another cluster graph $M' \in \mathcal{M}^{\leqslant 4}(V)$ such that $d(G, M') \leqslant d(G, M^*)$.

Denote by $\mathcal{Q}_4$ the collection of all Q_4-subgraphs in $\mathcal{O}$ and by $\mathcal{E}_i$ the collection of all i-cliques in $\mathcal{O}$, $i = 2, 3, 4$.

Let $N(\mathcal{O})$ and $N(\mathcal{A})$ be the number of edges in $\mathcal{E}_2 \cup \mathcal{E}_3 \cup \mathcal{Q}_4 \cup \mathcal{E}_4$ and $\mathcal{K}_2 \cup \mathcal{K}_3 \cup \mathcal{K}_4$, respectively. We have $N(\mathcal{O}) = 6|\mathcal{E}_4| + 5|\mathcal{Q}_4| + 3|\mathcal{E}_3| + |\mathcal{E}_2|$. Property 2 implies that

$$d(G, M^*) = |E \setminus E^*| + |E^* \setminus E| \geq \frac{4|\mathcal{E}_4| + 2|\mathcal{Q}_4| + 2|\mathcal{E}_3| + 2|\mathcal{E}_2|}{2} + |\mathcal{Q}_4|$$

$$= 2|\mathcal{E}_4| + 2|\mathcal{Q}_4| + |\mathcal{E}_3| + |\mathcal{E}_2|.$$

Here we add another term $|\mathcal{Q}_4|$ because each component $C = Q_4$ contributes 1 to the objective function.

Let k_i, $0 \leq i \leq 4$, be the number of 4-cliques of $\mathcal{E}_4$ that intersect precisely i vertices of $\mathcal{K}_3 \cup \mathcal{K}_4$. Let q_i, $0 \leq i \leq 4$, be the number of Q_4-subgraphs of $\mathcal{Q}_4$ that intersect precisely i vertices of $\mathcal{K}_3 \cup \mathcal{K}_4$. Let t_i, $0 \leq i \leq 3$, be the number of 3-cliques of $\mathcal{E}_3$ that intersect precisely i vertices of $\mathcal{K}_3 \cup \mathcal{K}_4$. At step 1, Algorithm 2 selects a maximal set of vertex-disjoint 4-cliques, so we have that $k_0 = 0$. Since Algorithm 2 selects a maximal set of vertex-disjoint triangles at step 3, we additionally have that $t_0 = 0$, $q_0 = 0$, and $k_1 = 0$.

We note that the number of vertices of $\mathcal{K}_3 \cup \mathcal{K}_4$ covered by $\mathcal{E}_3 \cup \mathcal{Q}_4 \cup \mathcal{E}_4$ is $u := 2k_2 + 3k_3 + 4k_4 + q_1 + 2q_2 + 3q_3 + 4q_4 + t_1 + 2t_2 + 3t_3$. Thus, the number of vertices of $\mathcal{K}_3 \cup \mathcal{K}_4$ not covered by $\mathcal{E}_3 \cup \mathcal{Q}_4 \cup \mathcal{E}_4$ is $w := 4|\mathcal{K}_4| + 3|\mathcal{K}_3| - u$. We obtain that the number of edges of $\mathcal{E}_2$ with at least one endpoint in $\mathcal{K}_3 \cup \mathcal{K}_4$ is at most w.

Let $z := |\mathcal{E}_2| - w$. We have that at least $\max\{0, z\}$ edges of $\mathcal{E}_2$ are disjoint from $\mathcal{K}_3 \cup \mathcal{K}_4$.

Next, every triangle of $\mathcal{E}_3$ or Q_4-subgraph of $\mathcal{Q}_4$ that intersects precisely one vertex of $\mathcal{K}_3 \cup \mathcal{K}_4$ has an edge that is disjoint from $\mathcal{K}_3 \cup \mathcal{K}_4$. Moreover, every 4-clique of $\mathcal{E}_4$ that intersects precisely 2 vertices of $\mathcal{K}_3 \cup \mathcal{K}_4$ contributes one such edge. Since $\mathcal{K}_2$ is a maximum matching of $G - (\mathcal{K}_3 \cup \mathcal{K}_4)$, we get

$$|\mathcal{K}_2| \geq k_2 + q_1 + t_1 + \max\{0, z\}. \tag{4}$$

Remind that $|\mathcal{E}_4| = k_2 + k_3 + k_4$, $|\mathcal{Q}_4| = q_1 + q_2 + q_3 + q_4$, and $|\mathcal{E}_3| = t_1 + t_2 + t_3$. We rewrite z as

$$z = |\mathcal{E}_2| - 4|\mathcal{K}_4| - 3|\mathcal{K}_3| + 2k_2 + 3k_3 + 4k_4 + q_1 + 2q_2 + 3q_3 + 4q_4 + t_1 + 2t_2 + 3t_3$$

$$= |\mathcal{E}_2| - 4|\mathcal{K}_4| - 3|\mathcal{K}_3| + 3|\mathcal{E}_4| - k_2 + k_4 + 2|\mathcal{Q}_4| - q_1 + q_3 + 2q_4 + 2|\mathcal{E}_3| - t_1 + t_3.$$

Since $N(\mathcal{A}) = 6|\mathcal{K}_4| + 3|\mathcal{K}_3| + |\mathcal{K}_2|$, taking into account (4) we obtain

$$N(\mathcal{A}) \geq 6|\mathcal{K}_4| + 3|\mathcal{K}_3| + k_2 + q_1 + t_1 + \max\{0, z\} \geq 6|\mathcal{K}_4| + 3|\mathcal{K}_3| + k_2 + q_1 + t_1 + z.$$

Substituting the value of z, we get

$$N(\mathcal{A}) \geq 2|\mathcal{K}_4| + 3|\mathcal{E}_4| + 2|\mathcal{Q}_4| + 2|\mathcal{E}_3| + |\mathcal{E}_2|.$$

By Remark 5, $|\mathcal{K}_4| \geq \frac{|\mathcal{E}_4|}{4}$. Thus, we obtain

$$N(\mathcal{A}) \geq 3.5|\mathcal{E}_4| + 2|\mathcal{Q}_4| + 2|\mathcal{E}_3| + |\mathcal{E}_2|.$$

Since M is the cluster subgraph of G, we have

$$d(G, M) \leq d(G, M^*) + N(\mathcal{O}) - N(\mathcal{A}) \leq d(G, M^*) + 2.5|\mathcal{E}_4| + 3|\mathcal{Q}_4| + |\mathcal{E}_3|,$$

and

$$\frac{d(G, M)}{d(G, M^*)} \leq 1 + \frac{2.5|\mathcal{E}_4| + 3|\mathcal{Q}_4| + |\mathcal{E}_3|}{d(G, M^*)}$$

$$\leq 1 + \frac{2.5|\mathcal{E}_4| + 3|\mathcal{Q}_4| + |\mathcal{E}_3|}{2|\mathcal{E}_4| + 2|\mathcal{Q}_4| + |\mathcal{E}_3| + |\mathcal{E}_2|} \leq \frac{5}{2}.$$

Theorem 3 is proved.

5   Conclusion

A version of the graph clustering problem is considered. In this version sizes of all clusters don't exceed a given positive integer s. This problem is NP-hard for every fixed $s \geqslant 3$. For $s = 3$, we prove that the problem is APX-hard and propose a new $\frac{5}{3}$-approximation algorithm. For the case $s = 4$, a similar $\frac{5}{2}$-approximation algorithm is proposed. Performance guarantees of these algorithms are better than those of earlier presented approximation algorithms.

Our algorithms are quite simple and have two stages. First, we introduce two simple data reduction rules that transform a given instance to simplified one. Second, for the simplified instance we use a heuristic that finds a maximal under inclusion collection of pairwise disjoint cliques of a given size.

We note that Hurkens and Schrijver [13] presented a more sophisticated heuristic that allows to find more accurate solutions to the problem of finding the maximum set of vertex-disjoint k-cliques in a graph G. However, the heuristic of Hurkens and Schrijver does not improve the worst-case performance of our algorithms.

Acknowledgements. The research of the first author was carried out within the framework of the state contract of the Sobolev Institute of Mathematics (project FWNF-2022-0019). The research of the second author was carried out within the framework of the state contract of the Sobolev Institute of Mathematics (project FWNF-2022-0020).

References

1. Ageev, A.A., Il'ev, V.P., Kononov, A.V., Talevnin, A.S.: Computational complexity of the graph approximation problem. Diskretnyi Analiz i Issledovanie Operatsii. Ser. 1. **13**(1), 3–11 (2006) (in Russian). English transl. in: J. of Applied and Industrial Math. 1(1), 1–8 (2007). https://doi.org/10.1134/s1990478907010012
2. Ailon, N., Charikar, M., Newman, A.: Aggregating inconsistent information: ranking and clustering. J. ACM. **55**(5), 1–27 (2008). https://doi.org/10.1145/1411509.1411513
3. Bansal, N., Blum, A., Chawla, S.: Correlation clustering. Mach. Learn. **56**, 89–113 (2004). https://doi.org/10.1023/B:MACH.0000033116.57574.95

4. Ben-Dor, A., Shamir, R., Yakhimi, Z.: Clustering gene expression patterns. J. Comput. Biol. **6**(3–4), 281–297 (1999). https://doi.org/10.1089/106652799318274

5. Charikar, M., Guruswami, V., Wirth, A.: Clustering with qualitative information. J. Comput. Syst. Sci. **71**(3), 360–383 (2005). https://doi.org/10.1016/j.jcss.2004.10.012

6. Chataigner, F., Manić, G., Wakabayashi, Y., Yuster, R.: Approximation algorithms and hardness results for the clique packing problem. Discrete Appl. Math. **157**(7), 1396–1406 (2009). https://doi.org/10.1016/j.dam.2008.10.017

7. Chawla, S., Makarychev, K., Schramm, T., Yaroslavtsev, G.: Near optimal LP algorithm for correlation clustering on complete and complete k-partite graphs. STOC '15 Symposium on Theory of Computing: ACM New York, pp. 219–228 (2015). https://doi.org/10.1145/2746539.2746604

8. Coleman, T., Saunderson, J., Wirth, A.: A local-search 2-approximation for 2-correlation-clustering. Lecture Notes in Computer Science. 5193, 308–319 (2008). https://doi.org/10.1007/978-3-540-87744-826

9. Demaine, E.D., Emanuel, D., Fiat, A, Immorlica, N.: Correlation clustering in general weighted graphs. Theor. Comput. Sci. **361**(2–3), 172–187 (2006). https://doi.org/10.1016/j.tcs.2006.05.008

10. Fridman, G.Š.: A graph approximation problem. Upravlyaemye Sistemy. Izd. Inst. Mat., Novosibirsk **8**, 73–75 (1971) (in Russian)

11. Fridman, G.Š.: Investigation of a classifying problem on graphs. Methods of Modelling and Data Processing (Nauka, Novosibirsk). 147–177 (1976). (in Russian)

12. Giotis, I., Guruswami, V.: Correlation clustering with a fixed number of clusters. Theory Comput. **2**(1), 249–266 (2006)

13. Hurkens, C.A.J., Schrijver, A.: On the size of systems of sets every t of which have an SDR, with an application to the worst-case ratio of heuristics for packing problems. SIAM J. Disc. Math. **2**(1), 66–72, (1989). https://doi.org/10.1137/0402008

14. Il'ev, V.P., Fridman, G.Š.: On the problem of approximation by graphs with a fixed number of components. Dokl. Akad. Nauk SSSR. **264**(3), 533–538 (1982) (in Russian). English transl. in: Sov. Math. Dokl. 25(3), 666–670 (1982)

15. Il'ev, V.P., Navrotskaya, A.A.: Computational complexity of the problem of approximation by graphs with connected components of bounded size. Prikl. Diskretn. Mat. **3**(13), 80–84 (2011) (in Russian)

16. Il'ev, V.P., Il'eva, S.D., Navrotskaya, A.A.: Graph clustering with a constraint on cluster sizes. Diskretn. Anal. Issled. Oper. **23**(3), 50–20 (2016) (in Russian). English transl. in: J. Appl. Indust. Math. 10(3), 341–348 (2016). https://doi.org/10.1134/S1990478916030042

17. Il'ev, V., Il'eva, S., Morshinin, A.: A 2-approximation algorithm for the graph 2-clustering problem. Lecture Notes in Comput. Sci. **11548**, 295–308 (2019). https://doi.org/10.1007/978-3-030-22629-9 21

18. Křivánek, M., Morávek, J.: NP-hard problems in hierarchical-tree clustering. Acta informatica. **23**, 311–323 (1986). https://doi.org/10.1007/BF00289116

19. Puleo, G.J., Milenkovic, O.: Correlation clustering with constrained cluster sizes and extended weights bounds. SIAM J. Optim. **25**(3), 1857–1872 (2015). https://doi.org/10.1137/140994198

20. Schaeffer, S.E.: Graph clustering. Comput. Sci. Rev. **1**(1), 27–64 (2005). https://doi.org/10.1016/j.cosrev.2007.05.001

21. Shamir, R., Sharan, R., Tsur, D.: Cluster graph modification problems. Discrete Appl. Math. **144**(1–2), 173–182 (2004). https://doi.org/10.1016/j.dam.2004.01.007

22. Tomescu, I.: La reduction minimale d'un graphe à une reunion de cliques. Discrete Math. **10**(1–2), 173–179 (1974)
23. Wahid, D.F., Hassini, E.: A literature review on correlation clustering: cross-disciplinary taxonomy with bibliometric analysis. Oper. Res. Forum **3**, 47, 1–42 (2020). https://doi.org/10.1007/s43069-022-00156-6
24. Zahn, C.T.: Approximating symmetric relations by equivalence relations. J. Soc. Industrial Appl. Math. **12**(4), 840–847 (1964)

Limiting the Search in Brute Force Method for Subsets Detection

B. V. Kuprijanov[(✉)] [iD] and A.A. Roschin[iD]

V. A. Trapeznikov Institute of Control Sciences of Russian Academy of Sciences,
Moscow, Russian Federation
kuprianovb@mail.ru, rochinaa@ipu.ru

Abstract. A set of methods for limiting the complete enumeration of the number of combinations C_n^k for $1 \leq k \leq n$ is proposed. The proposed methods can be used in combinatorial algorithms for calculating subsets of the original set. They are applicable if the calculation of subsets is carried out using a selection function with certain properties. An example of using an algorithm to calculate the set of minimal sections of a graph is considered. It is shown that these methods can be used to solve a number of different NP-complete problems. The results of numerical experiments are presented.

Keywords: NP-complete problem · Graph · Limiting the number of combinations to iterate through · Search for subsets

1 Introduction

The theory and practice of combinatorial analysis shows that the generality of the algorithm and its effectiveness are in antagonistic contradiction. It can be argued that if the general algorithm A solves some wide class of problems Z, then there is a more efficient (in terms of speed or memory usage) algorithm A' that solves a subset of $Z' \subset Z$ problems. When developing such an algorithm, it is important that the class of problems Z' is meaningful in an applied sense. This article discusses NP-complete problems that are solved by a complete search of the number of combinations of C_n^k when k changes from 1 to n. For large dimensions, such a solution has more theoretical than practical significance. The limiting dimensions of the problems solved by this method can be somewhat expanded by the methods proposed in this article to limit the number of iterations and increase the efficiency of calculations. The problem of splitting a set into a system of subsets satisfying certain properties is, in general, an NP-complete problem. Examples of such tasks are: calculation of a set of vertices or edges that are the graph's minimum section [1]; calculation of the minimum dominant set [2], etc. Let there be a set X and the task is to find subsets of this set that have

---

Supported by V. A. Trapeznikov Institute of Control Sciences of Russian Academy of Sciences.

N. Olenev et al. (Eds.): OPTIMA 2023, LNCS 14395, pp. 329–343, 2023.
https://doi.org/10.1007/978-3-031-47859-8_24

some common property. The presence of the property is checked by the function (algorithm) $F(S)$, where $S \subset X$ is the subset in question. $F(S) = 1$ if the subset S has the required property, and 0 otherwise. Comparing the proposed methods with the numerous existing ones in the field of solving NP-complete problems, first of all, it is necessary to limit consideration only to those problems whose solution is associated with iterating over the number of combinations. One of the well-known such problems is the calculation of the minimum dominant set on a graph. Among the publications on the topic of algorithm improvement, two directions are distinguished: exact algorithms and approximation algorithms. The closest to the proposed method is the branch and bound method. However, the tasks under consideration have two features:

– the objective function is a predicate;
– solutions to the problem can be either minimal or redundant.

Redundant solutions are not of interest. The branch and bound method in the traditional form does not solve the problem of finding minimal solutions.

Of the known exact algorithms, the fastest one running in time $O(1.5048^n)$ was found in [3]. In this case, the authors have not found any publications devoted to improving the efficiency of the algorithm based on the restriction of a full search. Since the algorithm proposed in the article finds all the exact solutions, its comparison with approximation algorithms would not be correct. The article shows an example of the applicability of this method for solving at least four known NP-complete graph theory problems. For the problem of calculating the minimum sections in a graph, the article presents the results of numerical experiments.

The article has the following structure. In Sect. 2 the basic definitions are given, in Sect. 3 the statement of the problem is carried out. Section 4 describes a limited brute force algorithm and describes methods for reducing the number of iterations. Section 5 contains the results of numerical experiments. Section 6 is devoted to the generalization of the method, and Sect. 7 is the conclusion.

2   Definitions

Definition 1. *Denote $N = \{1, 2, \ldots\}$ as the set of natural numbers.*

Definition 2. *An undirected graph $G = (V, E)$ is a structure consisting of a set of vertices V and a set of edges E, i.e. pairs of the form (v, w), where $v, w \in V$.*

Definition 3. *A simple chain connecting two vertices of a graph is called a path connecting these vertices.*

Definition 4. *Let $X = \{x_1, x_2, \ldots, x_n\}$ be a finite set and $F : X^+ \to \{1, 0\}$ is a function that selects some subsets of X^+. Here X^+ is the set of all subsets of X without an empty set. $F(S)$ is called a selection function such that $S \subset X$ is a selected subset if $F(S) = 1$.*

Definition 5. *The selected set $S \subset X$ is called minimal if, when removing any element from S, the set ceases to be selected. In other words, if S is the minimum selected set, then*

$$(F(S) = 1) \wedge (\forall x \in S \rightarrow F(S\backslash\{x\}) = 0).$$

Definition 6. *The selected set S is called redundant if it is not minimal.*

Since the article will provide examples from the problem of finding sections in a graph, we will give the following definition.

Definition 7. *Let the graph G have two vertices a and b. The edge section (hereinafter simply the section) of the graph G with respect to these vertices is such a set of edges $S \subset E$ that there are no paths from a to b in the graph $G' = (V, E\backslash S)$.*

Further in the article the vertices a and b are considered fixed. This precondition simplifies the presentation and does not reduce the generality of the problem statement. $|S|$ is the size of the section.

3   Problem Statement

Given a finite set $X = \{x_1, x_2, \ldots, x_n\}$ and a selection function $F : X^+ \rightarrow \{1, 0\}$. The task is to calculate all subsets of the original set X selected by the function F. Obviously, this problem can be solved only by a complete search of all combinations of C_n^k for $k = 1, 2, \ldots, n$ and the total number of options is

$$N = \sum_{k=1}^{n} C_n^k = 2^n - 1.$$

It is reasonable to replace the task of calculating all selected subsets of $S \subset X$ with the task of calculating all minimal selected subsets of S. Further, methods are proposed to limit the considered number of combinations and increase the efficiency of calculations.

4   The Algorithm of Limited Search

This section describes the algorithm of a brute force search, describes the properties of the selection functions and provides appropriate methods of limiting the search, which significantly reduce the number of options under consideration.

4.1   Description of the Brute Force Search

Let X ($|X| = n$) be some finite set, and all its elements have numbers from 1 to n. The original set X is matched with the combinatorial tree G, in which each vertex label of the combinatorial tree is matched with the element number of the original set. A combinatorial tree is a binary tree and is constructed in the following sequence. The root vertex has a label 1. Next, each vertex with some label $i = 1, 2, \ldots, n - 1$ is complemented by two adjacent vertices with a label $i + 1$ by the vertical and horizontal connections. The vertex $i + 1$ with a horizontal connection will be called *next* to the vertex i, and the vertex with a vertical connection *alternative* to the vertex i. An example of a combinatorial tree for an initial set containing 5 elements is shown in Fig. 1.

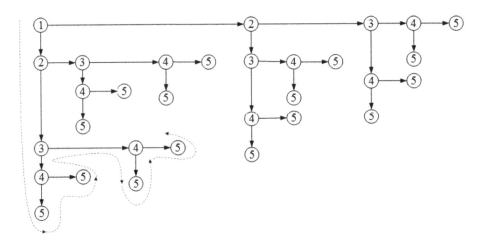

Fig. 1. An example of a combinatorial tree representation containing 5 edges.

Instead of sets of elements of the original set, we will consider sets of vertices of a combinatorial tree. There is a one-to-one correspondence between the vertex v_j of the combinatorial tree and the element x_j of the original graph. The tree G describes the iterations as follows. Each vertex of this tree uniquely corresponds to some path from the root vertex. The vertices traversed horizontally define the set of elements of the original set, and the vertices traversed vertically represent alternative values. The choice of a vertex on the vertical determines the subsequent horizon of the path. Figure 2 shows the examples of paths in a combinatorial tree are presented, corresponding sets of elements of the set (in curly brackets).

In the figure, the darkened vertices are alternative and do not generate the corresponding elements of subsets of X. Considering that the order of elements in the subset is not essential, we can make sure that all paths in the combinatorial tree describe all combinations of labels from one to five, i.e. all possible subsets of the set X.

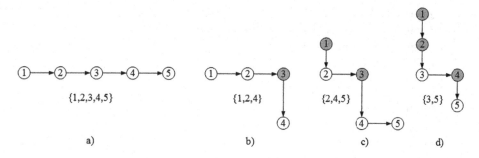

Fig. 2. Examples of paths of a combinatorial tree and subsets of their corresponding elements in the original set.

We can traverse the combinatorial tree by the rules "to the right and down" or "down and to the right". Further in the article, by default, the tree is supposed to be traversed down and to the right.

We have to define the function $F(S) : X^+ \to \{1,0\}$, which answers the question whether the set $S \subset X$ is a dedicated set.

Here we describe a general algorithm for generating a search for a given initial set X and the corresponding combinatorial tree $G = (V, E)$. Auxiliary variables are as follows:

T - combinatorial tree traversal stack;
i - stack pointer T;
S - stack of selected vertex numbers;
j - stack pointer S;
It is required to find all selected subsets of the set X.

The basic algorithm uses four recursive procedures: **Proc**, **Alter**, **Next** and **F**.

Algorithm 1. [(Algorithm for calculating selected subsets)]
Main procedure:
    Initialization: $i = 0$, $T(i) = 1$, $j = -1$.
   Proc
end.
Recursive procedure **Proc**:
  Alter
  F(S)
  Next
 end **Proc**.
Recursive procedure **Alter**:
   if $T(i) < n$ *then*
    $i = i + 1$
    $T(i) = T(i - 1) + 1$
    Proc
   end if
   $j = j + 1$

$$S(j) = T(i)$$
end **Alter**.
Recursive procedure **Next**:
    $if\ T(i) < n\ then$
        $i = i + 1$
        $T(i) = T(i-1) + 1$
        Proc
    $end\ if$
    $i = i - 1$
    $j = j - 1$
end **Next**.
End algorithm.

This algorithm, implemented through a set of recursive procedures, uses the stack T to traverse the vertices of the combinatorial tree and generates subsets using the stack S. The function $\mathbf{F}(S)$ calculates whether a subset of $\{x_{S(j)}, x_{S(j-1)}, \ldots x_{S(0)}\} \subset X$ is selected. If it is, then the algorithm registers the subset.

Next, the methods of limiting the number of iterations for this algorithm are considered.

4.2   Methods for Limiting the Number of Iterations and Improving the Efficiency of Calculations

In this section, we will consider the functions involved in the selection of subsets, each of which has a method associated with it that allows you to limit the search. These methods allow us to calculate all the minimum selected subsets and some of the redundant subsets. At the end of the section, a description of the algorithm of limited search is given.

Method 1. *Property* 1. Let the sets S_1 and S_2 be such that $S_1 \subset S_2 \subset X$ and the function $F(S)$ has the following property: if $F(S_1) = 1$ then $F(S_2) = 1$.

Consider the method of brute force reduction based on property 1. Let some path $\{v_1, v_2, \ldots, v_j\}$ be traversed in the combinatorial tree G and the corresponding set $S = \{x_1, x_2, \ldots, x_j\}$ is a selected subset in the set X, i.e. $F(S) = 1$. In this case, each further addition of vertices v_i for $i > j$ will generate subsets of the form $\{v_1, v_2, \ldots, v_j, v_{j+1}, \ldots v_q\}$, where $j < q \leq n$ i.e. find only redundant selected subsets. Therefore, traversing the tree G to the right can be stopped and the right subtree of the vertex v_j of the combinatorial tree can be cut off. Thus, the generation of redundant selected subsets of the set X will be stopped and the search will be reduced.

Method 2. *Property* 2. Define a function $g : X^+ \to N$ that has two properties:
1) if $S \subset X$, then $g(S) \leq g(S \cup \{x_i\})$, for any $x_i \in X$;
2) if $F(S) = 1$, then $g(S) = \max\limits_{D \subset (X \setminus S)} g(S \cup D)$.

The first property indicates that the function $g(S)$ is monotonous, and the second determines the relationship of the functions F and g. Such a relationship may look, for example, as follows:

$$F(S) = (g(S) = \max_{D \subset (X \setminus S)} g(S \cup D)).$$

We will assume that x_i is an empty complement to the set S if $g(S) = g(S \cup \{x_i\})$. For an empty complement, the following statement is true.

Statement 1. Given a set X and functions $F : X^+ \to \{0, 1\}$ and $g : X^+ \to N$ having the properties described above. If $S \subset X$ is such that $F(S) = 0$ and $x \in X$ is an empty complement to S $(g(S) = g(S \cup \{x\}))$, then $F(S \cup \{x\}) = 0$. The proof of this statement is simple and therefore not given.

Example 1. For example, consider the function g in the problem of finding a set of minimal sections in some graph $\mathbf{G} = (\mathbf{V}, \mathbf{E})$. Consider as a set X the set of edges $\mathbf{E}$ of this graph.

Deleting an edge in the graph $\mathbf{G}$ can cause blocking of one or more paths between two given vertices. Let the function g calculate the number of blocked paths.

If deleting a set of edges $S = \{e_1, e_2, \ldots, e_{j-1}\}$ blocks some set of paths and additional deletion of the edge e_j may lead to blocking new paths, then $g(S) < g(S \cup \{x_i\})$.

If deleting does not increase the set of blocked paths, then $g(S) = g(S \cup \{e_j\})$ and e_j is an empty complement to the set $\{e_1, e_2, \ldots, e_{j-1}\}$.

If all paths are blocked, then the set S is a section, the function $g(S)$ reaches a maximum, and $F(S) = 1$ defines the selected set.

Thus, it is possible to state: if $F(S) = 1$ then $g(S) = g(S \cup \{x_i\}) \ \forall x_i \in X$.

We formulate the following method in the form of a statement.

Statement 2. If the path $\{v_1, v_2, \ldots, v_j\}$ is traversed in the combinatorial tree G and the vertex v_j is an empty complement to $\{v_1, v_2, \ldots, v_{j-1}\}$, then the right subtree of the vertex v_j can be excluded from consideration and all the minimal selected subsets will still be calculated on the reduced tree. The number of redundant subsets may decrease in this case.

Proof. Suppose that in the process of traversing $\{v_1, v_2, \ldots, v_j\}$ the vertex v_j is an empty complement to the set $\{v_1, v_2, \ldots, v_{j-1}\}$. If the set $\{v_1, v_2, \ldots, v_{j-1}\}$ is not selected, then according to the statement 1 the set $\{v_1, v_2, \ldots, v_j\}$ is also not selected.

It is possible that there is such a set of vertices on the path $\{v_{j+1}, \ldots, v_q\}$ $(j < q \leq n)$ in the right subtree, which together with the initial set of vertices $\{v_1, v_2, \ldots, v_{j-1}, v_{j+1}, \ldots, v_q\}$ forms a selected set.

However, it is easy to note that in the alternative subtree v_j there will always be a path $\{v_{j+1}, \ldots, v_q\}$ $(j < q \leq n)$ and, therefore, when traversing the combinatorial tree down and to the right, this set will be calculated before discarding the right subtree of v_j.

Thus, no minimum selected set will be lost. The proof is complete.

4.3   The Algorithm of Limited Search

The algorithm is based on the following important property: on a combinatorial tree, when traversing it by the rule down and to the right, the minimum selected set is always found first.

Statement 3. Let the traversal of the combinatorial tree down and to the right reveals some set of selected subsets $\{S_i\}$ $(1 \leq i \leq q)$, such that

$$S_i = (\alpha_{i1}, Y_1, \alpha_{i2}, Y_2, \ldots, \alpha_{im}, Y_m, \alpha_{i(m+1)}),$$

where α_{ij} is a sequence of vertex labels (possibly empty) and Y_k are the vertex labels such that there is a minimal allocated set

$$S_j = (Y_1, Y_2, \ldots, Y_m).$$

In this case, the minimum allocated set S_j will always be the first detected in a given set of selected subsets.

Note the obvious fact that in any path in the combinatorial tree, the vertex numbers will be arranged in ascending order. Therefore, S_i will look on the graph like this – see Fig. 3.

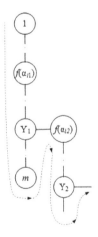

Fig. 3. A fragment of a combinatorial tree when traversing down and to the right.

We will denote $f(\alpha)$ as the first element of the chain α. In this case, because $f(\alpha_1) < Y_1$ and due to the traversal order, the vertex Y_1 will be considered first. When considering the set of alternatives following Y_1, Y_2 will also be considered before $f(\alpha_2)$, etc.

The statement is proven.

For example, if the minimum section is the set $\{3, 5\}$ (see Fig. 1), then the entire set of sections will be found in the following order:

$$\{3,5\}, \{3,4,5\}, \{2,3,5\}, \{2,3,4,5\}, \{1,3,5\}, \{1,3,4,5\}, \{1,2,3,5\}, \{1,2,3,4,5\}.$$

Taking into account the considered methods of search limitation, we describe the final complex algorithm for calculating the selected sets as a result of the modification of the algorithm 1.

To do this, we will introduce an additional object:

$\mathbf{g}(S(j))$ is a function with property 2, the argument of which is the set $\{S(j), S(j-1), \ldots, S(0)\}$, and the value is defined on the set N.

Algorithm 2. [(Algorithm for calculating selected sets)]
Main procedure:
    Initialization: $i = 0$, $T(i) = 1$, $j = -1$.
    Proc
end.
Recursive procedure **Proc**:
    Alter
    if $\mathbf{F}(S)$ *then*
        $i = i - 1$
        $j = j - 1$
        exit **Proc**
    end if
    if $(j > 0) \wedge (\mathbf{g}(S(j)) = \mathbf{g}(S(j-1)))$ *then*
        $i = i - 1$
        $j = j - 1$
        exit **Proc**
    end if
    Next
end **Proc**.
Recursive procedure **Alter**:
    if $T(i) < m$ *then*
        $i = i + 1$
        $T(i) = T(i-1) + 1$
        Proc
    end if
    $j = j + 1$
    $S(j) = T(i)$
end **Alter**.
Recursive procedure **Next**:
    if $T(i) < m$ *then*

$$i = i + 1$$
$$T(i) = T(i-1) + 1$$
Proc
end if
$$i = i - 1$$
$$j = j - 1$$
end **Next**.
End algorithm.

5   Results of Computational Experiments

During computational experiments, the problems of finding sections in several graphs, including graphs of transport and electrical networks, were solved. The two most representative graphs are shown in Fig. 4: graph of the IEEE 39 electrical network (left) and graph of the U22 transport network (right). The letters S and D indicate the source vertices and recipient vertices for which the sections were searched.

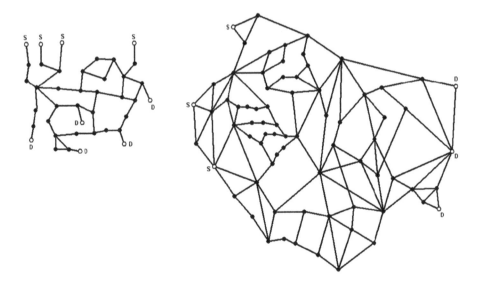

Fig. 4. Graphs IEE39 and U22 (from left to right).

Table 1 presents the results of computational experiments. During the experiments, the problems of searching for sections in the following graphs were solved:

- A conditional transport supply network (15 vertices, 17 edges) for two sources and three recipients without limiting the size of the section.

Table 1. Results of computational experiments.

Graph	Section size	Total combinations	Minimal sections	Numbering of edges	Combina- tions tested
Deliveries	≤ 17	131071	48	variant 1	15%
(15 nodes, 17 edges, 2 → 3)				variant 2	13%
Harary	≤ 16	16241060	209	variant 1	48%
(11 nodes, 24 edges, 1 → 1)				variant 2	43%
IEEE39	≤ 8	325374514	5214	variant 1	55%
(40 nodes, 46 edges, 4 → 5)				variant 2	43%
U22 (74 nodes, 131 edges,	≤ 6	6559349863	752	variant 1	29%
3 → 3, paths ≤ 9)				variant 2	53%

- The graph used as an example in [2] (11 vertices, 24 edges) for one source and one recipient with a section size of no more than 16 edges.
- The standard graph of the IEEE39 electrical network (40 vertices, 46 edges), in which 4 vertices were selected as sources and 5 as recipients, with a section size of no more than 8 edges.
- A real graph of a transport network (74 vertices, 131 edges) for three sources and three recipients with a limit of the maximum allowable path length to 9 transitions with a section size of no more than 6 edges.

The table shows the total number of possible combinations of edges that need to be checked for section when solving the problem by a brute force search, as well as the total number of minimum sections under given conditions of the problem. Since, when implementing a limited search, the construction of a combinatorial tree depends on the numbering order of the edges in the original graph, each experiment was conducted twice for two different ways of numbering the edges. The last column of the table shows the proportion of combinations of edges for which the function $F(S)$ was calculated (a check for the section was performed) during the limited search relative to the total number of combinations in full brute force search. The smaller this value is, the greater the gain is given by a limited search. It can be seen from the table that the gain increases when the restriction on the multiplicity (size) of the section is relaxed. This is due to the fact that the sizes of discarded subtrees in a combinatorial tree are the larger, the less this graph is bounded on the right by the allowable number of edges in the section.

With strong restrictions, the gain is smaller, but, on the other hand, the total number of combinations being checked also decreases.

It should also be noted that the gain may strongly depend on the numbering order of the edges – in the last experiment, there is almost a two-fold difference. If the edges of the original graph are numbered so that a sufficiently large number of sections are collected in the left part of the combinatorial tree, then the size of the discarded trees will be larger.

6   Generalization of the Applicability of Methods

This section will show the applicability of the described methods for solving some NP-complete problems. The methods are described in relation to sets, however, most of the known NP-complete problems are graph [4], therefore, graph problems are considered in the section and either the set of vertices or the set of edges of the graph act as the initial set.

Here are examples of tasks for which the proposed methods of reduction of brute force are applicable.

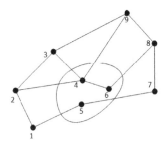

Fig. 5. Example of a graph and its dominating set.

Example 2. The problem of computing dominating sets [4]. In Fig. 5 an example of a graph and its smallest dominating set $S = \{4, 5, 6\}$ is given. If any vertex of the graph is added to the set S, then the set will remain dominating. The function $F(S)$, which checks whether the set belongs to the dominating one, will satisfy condition 1. We define $g(S)$ as a function that calculates the number of vertices of the graph that do not belong to S, but are adjacent to at least one vertex in S. In this case, the vertex $v \in V$ will be an empty complement to S if $g(S) = g(S \cup v)$. Therefore, methods 1 and 2 are applicable to this problem solution.

Example 3. One of the varieties of the problem of analyzing the vulnerability of the transport network [5] is reduced to the problem of maximizing the damage to the network described by the graph $\mathbf{G} = (\mathbf{V}, \mathbf{E})$. The formulation of this problem can be formulated as follows: find the set S for which

$$\max_{S \subset \mathbf{E}} (\sum_{i=1}^{n} \sum_{j=1}^{n} d_{ij}(c_{ij}^{S} - c_{ij}^{(0)})), \ |S| = q, \ \text{where}$$

S – the set of edges to be deleted,
$c_{ij}^{(0)}$ – the cost of travel along the shortest path (v_i, v_j) in the original graph,
c_{ij}^{S} - the cost of travel along the shortest path (v_i, v_j) in a graph with a set of deleted edges S,

d_{ij} – the need to travel along the path (v_i, v_j) which is an indicator of the amount of traffic flow along.

This problem can be reformulated as follows: find the minimum set S for which

$$\min_{S \subset \mathbf{E}} |S|$$

$$\sum_{i=1}^{n} \sum_{j=1}^{n} d_{ij} c_{ij}^{S} \geq \sum_{i=1}^{n} \sum_{j=1}^{n} d_{ij} c^{(0)} + D,$$

where D – some limit of exceeding the damage level.

In this case, the function $F(S)$ will look like this:

$$F(S) = \sum_{i=1}^{n} \sum_{j=1}^{n} d_{ij} c_{ij}^{S} \geq D_0,$$

where D_0 has the value

$$D_0 = \sum_{i=1}^{n} \sum_{j=1}^{n} d_{ij} c_{ij}^{(0)} + D$$

In general, such a problem is solved by a brute force search. Since removing an edge from the transport network increases damage, the function $F(S)$ will satisfy condition 1 and the function

$$g(S) = \sum_{i=1}^{n} \sum_{j=1}^{n} d_{ij} c_{ij}^{S}$$

will calculate the amount of damage, i.e. $F(S) = (g(S) \geq D_0)$.

Table 2 presents a list of known [4] NP-complete problems and the corresponding interpretations of the functions F and g.

Table 2. NP-complete problems and their corresponding functions F and g.

	Searching the graph for...	$F(S)$ detects that S is...	g calculates the count of...
1	a minimal section	a section	blocked paths
2	a dominating set	a dominating set	vertices adjacent to S
3	a minimal vertex cover	a vertex cover	edges incident to S
4	a maximum cut	a cut	edges in S

7   Conclusion

The methods considered in the article allow for a certain class of problems to limit the number of combinations when solving by a brute force search. The problem must meet the following conditions:

- the order of the objects in the combination is not important;
- redundant combinations are not of interest for the solution;
- the functions F and g with the properties described in the article are computable for any combination.

Such conditions often arise when solving various problems on graphs.

The recursive algorithms proposed in the article are described in accordance with the requirement of clarity and simplicity. In practice, the tree traversing can be implemented with a simpler and more efficient combination generator.

The method proposed in the article is based on traversing the combinatorial tree G in depth (DFS). To detect the selected minimal and redundant sets, the functions $F(S)$ and $g(S)$ must be calculated for each vertex in G.

To change this situation and after finding the minimum selected set $S = \{v_1, v_2, \ldots, v_j\}$ to calculate the remaining occurrences in the graph of the vertex v_j in an accelerated way, it is necessary to change the order of traversal of the combinatorial tree G. In this case, the graph should be traversed by width (BFS). It is known that such traversing is implemented using a queue (see Fig. 6).

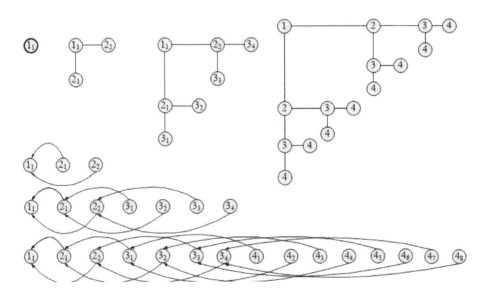

Fig. 6. Fragments of graph traversal steps by width and their corresponding queues.

It can be seen from the queue structure that all vertices of the same name lie in groups. In this case, if a section is found in some group at some position, then the remaining vertices of this group can be checked using an accelerated scheme. If the set $\{\alpha_1, v\}$ is a minimal selected set and there is a set $\{\alpha_2, v\}$ in the group such that $\{\alpha_1, v\} \subset \{\alpha_2, v\}$, then the set $\{\alpha_2, v\}$ is a redundant selected set.

With this method of traversal, the required memory grows at a rate of $c2^k - 1$, where k is the number of the vertex being processed and $c \leq 1$ is a certain queue reduction factor.

The proposed method of limiting the search is related to the subject set only by the number of elements of the set and the type of functions F and g. Choosing the numbering order of the elements of the set, for example edges, depending on the type of graph under study (numbering by the order of proximity to sources, by the number of paths containing an edge, etc.) will allow us to gain more – this issue requires further research.

References

1. Grebenuk, G., Nikishov, S.: Blocking of energy and resource supply of target objects in network infrastructures. Control Sci. **4**, 52–57 (2016)
2. Reinhard, D.: Graph Theory. Electronic Edition. Springer-Verlag, Heidelberg (2005) https://www.math.uni-hamburg.de/home/diestel/books/graph.theory/
3. van Rooij, J.M.M., Nederlof, J., van Dijk, T.C.: Inclusion/exclusion meets measure and conquer. In: Fiat, A., Sanders, P. (eds.) ESA 2009. LNCS, vol. 5757, pp. 554–565. Springer, Heidelberg (2009). https://doi.org/10.1007/978-3-642-04128-0_50
4. Garey, M., Johnson, D.: Computers and Intractability. W. H. Freeman and Company, United States (1979)
5. Taylor, M., D'Este, Glen M.: Transport network vulnerability: a method for diagnosis of critical locations in transport infrastructure systems. In: Critical infrastructure: Reliability and vulnerability, pp. 9–30. Springer, Berlin (2007). https://doi.org/10.1007/978-3-540-68056-7_2

Bicentered Interval Newton Operator for Robot's Workspace Approximation

Artem Maminov[1,2](✉) and Mikhail Posypkin[1,2]

[1] Federal Research Center "Computer Science and Control" of the Russian Academy of Sciences (FRC CSC RAS), 44/2, Vavilova street, 119333 Moscow, Russian Federation
artem_maminov@mail.ru

[2] HSE University, 20 Myasnitskaya Ulitsa, 101000 Moscow, Russian Federation

Abstract. The paper considers the approximation of the robot workspace. The developed method, based on interval Newton operator relies on Baumann bicentered theorem. We used this method for approximation of the solution sets of undetermined non-linear equation. This problem refers to the one of the most important problems in robotics: workspace approximation, since the robot kinematic systems are set with undetermined (usually non-linear) systems. We perform experiments for the DexTar robotic system and visualize the obtained approximations. As expected the bicentered modification provides tight approximation of the workspace compared with classical Newton method.

Keywords: Robot workspace · Parallel robot · Undetermined system of the equations · Newton method · Interval analysis

1 Introduction

Robotic systems have enabled significant advancements in various industries, including healthcare, manufacturing, and aerospace. However, these systems often face environmental uncertainties, such as changing operating conditions and unforeseen obstacles. Moreover, accurately determining the workspace of a robotic system is a critical step for motion planning and control, where the workspace refers to the set of all positions that the robot can reach while maintaining joint angle and velocity limits and avoiding obstacles.

A common approach to determining the workspace is to solve corresponding undetermined system of nonlinear equations where the kinematic constraints of the complicated robot are given. Traditional numerical methods, such as bisection and Newton's method, are often computationally expensive and time-consuming. Interval methods have emerged as a powerful alternative for approximating solutions of such nonlinear equations, which provides upper and lower bounds for the results besides the numerical approximation.

Interval methods offer several advantages, including guaranteed results, fast convergence rates, and robustness against uncertainties in the system's input parameters. Moreover, they can be used as a tool to optimize the robot workspace design. The Newton interval method, one of the most widely used interval root-finding algorithms, is commonly used in robotics to compute root approximations for workspace approximation by dividing the feasible roots into smaller intervals.

In this paper, we will focus on the Newton interval method and its bicentered modification for the approximation of the workspace of DexTar robotic systems. We will also explore the benefits of these methods and their potential for future developments in the field.

2   Related Works

The workspace of a robotic system is a fundamental challenge in robotics that involves finding the set of all positions that the robot can reach while avoiding obstacles, maintaining joint angle and velocity limits. Inaccurate approximation of the workspace can have profound consequences on robot functionality and safety.

Various techniques have been employed to approximate the workspace of robotic systems, including geometrical, discrete, and numerical methods. Geometrical methods generate the workspace through kinematic equations analysis and further geometrical interpretation of it [4]. Discrete methods, such as cell decomposition and visibility graphs, divide the configuration space into simple cells representing the workspace [3]. These methods may suffer from a lack of accuracy and are less efficient for high-dimensional systems.

Numerical methods for workspace approximation can be classified into deterministic and probabilistic methods. In Jo's approach [6], the constraints imposed by joint coordinates are carefully considered. By introducing additional variables, the author transformed the original inequality constraints into equality constraints. This technique is rooted in deterministic numerical methods that extensively analyze the Jacobi matrix and ensure the fulfillment of the robot's input coordinate limits.

Probabilistic methods, such as Monte Carlo simulations and Markov Chain Monte Carlo (MCMC) methods, offer a randomized approach to workspace approximation. In work [12] the authors employed MCMC methods to generate a dense sample of configuration space that can be used to approximate the workspace of robotic systems.

Interval methods, which use interval analysis to provide both numerical approximations and reliable bounds for solution sets, have gained significant attention in the field of robotics. These methods have proven to be a viable alternative to traditional numerical techniques for finding the workspace of robotic systems.

One of the most widely studied interval root-finding algorithms is the Newton interval method [14,15]. By subdividing the feasible roots for the system of nonlinear equations, it applies Newton's method iteratively within smaller intervals to improve the approximation's accuracy.

The use of interval methods in robotic workspace approximation has attracted significant research interest. Caro et al. [2] employed the interval iteration method to determine the manipulation workspace of parallel robots. They used a bounding process that computed the precise boundaries of the workspace. Additionally, Kumar et al. [9] proposed an algorithm based on the interval Newton method to find the reachable workspace of 3-degrees-of-freedom serial link redundant manipulator. The method's efficiency and accuracy were demonstrated by computing the reachable workspace of a prototype robotic arm.

Another workspace approximation method used interval analysis is based on nonuniform covering method [5]. This method was successfully applied for approximation of different robots working spaces [10, 16, 17]. But researches have to transform initial kinematic system into the inequalities, which requires additional efforts.

In our previous researches the Krawczyk method and its bicentered version [13] with different modifications [11] were used for 2-RPR and passive orthosis workspace approximation.

In this work we study the interval Newton method and its bicentered modification for workspace approximation of the new robotic system.

3   Problem Statement

The accurate approximation of the robot workspace is crucial for optimizing the performance of robotic systems. In this study, we focus on the DexTar robot, a versatile and high-performance robotic platform utilized in various industries and applications. To achieve accurate workspace approximation, it is imperative to understand and model the kinematic system of equations governing the DexTar robot's movement.

The DexTar robot (also can be called RRRRR) is a state-of-the-art manipulator designed for complex tasks, such as assembly, pick-and-place operations, and other industrial applications. Featuring multiple degrees of freedom, the DexTar robot provides exceptional dexterity and flexibility in its movements (see Fig. 1).

In Fig. 2 the scheme of the robot is presented. The structure of the robot includes 4 rods of constant length, as usually pairwise equal, i.e. $l_a = l_d$ and $l_b = l_c$. We consider the case, when the engines are located above the plane of the workspace area and do not affect it. Distance between anchor points bar is equal to d. The letters R in the name of the robot denote 5 rotational pairs (C, D, P, B, A), two of which are underlined, which means 2 drives (C and A), whose rotation determines the translational movement of the working body along the X and Y axes.

To accurately model and approximate the robot's workspace, it is necessary to define the kinematic system of equations that govern the robot's motion. These equations establish the relationship between the joint angles or joint displacements and the position and orientation of the robot's end-effector.

Fig. 1. Model of the DexTar robot

The original kinematic system of equations for the DexTar robot is represented as follows:

$$\begin{cases} x_p = l_b \cdot \cos(q_B) + l_a \cdot \cos(q_A) + \frac{d}{2}, \\ x_p = l_d \cdot \cos(q_D) + l_c \cdot \cos(q_C) - \frac{d}{2}, \\ y_p = l_a \cdot \sin(q_a) + l_b \cdot \sin(q_B), \\ y_p = l_d \cdot \sin(q_D) + l_c \cdot \sin(q_C), \end{cases} \tag{1}$$

With basic geometric transformations we can rewrite the system (1), since there are only two active angles, which control the working tool movement. The simplified System is the following:

$$F(x_p, y_p, q_A, q_D) = \begin{cases} (x_p + \frac{d}{2} - l_a \cdot \cos(q_A))^2 + (y_p - l_a \cdot \sin(q_A))^2 - l_b{}^2 \\ (x_p - \frac{d}{2} - l_d \cdot \cos(q_D))^2 + (y_p - l_d \cdot \sin(q_D))^2 - l_c{}^2 \end{cases} = 0 \tag{2}$$

The kinematic equations capture the complex transformations and constraints involved in the DexTar robot's motion. These equations take into account the geometric parameters, joint limits, and mechanical properties of the robot's structure. By solving the kinematic equations, we can determine the feasible joint configurations that correspond to different end-effector positions and orientations.

As you can see the system (2) is undetermined (the number of variables is more then number of equations). Thus, the solution for such system is not unique

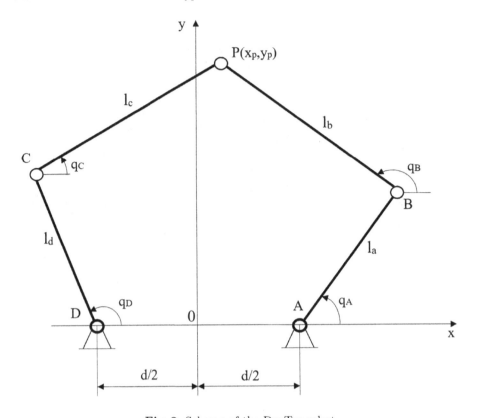

Fig. 2. Scheme of the DexTar robot

and formed by sets. The workspace of the robotic system is formed by such set and in fact all points, which satisfies system (2).

For the solution of the such systems we use interval analysis methods. Thus, we consider the problem of solving undetermined system of interval nonlinear equations. Let us rewrite in more appropriate form:

$$F(u, v) = 0, \tag{3}$$

where u, v are vectors in the spaces $\mathbb{R}^m$, $\mathbb{R}^n$, respectively, and $F : \mathbb{R}^{m+n} \to \mathbb{R}^n$ is a continuously differentiable mapping. Vector u denotes the passive variables of the system (in our case the position of the point P), and vector v denotes the active variables of the system (in our case the rotational angles q_A and q_D). Given a box $U = [\underline{u}_1, \overline{u}_1] \times \cdots \times [\underline{u}_m, \overline{u}_m] \subseteq \mathbb{R}^m$ and a box $V = [\underline{v}_1, \overline{v}_1] \times \cdots \times [\underline{v}_n, \overline{v}_n] \subseteq \mathbb{R}^n$, let us define the *solution set* Ω as

$$\Omega = \{u \in U \subseteq \mathbb{R}^m \mid \exists v \in V \subseteq \mathbb{R}^n \text{ such that } F(u, v) = 0\}, \tag{4}$$

i.e. as the set formed by all such parameters $u \in U$ that there exists a solution v for the equations system $F(u, v) = 0$.

For our problem we have box $U = [\underline{u}_1, \overline{u}_1] \times [\underline{u}_2, \overline{u}_2]$, which limits the workspace of the mechanism. The box $V = [\underline{v}_1, \overline{v}_1] \times [\underline{v}_2, \overline{v}_2]$ limits the angles q_A and q_D accordingly:

$$F(u, v) = \begin{cases} (u_1 + \frac{d}{2} - l_a \cdot \cos(v_1))^2 + (u_2 - l_a \cdot \sin(v_1))^2 - l_b{}^2 \\ (u_1 - \frac{d}{2} - l_d \cdot \cos(v_2))^2 + (u_2 - l_d \cdot \sin(v_2))^2 - l_c{}^2 \end{cases} = 0 \quad (5)$$

4   Theoretical Results and Algorithms

4.1   Preliminaries and Notation

In this work we use interval analysis technique [8,14,15]. Interval analysis is a mathematical tool that plays a pivotal role in the computation and manipulation of uncertainty in numerical computing. It offers a rigorous framework for working with intervals, which are sets of real numbers bounded by lower and upper bounds.

Interval $u \subseteq \mathbb{R}$ is a set of real numbers enclosed between its left and right endpoints, that is, $u = \{u \in \mathbb{R} \mid \underline{u} \leq u \leq \overline{u}\}$. The set of all intervals is denoted as $\mathbb{IR}$. The direct Cartesian product of n intervals is defined by an n-dimensional *box*, and the set of such boxes is denoted by $\mathbb{IR}^n$. In our work, we use boldface font to denote intervals and boxes, while normal font denotes real numbers and vectors.

Arithmetic operations can be defined between intervals, and thus we get various interval arithmetic. In our work, we use the extended Kahan interval arithmetic [7] . This type of interval computations provides the rules for division by any interval-number containing zero.

The four fundamental interval mathematical operations include addition, subtraction, multiplication, and division. For two intervals $x = [\underline{x}, \overline{x}]$ and $y = [\underline{y}, \overline{y}]$, the basic interval operations can be defined as follows:

$$x + y = [\underline{x} + \underline{y}, \ \overline{x} + \overline{y}], \qquad x - y = [\underline{x} - \overline{y}, \ \overline{x} - \underline{y}],$$

$$x \cdot y = [\min\{\underline{x}\,\underline{y}, \underline{x}\,\overline{y}, \overline{x}\,\underline{y}, \overline{x}\,\overline{y}\}, \ \max\{\underline{x}\,\underline{y}, \underline{x}\,\overline{y}, \overline{x}\,\underline{y}, \overline{x}\,\overline{y}\}],$$

$$x/y = x \cdot [1/\overline{y}, \ 1/\underline{y}] \qquad \text{for } y \not\ni 0,$$

$$x/y = \begin{cases} [\overline{x}/\underline{y}, \inf], & \text{if } \overline{y} = 0 \text{ and } \overline{x} \leq 0 \\ [[-\inf, \overline{x}/\overline{y}], [\overline{x}/\underline{y}, \inf]], & \text{if } \overline{x} < 0 \text{ and } y \ni 0 \\ [-\inf, \overline{x}/\overline{y}], & \text{if } \overline{x} \leq 0 \text{ and } \underline{y} = 0 \\ [[-\inf, \underline{x}/\underline{y}], [\underline{x}/\overline{y}, \inf]], & \text{if } \underline{x} > 0 \text{ and } y \ni 0 \\ [\underline{x}/\overline{y}, \inf], & \text{if } \underline{x} \geq 0 \text{ and } \underline{y} = 0 \\ [-\inf, \inf], & \text{if } x \ni 0 \text{ and } y \ni 0 \end{cases}$$

These operations provide the means to compute with intervals, allowing for the encapsulation of uncertainties and the determination of robust bounds on calculations. According to Kahan arithmetic we also define the special cases of interval division by interval containing zeros. In most cases we get default interval, but we also can get two intervals if the dividend is fully positive or fully negative interval. Such intervals are called "multi-intervals".

In addition to basic interval mathematical operations, interval analysis also facilitates the evaluation of functions over intervals. The extension of functions to intervals provides valuable insights into the behavior and characteristics of these functions over a range of inputs.

To utilize basic functions, such as the power function or logarithmic function, in interval calculations, it is necessary to redefine them by considering their monotonicity properties. By understanding the behavior of these functions over intervals, we can determine their range of values in a straightforward manner.

For monotone functions of a single variable, determining the range of values is relatively simple. If we have a function $f : \mathbb{R} \to \mathbb{R}$ that is monotonic over an interval $a = [\underline{a}, \overline{a}]$, we can obtain the range of values by computing the function's values at the endpoints of the interval. Specifically, for a monotonically increasing function, the range of values can be expressed as $f(a) = [f(\underline{a}), f(\overline{a})]$. Similarly, for a monotonically decreasing function, the range of values is given by $f(a) = [f(\overline{a}), f(\underline{a})]$.

In our case we use power and trigonometric functions in system (5). For power function we can distinguish two cases: odd powers and even powers. For odd powers the function is monotonically increasing on the whole domain of the definition. But in case of even powers we should check if the interval includes negative numbers.

Similarly, trigonometric sine and cosine functions do not exhibit monotonic behavior over the entire real number axis. Therefore, it becomes necessary to determine if the interval under consideration contains the minimum and maximum values of the sine and cosine functions. These critical values occur at $-\frac{\pi}{2}, \frac{\pi}{2}$ and $0, \pi$ for the sine and cosine functions respectively.

By taking into account the monotonicity properties and critical values of these functions, we can accurately redefine them for interval calculations. Through this approach, we ensure that the interval arithmetic and computations involving these functions yield valid and meaningful results.

For a given function $f(x)$, where x represents a real variable, the extension of f to intervals can be denoted as $f(X)$, where $X = [a, b]$ is an interval. The interval extension $f(X)$ of the function $f(x)$ over the interval X is defined as:

$$f(X) = \{f(x) : x \in X\}$$

The interval extension of a function enables the computation of rigorous bounds on the function values over the interval domain. By evaluating the function at the endpoints of the interval and considering the possible variation within the interval, a tight enclosure of the function's range can be obtained.

Interval extensions are techniques employed in interval analysis to approximate functions that are not explicitly defined over intervals. These extensions aim to provide a conservative, yet reasonably tight approximation of the function values within the interval domain.

4.2   Interval Newton Method

The Newton method is well-known method for finding zeros of the equations. Lets consider the function $f : R \supseteq X \rightarrow R$, where f has some solution x^* on the interval X and differentiable on it. The classical Newton method can be written as follows:

$$x^* = \hat{x} - \frac{f(\hat{x})}{f'(\epsilon)},$$

where $\hat{x} \in X$ and ϵ is the point between $\hat{x}$ and x^*.

If $f'(X)$ is some interval estimation of the derivative of the function $f(x)$ on X, then $f(\epsilon) \in f(X)$ and, by intervalizing the written equality, we obtain the inclusion:

$$x^* \in \hat{x} - \frac{f(\hat{x})}{f'(X)} \tag{6}$$

Taking the right part of the Eq. (6) we obtain the interval Newton operator:

$$N(X, \hat{x}) = \hat{x} - \frac{f(\hat{x})}{f'(X)} \tag{7}$$

4.3   Bicentered Interval Newton Method

The modification of Newton operator based on the Baumann theorem [1] can give tighter bounds. Such modification was proposed in [18] for Krawczyk method and used for 2-RPR robot [13] and passive part of lower limb rehabilitation system [11]. Denote $d^{(i)} = \natural f'(X)$, $d^{(i)} \in \mathbb{IR}$. According to the Baumann theorem,

$$f(X) \subseteq \tilde{f}\left(X, \hat{c}^{(i)}\right) \cap \tilde{f}\left(X, \check{c}^{(i)}\right), \tag{8}$$

$$\hat{c}^{(i)} = \begin{cases} \overline{X}, & \text{if } \overline{d}^{(i)} \leq 0, \\ \underline{X}, & \text{if } \underline{d}^{(i)} \geq 0, \\ \dfrac{\underline{d}^{(i)} \underline{X} - \overline{d}^{(i)} \overline{X}}{\underline{d}^{(i)} - \overline{d}^{(i)}}, & \text{if } \underline{d}^{(i)} < 0 < \overline{d}^{(i)}, \end{cases}$$

$$\check{c}^{(i)} = \begin{cases} \overline{X}, & \text{if } \overline{d}^{(i)} \leq 0, \\ \underline{X}, & \text{if } \underline{d}^{(i)} \geq 0, \\ \dfrac{\overline{d}^{(i)} \underline{X} - \underline{d}^{(i)} \overline{v}}{\overline{d}^{(i)} - \underline{d}^{(i)}}, & \text{if } \underline{d}^{(i)} < 0 < \overline{d}^{(i)}. \end{cases}$$

We define the *bicentered Newton operator* $N_{bic}(u, v)$ as follows:

$$N_{bic}(\boldsymbol{X}) = \boldsymbol{N}\left(\boldsymbol{X}, \hat{c}^{(i)}\right) \cap \boldsymbol{N}\left(\boldsymbol{X}, \check{c}^{(i)}\right) \qquad (9)$$

where $\boldsymbol{N}\left(\boldsymbol{X}, c^{(i)}\right)$ is the classical Newton operator.

4.4 Algorithms Description

In the Algorithm 1, the box $\mathbf{U}$ is divided into smaller boxes using a uniform grid with k nodes in each dimension. This partitioning results in the formation of $\mathbf{U}$ as the union of smaller boxes, denoted as $\mathbf{U}^{(i)}$ for i ranging from 1 to m^k. These computed boxes are stored in a list L. For every element $\mathbf{U}^{(i)}$ in the list L, the procedure checkBox is executed (lines 3–4). Depending on the return value of this procedure, which can be IN or $INDET$, the corresponding box is added to either the list S_{in} or the list S_b respectively.

Lists S_{in} and S_b constructed by the algorithm satisfy the following condition: $\cup_{u \in S_{in}} \subseteq \Omega \subseteq \cup_{u \in S_{in} \cup S_b} \subseteq U$ providing that procedure checkBox correctly classifies the box, where Ω is the solution set: $\Omega = \{u \in U \subseteq \mathbb{R}^m \mid \exists v \in V \subseteq \mathbb{R}^n$ such that $F(u,v) = 0\}$. However the algorithm does not guarantee the tightness of the constructed approximation. The collection S_b can contain boxes having no points from the boundary of a set Ω. Let us consider checkBox in detail.

Algorithm 1. Cover algorithm with enlargement input box

U, V — bounding boxes for u and v, max_depth — the maximum depth of the recursion

1: Construct a uniform meshes $L_u = \left\{\boldsymbol{U}^{(i)}\right\}$, $\boldsymbol{U} = \cup_{i=1}^{i=m^k} \boldsymbol{U}^{(i)}$
2: Initialize lists $S_{in} = \{\}, S_b = \{\}$
3: **for all** $\boldsymbol{u} \in L_u$ **do**
4:      $r := \text{checkBox}(\boldsymbol{u}, \boldsymbol{V}, max\_depth)$
5:      **if** $r = IN$ **then**
6:          $S_{in} := S_{in} \cup \boldsymbol{u}$
7:      **else if** $r = INDET$ **then**
8:          $S_b := S_b \cup \boldsymbol{u}$
9:      **end if**
10: **end for**
11: **return** S_{in}, S_b

The Algorithm 2 is based on intermediate value theorem: if continuous function has values of opposite signs inside an interval, then it has a root in that interval. Note, that we need to proof only existence of the root on the interval v. Thus, we check signs on the ends of the input interval (lines 3–4) and if it is opposite, we can conclude, that box u has root on v. Obviously, if the natural interval extension does not contain zero, we can conclude, that there is no roots (lines 5–6).

Algorithm 2. Check box

---

u, v — boxes, $depth$ — the depth of the recursion
1: **if** $depth = 0$ **then**
2:     **return** $INDET$
3: **else if** $f(u, [\overline{v}, \overline{v}]) \leq 0 \leq f(u, [\underline{v}, \underline{v}])$ **or** $f(u, [\underline{v}, \underline{v}]) \geq 0 \geq f(u, [\overline{v}, \overline{v}])$ **then**
4:     **return** IN
5: **else if** $\underline{f(u, v)} > 0$ **or** $\overline{f(u, v)} < 0$ **then**
6:     **return** $\overline{OUT}$
7: **else**
8:     $v' := G(u, v)$
9:     $v'' = v' \cap V$
10:     **if** $v'' = \varnothing$ **then**
11:         **return** OUT
12:     **else**
13:         **return** checkBox$(u, v'', depth - 1)$
14:     **end if**
15: **end if**

If none of these conditions are satisfied, we compute an interval extension $G(u, v)$, which results in a box v'. In our work we test two methods: classical interval Newton extension and bicentered Newton extension. Then, we compute the intersection $v'' = v' \cap v$. If it is empty, then the box u lies outside the solution set Ω. At the final step, the procedure is called recursively (line 13) with a changed box v''. When we achieve the maximal depth of the recursion, we can not classify box into outside or inside boxes, so we return $INDET$, what in fact means, that the box is boundary.

5   Experimental Results

For the computational experiments we used system (5), constant parameters $l_a = l_d = 8$, $l_b = l_c = 5$, $d = 9$, and the whole trigonometric axe for input angles limits: $v_1 = v_2 = [-\pi, \pi]$. It's obvious, that for given ranges the workspace lies within the region $U = [-20, 20] \times [-20, 20]$. In the Fig. 3 the workspace approximation of the DexTar robot for classical interval method (a) and for bicentered Newton method (b) is presented. The red rectangular denotes initial box U, the analytical workspace is shown with black lines. The approximation is illustrated with green and yellow boxes, where green boxes — inside boxes, and yellow boxes — border boxes. The experiment was done for two different uniform grids (32×32 (top figures) and 64×64 (bottom figures)). The bicentered Newton method generates tight approximations, while those produced by the classical Newton method contain redundant boxes.

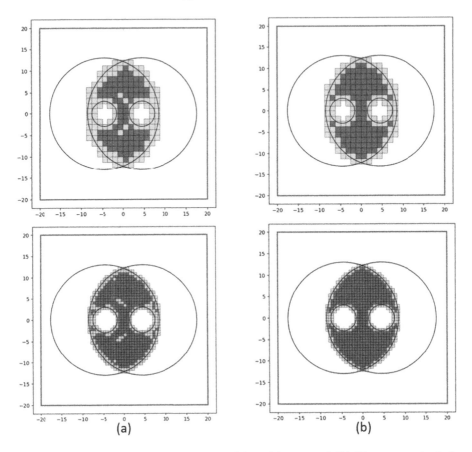

Fig. 3. The results produced by classical (a) and bicentered (b) Newton methods for uniform grids of sizes 32×32 (top figures) and 64×64 (bottom figures) for DexTar robot

6   Conclusions

This paper presents a method for constructing inner and outer approximations of solution sets for indeterminate systems of non-linear equations. Specifically, we compare two approaches: the classical Newton method and a novel technique based on the Baumann bicentered form. The experiments were done on DexTar robotic system, where the modificated bicentered Newton method showed its superiority.

Currently, the computational cost associated with our technique is relatively high. However, this can be mitigated by utilizing high-performance computing methods, such as parallel computing, to enhance efficiency. Additionally, there is potential for further research on various robotic systems, exploring different applications of our modification technique.

Acknowledgment. The research was carried out using the infrastructure of the Shared Research Facilities "High Performance Computing and Big Data" (CKP "Informatics") of FRC CSC RAS (Moscow).

References

1. Baumann, E.: Optimal centered forms. BIT Numer. Math. **28**(1), 80–87 (1988)
2. Caro, S., Chablat, D., Goldsztejn, A., Ishii, D., Jermann, C.: A branch and prune algorithm for the computation of generalized aspects of parallel robots. Artif. Intell. **211**, 34–50 (2014)
3. Chablat, D., Wenger, P.: Moveability and collision analysis for fully-parallel manipulators. arXiv preprint arXiv:0707.1957 (2007)
4. Chen, Y., Han, X., Gao, F., Wei, Z., Zhang, Y.: Workspace analysis of a 2-dof planar parallel mechanism. In: International Conference of Electrical, Automation and Mechanical Engineering, pp. 192–195 (2015)
5. Evtushenko, Y., Posypkin, M., Rybak, L., Turkin, A.: Approximating a solution set of nonlinear inequalities. J. Global Optim. **71**(1), 129–145 (2018)
6. Jo, D.Y., Haug, E.J.: Workspace analysis of closed loop mechanisms with unilateral constraints. In: International Design Engineering Technical Conferences and Computers and Information in Engineering Conference. vol. 3691, pp. 53–60. American Society of Mechanical Engineers (1989)
7. Kahan, W.M.: A more complete interval arithmetic. Lecture notes for an engineering summer course in numerical analysis, University of Michigan **4**, 31 (1968)
8. Kearfott, R.B.: Rigorous Global Search: Continuous Problems. Springer, Cham (2013)
9. Kumar, V., Sen, S., Roy, S., Das, S., Shome, S.: Inverse kinematics of redundant manipulator using interval newton method. Int. J. Eng. Manuf. **5**(2), 19–29 (2015)
10. Malyshev, D., Nozdracheva, A., Dubrovin, G., Rybak, L., Mohan, S.: A numerical method for determining the workspace of a passive orthosis based on the RRRR mechanism in the lower limb rehabilitation system. In: Pisla, D., Corves, B., Vaida, C. (eds.) EuCoMeS 2020. MMS, vol. 89, pp. 138–145. Springer, Cham (2020). https://doi.org/10.1007/978-3-030-55061-5_17
11. Maminov, A., Posypkin, M.: Robot workspace approximation with modified bicenetred krawczyk method. In: Olenev, N., et al. (eds.) OPTIMA 2022. LNCS, vol. 13781, pp. 238–249. Springer, Cham (2022). https://doi.org/10.1007/978-3-031-22543-7_17
12. Maminov, A.D., Posypkin, M.A.: Research and developing methods of solving engineering optimization problems for parallel structure robots. Int. J. Open Inf. Technol. **7**(11), 1–7 (2019)
13. Maminov, A.D., Posypkin, M.A., Shary, S.P.: Reliable bounding of the implicitly defined sets with applications to robotics. Procedia Comput. Sci. **186**, 227–234 (2021)
14. Moore, R.E., Kearfott, R.B., Cloud, M.J.: Introduction to interval analysis. SIAM (2009)
15. Neumaier, A.: Interval Methods for Systems of Equations. Cambridge University Press, Cambridge (1990)
16. Posypkin, M.: Automated robot's workspace approximation. J. Phys. Conf. Ser. **1163**, 012050 (2019)

17. Rybak, L., Gaponenko, E., Malyshev, D.: Approximation of the workspace of a cable-driven parallel robot with a movable gripper. In: Hernandez, E.E., Keshtkar, S., Valdez, S.I. (eds.) LASIRS 2019. MMS, vol. 86, pp. 36–43. Springer, Cham (2020). https://doi.org/10.1007/978-3-030-45402-9_5
18. Shary, S.P.: Krawczyk operator revised. In: Proceedings of International Conference on Computational Mathematics ICCM-2004, Novosibirsk, Russia, June 21–25, 2004. pp. 307–313. Institute of Computational Mathematics and Mathematical Geophysics (ICM&MG) (2004)

Statistical Performance of Subgradient Step-Size Update Rules in Lagrangian Relaxations of Chance-Constrained Optimization Models

Charlotte Ritter[1] and Bismark Singh[2](✉)

[1] Department of Mathematics, Friedrich-Alexander-Universität Erlangen-Nürnberg, Erlangen 91058, Germany
[2] School of Mathematical Sciences, University of Southampton, Southampton SO17 1BJ, UK
b.singh@southampton.ac.uk

Abstract. Lagrangian relaxation schemes, coupled with a subgradient procedure, are frequently employed to solve chance-constrained optimization models. Subgradient procedures typically rely on step-size update rules. Although there is extensive research on the properties of these step-size update rules, there is little consensus on which rules are most suitable practically; especially, when the underlying model is a computationally challenging instance of a chance-constrained program. To close this gap, we seek to determine whether a single step-size rule can be statistically guaranteed to perform better than others. We couple the Lagrangian procedure with three strategies to identify lower bounds for two-stage chance-constrained programs. We consider two instances of such models that differ in the presence of binary variables in the second-stage. With a series of computational experiments, we demonstrate—in marked contrast to existing theoretical results—that no significant statistical differences in terms of optimality gaps is detected between six well-known step-size update rules. Despite this, our results demonstrate that a Lagrangian procedure provides computational benefit over a naive solution method—regardless of the underlying step-size update rule.

Keywords: Chance constraints · Statistical guarantees · Lagrangian decomposition · Subgradient · Progressive hedging

Supplementary Information The online version contains supplementary material available at https://doi.org/10.1007/978-3-031-47859-8_26.

1   Introduction

Consider the generic formulation of a two-stage stochastic linear program with chance constraints:

$$\mathcal{Z} = \max_{x,y} \sum_{t \in T} R_t x_t - \mathbb{E}\left[C_t^\omega y_t^\omega\right] \tag{1a}$$

$$\text{s.t. } \mathbb{P}\left(x_t \le y_t^\omega + u_t^\omega, \forall t \in T\right) \ge 1 - \varepsilon \tag{1b}$$

$$x_t \ge 0, y^\omega \in Y^\omega, \forall t \in T, \omega \in \Omega. \tag{1c}$$

Here, x_t is a first-stage decision that provides a reward $R_t, \forall t \in T$. Following this decision, the uncertainty is realized via a scenario $\omega \in \Omega$. Next, we make a so-called recourse, or second-stage, decision y_t^ω while paying a cost C_t^ω, respectively $\forall t \in T, \omega \in \Omega$. Then, the objective function in Eq. (1a) seeks to maximize the overall profit by rewarding first-stage decisions and penalizing the expected cost of second-stage decisions. Constraint (1b) is a typical joint chance constraint (JCC); here, u_t^ω is data that is revealed after the uncertainty ω is realized. The JCC ensures that the joint probability of satisfying this inequality is no less than $1 - \varepsilon$, where $\varepsilon > 0$ is a risk threshold. Constraint (1c) enforces a non-negativity restriction for the x variables, and that the y variables belong to some set Y, $\forall \omega \in \Omega$.

Chance constraints were first studied by Charnes and Cooper [6], and have been actively investigated in the past few decades. One-stage chance-constrained programs, that allow only non-adaptive decisions, are relatively easier to solve as strong integer programming reformulations resulting from strong linear programming (LP) relaxations are available, see, e.g., [2,13]. Two-stage chance-constrained programs, where the y variables adapt to the scenario that is realized, require specialized algorithms for their solution since direct methods are often computationally intractable. Examples of such methods include decomposition methods [15], heuristics [30], or approximations [27]. Model (1) is an example of a typical two-stage chance-constrained model that requires the use of similarly tailored solution methods. Further, the set Y can include integer variables, and coupling this set with a JCC leads to an even greater requirement of computational effort [26].

The JCC in constraint (1b) can be reformulated with the following non-convex feasible region.

$$x_t - y_t^\omega \le u_t^\omega + M_t^\omega z^\omega, \forall t \in T, \omega \in \Omega \tag{2a}$$

$$\sum_{\omega \in \Omega} p^\omega z^\omega \le \varepsilon \tag{2b}$$

$$z^\omega \in \{0,1\}, \forall \omega \in \Omega. \tag{2c}$$

Here, M_t^ω is large enough such that Eq. (2a) is vacuous when $z^\omega = 1$. Constraint (2b) is a *knapsack* constraint. For equally likely scenarios, this constraint reduces to $\sum_{\omega \in \Omega} z^\omega \le \lfloor |\Omega|\varepsilon \rfloor$, where $\lfloor \cdot \rfloor$ rounds its argument down to the nearest integer. With the reformulation of the JCC as in Eq. (2), model (1) can be directly solved using a commercial mixed-integer programming (MIP) solver; however, the computational effort can be prohibitive. Due to the binary z variables, model (1) is non-convex.

This work investigates the computational performance of a series of procedures that provide lower and upper bounds for model (1) with the JCC represented as Eq. (2). For lower bounds of model (1), we summarize, extend, and compare three methods based on *regularization, aggregation,* and *sorting* techniques, respectively. For upper bounds of model (1), we use a Lagrangian relaxation procedure that we solve with a subgradient algorithm using six step-size update rules. The theory of these procedures is largely developed; however, as we show later in this work, this theory is not always reflected consistently in practice. Specifically, differences arise since in practice we solve models with a given time limit; and, the theoretical "best" bounds are not always achieved in this time limit. To this end, we are interested in the following two questions:

Q1 Given a computationally intractable chance-constrained program, is there any merit for a bounding procedure that is also computationally challenging?
Q2 Which bounding procedures are the most effective in practice and when?

Questions Q1 and Q2 are fundamental when facing a choice of bounding procedures for chance-constrained optimization models. We provide a set of guidelines that assist in this choice. The following are the key contributions of this work.

(i) Using six well-known step-size rules, we empirically demonstrate that a single rule cannot be statistically validated as the best performer for a chance-constrained optimization model within a sufficiently large time limit.
(ii) Despite only conservative upper bounds for model (1) resulting from Lagrangian procedures, there is improvement over a naive solution method no matter which step-size rule is employed. For example, for one instance class, we reduce optimality gaps up to 20.9% on average.
(iii) For lower bounds of model (1), we investigate three methods based on regularization, aggregation, and sorting techniques, that improve the objective function obtained naively by 16.2%, 0.5%, and 10.3%, respectively, even when using only half or less of the time consumed naively.
(iv) We demonstrate that regularization with aggregated scenarios does not always benefit over regularization with independent identically distributed (i.i.d) scenarios. In the absence of customized aggregation procedures, choosing i.i.d. scenarios for the SAA of model (1) is advisable.
(v) Finally, we show that good lower and upper bounds used together—even when the bounding procedures are terminated prematurely—result in smaller optimality gaps than those obtained naively. In our instances, average gaps reduce to 30.3% from 37.0% obtained naively.

In the interest of space, we reserve results of all our computational experiments for two appendices that are available from our GitHub page that we mention below. For a better understanding of this work, we recommend reading the appendix together with the article.

2 Upper Bounds for Model (1)

2.1 Lagrangian Relaxation

Relaxing constraint (1b) via a Lagrangian dual procedure results in the following model that we call as the dual model for model (1).

$$\mathcal{Z}_D(\lambda) = \max_{x,y,z} \sum_{t \in T} R_t x_t - \mathbb{E}[C_t^\omega y_t^\omega] + \lambda\big(\varepsilon - \sum_{\omega \in \Omega} p^\omega z^\omega\big) \tag{3a}$$

$$\text{s.t. } (1c), (2a), (2c). \tag{3b}$$

Although model (3) is still non-convex, we require just a single dual variable λ. We have the following basic bounding principle (BBP): $\mathcal{Z} \leq \mathcal{Z}_D(\lambda), \forall \lambda \geq 0$. The BBP follows since an optimal solution for model (1) is feasible for model (3), and with a non-negative λ the objective function value in equation (3a), is more than that of equation (1a); see, e.g., [5]. Given an optimal solution (x, y, z) of model (3) for a $\lambda \geq 0$, the BBP holds with equality, $\mathcal{Z} = \mathcal{Z}_D(\lambda)$, if the following two sufficient conditions are met: (i) $\varepsilon \geq \sum_{\omega \in \Omega} p^\omega z^\omega$, and (ii) $\lambda\big(\varepsilon - \sum_{\omega \in \Omega} p^\omega z^\omega\big) = 0$. Then, the Lagrangian dual solves model (1).

However, even for LP models, these conditions rarely hold true in practice. For a numerical solution of MIP models—such as model (1) and model (3)—we require the use of an optimization solver, e.g., CPLEX or Gurobi, that works by branching. The obtained solutions are either provably optimal up to a tolerance, or are the best known feasible solutions in a given amount of time. In this time limit, if the dual problem is solved to optimality, the BBP holds (although not necessarily with equality). Further, this is true in the presence of integer variables as well. If we have a lower bound for model (1) available, we can construct a range for the optimal objective function value of model (1). This is true even if the optimization models for the lower bounding procedure are solved suboptimally. However, if the dual problem is solved suboptimally, only a weak, or conservative, bounding procedure results; see Example S1 in Appendix A for an illustration.

Let $\underline{\mathcal{Z}_D}(\lambda)$ and $\overline{\mathcal{Z}_D}(\lambda)$ denote the lower and upper bounds of the dual model reported by the optimization solver. Thus, we modify the BBP conservatively as $\mathcal{Z} \leq \overline{\mathcal{Z}_D}(\lambda), \forall \lambda \geq 0$, and denote it as the modified basic bounding procedure (mBBP). Then, smaller values of $\overline{\mathcal{Z}_D}(\lambda)$, or $\mathcal{Z}_D(\lambda)$, are of particular interest to us. Solving the one-dimensional non-convex optimization problem

$$\overline{\mathcal{Z}} = \min_{\lambda:\lambda \geq 0} \mathcal{Z}_D(\lambda) \tag{4}$$

achieves this, and we do so via the classical subgradient algorithm of Shor [24]. Let γ denote the one-dimensional *subgradient*, $\varepsilon - \sum_{\omega \in \Omega} p^\omega z^\omega$, of the function $\mathcal{Z}_D(\lambda)$ for any $\lambda \geq 0$ at which z is an optimal solution for model (3). Let Δ denote the step size that determines the magnitude of decrease in the descent direction. A number of strategies to determine good step sizes exist, see, e.g., a series of works by Polyak [18, 19] Then, at iteration k we update λ by taking a step in a direction opposite to the subgradient: $\lambda_{k+1} \leftarrow \max\{\lambda_k - \Delta_k \gamma_k, 0\}$.

This straightforward Lagrangian relaxation procedure of a chance-constrained optimization model is well-studied. The progressive hedging algorithm of Rockafellar and Wets [21] is also based on this scheme as follows. In the absence of the first-stage variables x, model (3) separates across the scenarios. Thus, we include a so-called non-anticipativity constraint that creates copies of the x variables: $x_t = \bar{x}_t^\omega, \forall t \in T, \omega \in \Omega$. Then, relaxing both the non-anticipativity and the knapsack constraints completely splits the problem across the scenarios; these separable problems are solvable in parallel to further reduce computational effort. Similar ideas are employed in work on decomposition algorithms for chance-constrained optimization models [3,8]. Such schemes find particular application in the power systems literature, see, e.g., [17,25].

The above cited works bound stochastic optimization models with the following motivation. The original optimization model—such as model (1)—is difficult to solve naively. Relaxing a set of constraints by a dual procedure results in a relatively easier optimization model—such as model (3)—that is computationally more tractable. Then, these works iteratively determine the best bounds from the relaxation by solving model (4). Even when $\arg\min_\lambda \mathcal{Z}_D(\lambda)$ cannot be computed, such schemes provide significant value that is especially evident when the Lagrangian models are completely [30] or nearly-completely [28] decomposed into independent subproblems.

As we mention in Sect. 1, we are interested in a different setting than these works. Specifically, we study the situation where the relaxed model is *also* computationally challenging to solve, such that not only $\arg\min_\lambda \mathcal{Z}_D(\lambda)$ is difficult to compute but also the individual models in the sequence of Lagrangian problems are only solved suboptimally. To this end, our work on upper bounds specifically focuses on the gap between theory and practice in relation to the parameter Δ for the two-stage chance-constrained program given by model (1). Below are six well-known strategies to update this step-size parameter that we consider in this work.

2.2 Subgradient Methods

We consider the following six strategies to update the step-size parameter Δ.

(i) Rule I: First, we consider a constant step size; i.e., $\Delta_k = \Delta > 0$. In Sect. 5.4, based on a preliminary analysis we use $\Delta_k = 0.58$.

(ii) Rule II: Second, we consider a constant *step length*. The quantity $|\gamma_k|\Delta_k$ provides the step length or the total amount we change λ by. We then have $\Delta_k = \alpha/|\gamma_k|$, with $\alpha > 0$. In Sect. 5.4, we use $\Delta_k = 200/|\gamma_k|$.

(iii) Rule III: Third, we consider a sequence Δ_k where the sum diverges, but the squared sum is finite; i.e., $\Delta_k \geq 0$, with $\sum_{k=1}^\infty \Delta_k^2 < \infty, \sum_{k=1}^\infty \Delta_k = \infty$. An example of a sequence that satisfies this criterion is $\Delta_k = a/(b+k)$, where $a, b > 0$. In Sect. 5.4, we use $\Delta_k = 3/(2+k)$.

(iv) Rule IV: Fourth, we consider a sequence Δ_k where the sum diverges as in Rule III, however the step size shrinks to zero; i.e., $\Delta_k \geq 0$, with $\lim_{k\to\infty} \Delta_k =$

$0, \sum_{k=1}^{\infty} \Delta_k = \infty$. An example of a sequence that satisfies this criterion is $\Delta_k = a/k^n$, where $a > 0$ and $0 < n \leq 1$. In Sect. 5.4, we use $\Delta_k = 1/\sqrt{k}$.

(v) Rule V: Fifth, analogous to Rule IV, we consider step lengths that diminish but are not summable; i.e., $\Delta_k = \alpha_k/|\gamma_k|$, with $\alpha_k \geq 0$, $\lim_{k \to \infty} \alpha_k = 0$, $\sum_{k=1}^{\infty} \alpha_k = \infty$. In Sect. 5.4, we use $\Delta_k = 300/k/|\gamma_k|$.

(vi) Rule VI: Finally, we consider the step size proposed by Polyak [20, Chapter 5]: $\Delta_k = \theta_k(\overline{\mathcal{Z}_D}_k - \underline{\mathcal{Z}})/\gamma_k^2$, where $\overline{\mathcal{Z}_D}_k$ is an upper bound for the optimal objective function value for model (3) at iteration k given a $\lambda \geq 0$, $\underline{\mathcal{Z}}$ is any lower bound for model (1), and $\theta_k > 0$ is a scalar. From the mBBP we are assured that $\Delta_k \geq 0$ for any lower bound $\underline{\mathcal{Z}}$. In Sect. 3, we provide three ways to compute lower bounds, $\underline{\mathcal{Z}}$. Following the suggestion by Held et al. [9], in Sect. 5.4 we use $\theta_1 = 2$ and progressively half this value if there is no improvement in $\mathcal{Z}_D(\lambda)$ in two iterations.

We summarize the above discussion in Algorithm S1 in Appendix A. Here, we begin with the LP relaxation of model (1) to provide an initial upper bound for model (1). We terminate if one of the following happens: (i) the gap between the estimated upper bound and the input lower bound is less than a threshold, δ, (ii) a time limit, *time*, is reached, or (iii) the values of λ do not change by more than a threshold, ζ, between two iterations. In the last two cases, the algorithm converges without finding the optimal objective function value $\overline{\mathcal{Z}}$ of model (4). Then, we report the available optimality gap and the conservative upper bound Z_D. As we mention in Sect. 2.1, we are especially interested in the second of these three cases.

2.3 Statistical Validation

In practice, fast high quality bounds are of greater interest than provably optimal bounds that require significant computational effort. For example, although Polyak's works [18,19] provide strategies guaranteed to converge to the optimal λ for convex optimization models, they often require improvisation in practice [9]. To this end, we employ 20 batches of two computationally expensive instances of chance-constrained programs from existing works to develop statistical guarantees; i.e., for both instances of model (1), given the parameters, we sample the uncertainties 20 times each. Both of these instances could not be solved to optimality in their existing works.

Let μ_r denote the expected gap computed by Algorithm S1 (denoted by δ) for step-size rule r. Consider the following null and alternate hypotheses: $H_0 : \mu_r \geq \mu_s$, and $H_1 : \mu_r < \mu_s$. We employ a Welch t-test, due to unequal variances and the small sample size, with a test statistic T that has a Student-t distribution with ν degrees of freedom as follows:

$$T = \frac{\bar{r} - \bar{s}}{\sqrt{\frac{s_r^2}{20} + \frac{s_s^2}{20}}}, \quad \text{where } T \sim t_\nu, \quad \text{with } \nu = \left\lfloor \frac{\left(\frac{s_r^2}{20} + \frac{s_s^2}{20}\right)^2}{\frac{1}{19}\left(\frac{s_r^2}{20}\right)^2 + \frac{1}{19}\left(\frac{s_s^2}{20}\right)^2} \right\rfloor. \tag{5}$$

In Eq. (5), $\bar{r}$ and $\bar{s}$ denote the average gap over the 20 batches for step-size rules r and s, respectively. Analogously, s_r and s_s are the corresponding sample standard deviations of the gaps. For a 5% level of significance, we reject the null hypothesis if $T < Q_{0.05,\nu}$, where $Q_{0.05,\nu}$ is the 5-percentile quantile of a t-distribution with ν degrees of freedom. Rejecting the null hypothesis suggests rule r performs better than rule s. We present results of this test in Sect. 5.4.

3   Lower Bounds for Model (1)

Although a lower bound, $\underline{Z}$, for model (1) is available via any feasible solution, better lower bounds are of interest for at least two reasons. First, they help determine an optimality gap from Sect. 2. Second, step-size Rule VI requires a lower bound. We now summarize three lower bounding methods with their merits and shortcomings.

3.1   An Iterative Regularization Bound

In [26], the authors provide an iterative regularization-based (IR) heuristic that achieves statistically validated lower bounds for a chance-constrained optimization model. This regularization method is motivated by classical descent algorithms in the continuous optimization literature (see, e.g., [5]), with a notable difference of the use of independent samples of scenarios at each iteration. Traditional regularization methods work by modifying the objective function via a proximal term, see, e.g., [29]. Singh et al. [26] exploit this idea to break symmetries in MIP models and achieve good quality feasible solutions quickly. Model (6) is the regularized model, where $\rho > 0$ is a regularization parameter and $\hat{x}$ is a proximal term.

$$\max_{x,y} \sum_{t \in T} \left(R_t x_t - \mathbb{E}[C_t^\omega y_t^\omega] \right) - \sum_{t \in T} \rho |x_t - \hat{x}_t| \tag{6a}$$

$$\text{s.t. } (1c), (2a) - (2c). \tag{6b}$$

The IR iteratively solves SAAs of model (6) of progressively increasing size, m; i.e., all the considered scenarios have a probability of $1/m$. Algorithm S2 in Appendix A summarizes this scheme. In [26], the authors achieve lower bounds nearly double of those obtained naively. Thus, the method is promising; however, to the best of our knowledge, it is untested on other instances of chance-constrained models. We test the computational performance of this method in Sect. 5.3.

3.2   An Aggregation Bound

A large number of scenarios leads to computational difficulties in solving model (1). Choosing a new, smaller, set of scenarios that "represents" Ω is one way to handle this computational difficulty. This idea again relates to the classical scenario aggregation technique proposed in [21]; although clustering techniques for general data points are well-studied in other disciplines, see, e.g., [10,22]. We develop a second lower bound by integrating a clustering procedure within the IR bound.

We begin with a large pool of scenarios that we aggregate (or, cluster) into a set of a few representative scenarios. Several methods are available to determine these representative scenarios, see, e.g., [12] for a comparison of these methods and their respective performance on energy system models. We then solve this relatively smaller problem to obtain an optimal solution for the first-stage x variables. Next, we solve the original problem with the x variables fixed to this value; the resulting objective function value provides a lower bound for $\mathcal{Z}$. Ideally, both of these problems are solved quickly. However, the procedure fails if the second problem is infeasible. For models with relatively complete recourse—where the second-stage problem is feasible for any choice of the first-stage variables—this procedure guarantees a feasible solution.

We further improve the above described scheme in Algorithm S3 in Appendix A. We use the IR procedure of Algorithm S2 but instead of sampling m identically distributed samples we use the above-described clustering procedure to generate m samples. In Sect. 5.3, we use the so-called hierarchical aggregation method presented by Hoffmann et al. [11] for the procedure cluster(M, m) in Algorithm S3. Given a pool of samples of size M, this procedure returns m aggregated scenarios with respective weights for each scenario. These weights determine the probability, or relevance, of each of these m scenarios; i.e., the scenarios are no longer equally likely. We denote this lower bound as the aggregation procedure (AP) bound.

3.3   A Quantile Bound

Another lower bound is available from a procedure described and implemented in [25]. Here, Singh and Knueven [25] modify the quantile-based bound of Ahmed et al. [1]—originally used to generate upper bounds—to generate lower bounds from feasible solutions. We first note that a feasible solution for the z variables for model (1) is available by choosing an arbitrary number, k, of scenarios ω as 1 and the rest as 0, where k is the cardinality of the set $\{\omega : \sum_{\omega \in \Omega} p^\omega \leq \varepsilon\}$. For equally likely scenarios, we have $k = \lfloor |\Omega| \varepsilon \rfloor$.

In [25], the authors consider equally likely scenarios and first solve model (1) separately for each scenario ω by fixing $z^\omega = 0$; i.e., a set of $|\Omega|$ subproblems. Similar to the progressive hedging subproblems that we discuss in Sect. 2, each of these subproblems are computationally cheap to solve and also solvable in parallel. Then, the procedure chooses $z^\omega = 1$ for the worst performing $\lfloor |\Omega| \varepsilon \rfloor$ scenarios; i.e., those with the lowest objective function values. Solving model (1) for this fixed value of z provides a lower bound, $\underline{\mathcal{Z}}$. On the test instances defined in [25], the authors complete this procedure for $|\Omega| = 1200$ in about half a minute. However, if this final problem is computationally intensive—despite the z variables being fixed—this heuristic is not effective. As we demonstrate later in our work, this can indeed happen. We denote this bound as the quantile procedure (QP) bound, and summarize this discussion in Algorithm S4 the Appendix A.

4   Two Instances of Model (1)

We first consider the chance-constrained optimization model with equally likely scenarios presented in [25]. This model maximizes profit from a day-ahead promise of power minus the cost of operating a battery. Energy is produced only via a photovoltaic power station, and excess production is stored in the battery. The JCC ensures highly reliable operations despite the limited knowledge of future solar power forecasts. The set Y determines the standard operating regime for the battery—ramping rates and storage limits. This set does not contain any binary variables. We denote this instance of model (1) as Model I.

Next, we consider the adaptive chance-constrained optimization model with equally likely scenarios presented in [26]. This model maximizes profit given by revenue from a day-ahead bid for power minus the expected generation costs. Energy is generated via a natural gas generator. The JCC again ensures highly reliable operations for the day-ahead market. The set Y consists of nearly-standard operating constraints for a generator—ramping, startup and shutdown rates, and minimum and maximum power generation limits. These constraints form part of so-called unit commitment models. The set Y now contains binary variables; i.e., model (1) contains binary variables additional to the binary z variables of equation (2). Thus, this instance of model (1) is more challenging than Model I. We denote this instance as Model II.

5   Computational Results

5.1   Computational Setup

All computational experiments are carried out on two high performance computing clusters at the Regionales Rechenzentrum Erlangen with Intel Xeon E3-1240 v5 processors with 32 GB of RAM with GAMS version 24.8.5. We use Gurobi version 7.0.2 for solving all the optimization models. We use four sets of scenario-size regimes for both Model I and Model II: $|\Omega| = \{100, 600, 900, 1500\}$; for each, we sample scenarios independently using procedures detailed in [26] and [25] for Model I and Model II, respectively. The parameters in the set Y are unchanged from those in the original works. We use four regimes of the reliability threshold: $\varepsilon = \{0.01, 0.03, 0.05, 0.07\}$; i.e., we have 16 instances for both Model I and Model II.

5.2   Analysis: Naive Solve

First, we present results using a direct naive solution method of Model I and Model II. Table S1 in Appendix B summarizes our computational experiments. Naturally, instances with a larger number of scenarios are computationally more challenging to solve. Increasing the reliability threshold, ε, leads to a greater number of combinatorics that is further reflected in an increased computational effort. Except for the smallest instances of Model I with $|\Omega| = 100$, none of the instances are solved to optimality. As we mention before, Model II is computationally more challenging to solve than Model I. For each of the 16 instances we consider, the optimality gap—see fifth column of Table S1 in Appendix B—is larger for Model II as compared to Model I. The average gap for the naive solution method over the 16 instances for Model I and Model II is 12.9% and 61.0%, respectively.

The set of scenarios we use, as well as our computational setup, is different than those used in [25, 26], and our gaps in Table S1 are larger than those reported previously. Thus, to further validate the trend we report, we repeat this experiment using 20 independent batches of scenarios for both Model I and Model II. We do so for all four scenario-size regimes with $\varepsilon = 0.05$. Table S2 in Appendix B presents the 95% confidence intervals for the objective function value, the time, and the optimality gap. The observations we report above again follow suggesting that the computational difficulty is not biased by the particular scenarios sampled.

5.3 Analysis: Lower Bounds

We now compare the three lower bounding heuristics of Sect. 3 for Model I and Model II. Table S3 in Appendix B summarizes our results.

Similar to the naive solution method, the IR and AP bounds require greater computational effort for instances with a larger number of scenarios and regimes with a higher tolerance threshold. This is because Step 7 of Algorithm S2 and Step 8 of Algorithm S3, require solving model (1) which can be challenging despite the fixed first-stage variables. The largest instance—see last row of Table S1—requires over 2100 s for the IR bound. Still, even for this particular instance, this computational effort is half of what we use as the time limit for the naive solution, and, yet, we achieve an improvement over 10%. Further, for the IR bound, 10 and 12 of the 16 instances obtain bounds at least as good as that obtained naively in Table S1 for Model I and Model II, respectively. The average improvement over the 16 instances, for Model I and Model II using the IR bound is 2.8% and 29.6%, respectively. The larger instances of model (1) are particularly suited for the IR bound. For the eight instances of the $|\Omega| = 900$ and $|\Omega| = 1500$ regimes, the average improvements for Model I and Model II are 7.4% and 54.5%, respectively. We emphasize that this improvement is achieved in less than half the time required by the naive solution method.

In terms of the computational effort required, the AP bound behaves similar to the IR bound. However, as we observe from Table S3 with the blanks, the AP bound is often unsuccessful in obtaining a feasible solution. For Model I, the AP bound is only successful for four of the 16 instances and is always worse than the naively obtained objective function value. Yet, analogous to the IR bound, the AP bound is useful for the more challenging instances of Model II. Although four of the 16 instances are infeasible within the give time limit, the average improvement over the other 12 is 14.3%. For the six feasible instances of the $|\Omega| = 900$ and $|\Omega| = 1500$ regimes, the average improvements for Model II increase to 35.6%.

The QP bound is obtained fast for Model I. For Model I, 12 of the 16 instances obtain bounds at least as good as that obtained naively in Table S1a. Except for two instances, all instances are solved in less than about a minute. Although the QP bound is not obtained as quickly as for Model I for Model II, it still manages to achieve solutions that are comparable or better than that obtained by the naive solution method. For Model II, 14 of the 16 instances obtain bounds at least as good as that obtained naively in Table S1b. The average improvement over the 16 instances for Model I and Model II using the QP bound is 3.9% and 16.7%, respectively.

To get more representative results, we repeat the calculation of the three different lower bounds on the same 20 independent batches of scenarios that we use in Sect. 5.2 for all four scenario-size regimes for Model I and Model II. We do this for $\varepsilon = 0.05$; see, Table S4 in Appendix B. These results qualitatively mirror the trend we report here.

Summarizing, the computational difficulties associated with the challenging structure of Model II are reflected in the three heuristics we study as well. We now revisit Q1 and Q2 from Sect. 1. All instances of Model II consume the entire time limit for each of the three heuristics; yet, to answer Q1, they show significant merit compared to a naive solution method. To answer Q2 for the lower bounding procedures, we observe that the IR bound is the most effective in terms of obtaining large lower bounds for both Model I and Model II. The AP bound is not effective for Model I, however it has value for Model II. For Model I, the QP bound is also efficient in the sense of achieving good quality bounds quickly in only a fraction of the time required by the IR bound.

We further conclude that regularization with aggregated scenarios, at least using the hierarchical aggregation procedures described in [11], has no significant advantage compared to regularization using i.i.d. scenarios. Future work could examine specialized aggregation techniques that assist in regularization methods for chance-constrained programs. Quantile-based bounds are particularly effective for models that are easy to solve when the binary z variables are fixed. On the other hand, regularization-based heuristics offer advantage when the second-stage model is easy to solve given the first-stage variables. Future work could also study lower bounding procedures that begin with a quantile-based bound and then proceed to regularization-based heuristics.

5.4 Analysis: Comparison of the Six Step-Size Rules

In this section, we statistically validate the performance of the six step-size rules of Sect. 2.2. For each of the four scenario-size regimes, and for the two tolerance threshold regimes of $\varepsilon = 0.01, 0.05$, for both Model I and Model II, we compare the gaps of the Lagrangian relaxation procedure given by Algorithm S1; here, we use $\underline{Z}$ as the maximum of the three lower bounds computed in Sect. 5.3 for the corresponding instance. We are interested in determining step-size rules that reject the null hypothesis of the Welch-t test of Sect. 2.3. Table S5 of Appendix B summarizes the results of this test for Model I and Model II using the same 20 batches of scenarios reported in Sect. 5.2 for each of the four scenario-size regimes.

First, we analyze the results for $\varepsilon = 0.05$ that we report on the right side of the entries of Table S5. For Model I, Rule VI has statistically significant evidence to perform better than all other rules for the smallest regime of $|\Omega| = 100$. Rule VI—given by Polyak [19] and further modified by Held et al. [9]—is one of the most employed subgradient step-size rules with a rich history of extensions [7, 14, 16]. However, for Model II, the computational evidence in favor of any particular rule is not convincing; e.g., Rule I, Rule II, Rule III, Rule IV, Rule V, and Rule VI have only three, zero, three, two, two, and zero instances, respectively, that perform better than any of the other five rules (over the 32 instances). Thus, overall, the only statistically validated conclusion we can draw is that Rule V and Rule VI perform well on instances that are the cheapest computationally; i.e., those for Model I with $|\Omega| = 100$. For all of the other instances, however, we have few rejections and thus little evidence for which rule performs the best.

The above results bring us to the two questions we raise in Sect. 1. For computationally challenging instances of Model I and Model II, iterations of model (3) are often not solved to optimality in the given time limit. Depending on the step size, this suboptimality happens at different iterations of Algorithm S1. This observation is especially visible for Rule IV where for all the scenario regimes, except $|\Omega| = 100$, even the first iteration of Algorithm S1 is not solved to optimality. Although our computational setup ensures that at least two iterations of Algorithm S1 are completed in the given time limit (albeit, suboptimally), we observe that the smallest upper bound is almost always obtained at the first iteration itself. Thus, the upper bound is often the same for all six rules. The first iteration is likely assisted by the dual values of λ from the LP relaxation in Step 1 of Algorithm S1. Hence, there is inconclusive evidence to determine a single rule that performs better than others. To further validate this observation, we again conduct additional computational experiments.

(i) First, we repeat this experiment for Model I and Model II at a 90% significance level (not shown); however, the results are practically unchanged.

(ii) Second, we perform the same experiments and tests for a tolerance of $\varepsilon = 0.01$. These results are listed as the left entries in Table S5 of Appendix B. However, this new experimental design does not significantly change the results we reported previously. For Model I, there are only five rejections over all the sets of instances—and all of these are for the $|\Omega| = 100$ regime. For Model II, there is not even a single rejection. This again provides a validation to our observation that no single rule performs better than the others.

(iii) Third, we repeat these experiments using a time limit that is an order of magnitude larger than our previous limit; i.e., in Algorithm S1 we set $time = 41000\,\text{s}$. Further, in this setup, we solve the first two iterations of

Algorithm S1 with a time limit of 5000 s each, and allow all other iterations a time limit of 15000 s. Thus, we are ensured that at least five iterations of Algorithm S1 are completed for each of the six rules. We summarize these results in Table S6 in Appendix B, and provide a corresponding discussion. As opposed to the larger instances where the first iteration typically provides the best bound, here the best iteration varies. This behavior is expected from smaller-sized Lagrangian models as the bounds can improve in later iterations, especially when iterations are solved optimally. Observing the sets of instances for this regime, we find different rules performing better; thus, again there is no statistical best performer for this regime.

Summarizing, the above results suggest that statistical evidence in favor of a step-size rule only appears when given a sufficiently high time limit. Even then, given differences in the structure of the underlying optimization models, different rules can perform better. Revisiting Q1, we can say that for applications where this large time limit is not a concern, our results suggest investing in the determination of a good step-size rule. Then, the answer to Q2 is that Rule II and Rule VI—although neither can be declared as the best—are strong candidates. Indeed, the constant step-length rule given by Rule II is widely used in several applications; see, e.g., [23,31]. Polyak's step-size Rule VI, as we mention before, is also widely used, as well as studied; see, e.g., [4,19]. However, for several applications, the nearly half-a-day time limit we consider could be impractical. Then, our answer to Q2 changes. Here, our results suggest that practitioners could well choose any of the step-size rules and achieve results that have no statistical differences from others.

5.5 Analysis: Comparison of a Naive Solution Method and the Lagrangian Procedure

As we mention in Sect. 1, a goal of this work is to investigate whether a Lagrangian relaxation procedure using an appropriate step-size rule improves the computational tractability of model (1). In this section, we bring together results of the previous sections to analyze the output of Algorithm S1. We use the same batch of scenarios as in Table S1 and choose $r = $ Rule VI for the experiments in this section. Similar to Sect. 5.4, we run Algorithm S1 with $\underline{z}$ set to the largest of three lower bounds of Table S3. Table S7 in Appendix B summarizes our results.

For Model I, except for the four instances in the first four rows of Table S7a and one additional instance of $|\Omega| = 1500$, $\varepsilon = 0.01$, Algorithm S1 always presents an improvement. For Model II there are a total of three such instances; all are in the $|\Omega| = 100$ regime. The instances of $|\Omega| = 100$ are the computationally least demanding instances; for Model I, all four of these instances are solved to optimality naively, see Table S1. Excluding these instances, the value of the algorithm is immediately visible in the larger and computationally more challenging instances. For Model I—see Table S7a—the average improvement over the 16 instances is 29.4%. For Model II—see Table S7b—the average improvement over the 16 instances is 12.3%.

Next, we revisit the sufficient conditions that we discussed in Sect. 2 for the BBP to hold with equality. These conditions ensure that the Lagrangian procedure identifies an optimal solution for model (1). They hold for only one of the 32 instances, which is also computationally the cheapest: the first row in Table S7a and Table S1a with an objective of 278.54. In this particular instance, the lower bound and the upper bounds from Algorithm S1 are identical (subject to a tolerance). Thus, there is no additional contribution from the BBP. Instead, if the lower bound was not tight, this contribution would have been immediately apparent. Then, we could conclude that the optimal objective function value of this instance is indeed that reported by Algorithm S1 even without knowledge of the lower bound. That being said, as we mention in Sect. 2 and as our computational experiments show, in practice, the BBP rarely holds with an equality.

Next, we address question Q1 from Sect. 1 that we reformulate as: does the Lagrangian procedure coupled with a suitable step-size rule and a suitable lower bounding heuristic, improve the optimality gaps as opposed to a naive solution method? In Table S7a we already provided evidence and conditions to answer this question with a "yes" for $r =$ Rule VI. These results include a single batch of scenarios for the 16 instances each for Model I and Model II; thus, they are insufficient to make a statistical conclusion. We now repeat this experiment, as we did in Sect. 5.3 and Sect. 5.4, using the same 20 batches of scenarios for each of the four scenario-size regimes. We use $\varepsilon = 0.05$, the computational setup for Algorithm S1 as described in Sect. 5.1, and again set $\underline{Z}$ as the maximum value obtained from the three lower bounding procedures of Sect. 3. We then apply a Welch t-test similar to that in Sect. 2.3 but with the null and alternate hypotheses modified for step-size rule r as follows: $H_0 : \mu_r \geq \mu_{\text{naive}}$ and $H_1 : \mu_r < \mu_{\text{naive}}$. Here μ_r is the optimality gap reported by Algorithm S1 for step-size rule r and μ_{naive} denotes the optimality gap from the naive solution method. Rejecting the null hypothesis suggests that there is sufficient evidence to conclude that Algorithm S1, with step size r, performs better than the naive solution method.

As in Eq. (5), the test statistic T has a Student-t distribution with ν degrees of freedom, where $T = (\bar{r} - \overline{\text{naive}}) / \sqrt{s_r^2/20 + s_{\text{naive}}^2/20}$, with $T \sim t_\nu$, and $\nu = \left\lfloor \left(s_r^2/20 + s_{\text{naive}}^2/20 \right)^2 / \left(1/19 \left(s_r^2/20 \right)^2 + 1/19 \left(s_{\text{naive}}^2/20 \right)^2 \right) \right\rfloor$. Here, $\bar{r}$ and $\overline{\text{naive}}$ denote the average gap over the 20 batches for step-size rule r and the naive solution method, respectively. Analogously, s_r and s_{naive} are the corresponding sample standard deviations of the gaps. For a 5% level of significance, we reject the null hypothesis if $T < Q_{0.05,\nu}$, where $Q_{0.05,\nu}$ denotes the 5-percentile quantile of the t-distribution with ν degrees of freedom. In Table S5 of the Appendix, we present the results of this statistical test; i.e., every entry results from the solution of 20 sets of optimization models.

Table S5a and Table S5b present consistent results for both Model I and Model II, respectively. For all scenario regimes, except $|\Omega| = 100$, for both models we have statistically sufficient evidence to reject the null hypothesis. These results provide further validation for our conclusions from Sect. 5.4, that the improvement of Algorithm S1 over the naive solution method is particularly noticeable in the computationally more intensive regimes; i.e., for these upper bounds, we answer Q1 with a convincing "yes". For computationally less intensive regimes, our results suggest there is no significant benefit to use a Lagrangian procedure; this is again intuitive. Further, in line with our conclusions in Sect. 5.4, our results suggest that there is an improvement regardless of the chosen step-size rule. This answers Q2. For computationally intensive regimes, a Lagrangian relaxation procedure is always effective over a naive solution, and that too for any step-size rule. However, we note with caution, as our results in Sect. 5.4 suggest, that the extent of this improvement is dependent not only on the chosen step-size rule but also on the underlying structure of the model.

6   Summary

We study a Lagrangian relaxation of a general two-stage chance-constrained optimization model. We show that there exist considerable differences between theory and practice, especially for computationally challenging instances of such models. Specifically, well-studied schemes to update step-sizes in the classical subgradient method do not always perform well in practice. To further validate this observation, we employ a series of statistical tests by solving batches of scenarios. In marked contrast to rules that are known to perform better than others in theory—for example in rates of convergence—our results show that no single rule can be uniformly declared as the best performer in practice. Nonetheless, our results demonstrate that such schemes show significant improvements over naive methods even when iterations in the Lagrangian procedure are solved suboptimally; i.e., our results are highly conservative. Future work could examine necessary or sufficient conditions to use the optimistic bound, rather than the conservative bound, resulting from the Lagrangian procedure.

We also study three lower bounds for such chance-constrained models. We find a quantile-based bound to be highly effective in achieving fast solutions, however regularization-based bounds are better when more compute time is available. We also find no benefit in constructing representative scenarios using hierarchical clustering methods over i.i.d. scenarios. This suggests another line of future work where customized aggregation schemes for chance-constrained programs could be developed. A third direction of future work is the development of iterative algorithms that compute lower bounds at each iteration of the Lagrangian procedure.

Acknowledgments. We gratefully acknowledge the compute resources and support provided by the Erlangen Regional Computing Center (RRZE). The authors acknowledge the financial support by the Federal Ministry for Economic Affairs and Energy of Germany in the project METIS (project number 03ET4064).

Data Availability. All our codes and data, as well as Appendix A and B containing our algorithms and computational results, are publicly available at: https://github.com/charlotteritter/ArticleSubgradient.

References

1. Ahmed, S., Luedtke, J., Song, Y., Xie, W.: Nonanticipative duality, relaxations, and formulations for chance-constrained stochastic programs. Math. Program. **162**(1–2), 51–81 (2017). https://doi.org/10.1007/s10107-016-1029-z
2. Ahmed, S., Shapiro, A.: Solving chance-constrained stochastic programs via sampling and integer programming. In: State-of-the-Art Decision-Making Tools in the Information-Intensive Age, pp. 261–269. INFORMS (2008). https://doi.org/10.1287/educ.1080.0048
3. Ahmed, S., Xie, W.: Relaxations and approximations of chance constraints under finite distributions. Math. Program. **170**(1), 43–65 (2018). https://doi.org/10.1007/s10107-018-1295-z
4. Allen, E., Helgason, R., Kennington, J., Shetty, B.: A generalization of Polyak's convergence result for subgradient optimization. Math. Program. **37**(3), 309–317 (1987). https://doi.org/10.1007/BF02591740
5. Boyd, S., Boyd, S.P., Vandenberghe, L.: Convex optimization. Cambridge University Press, Cambridge (2004). https://doi.org/10.1017/CBO9780511804441
6. Charnes, A., Cooper, W.W.: Chance-constrained programming. Manage. Sci. **6**(1), 73–79 (1959). https://doi.org/10.1287/mnsc.6.1.73
7. d'Antonio, G., Frangioni, A.: Convergence analysis of deflected conditional approximate subgradient methods. SIAM J. Optim. **20**(1), 357–386 (2009). https://doi.org/10.1137/080718814
8. Deng, Y., Shen, S.: Decomposition algorithms for optimizing multi-server appointment scheduling with chance constraints. Math. Program. **157**(1), 245–276 (2016). https://doi.org/10.1007/s10107-016-0990-x
9. Held, M., Wolfe, P., Crowder, H.P.: Validation of subgradient optimization. Math. Program. **6**(1), 62–88 (1974). https://doi.org/10.1007/BF01580223
10. Hoffmann, M., Kotzur, L., Stolten, D., Robinius, M.: A review on time series aggregation methods for energy system models. Energies **13**(3), 641 (2020). https://doi.org/10.3390/en13030641
11. Hoffmann, M., Priesmann, J., Nolting, L., Praktiknjo, A., Kotzur, L., Stolten, D.: Typical periods or typical time steps? A multi-model analysis to determine the optimal temporal aggregation for energy system models. Appl. Energy **304**, 117825 (2021). https://doi.org/10.1016/j.apenergy.2021.117825
12. Kotzur, L., Markewitz, P., Robinius, M., Stolten, D.: Impact of different time series aggregation methods on optimal energy system design. Renewable Energy **117**, 474–487 (2018). https://doi.org/10.1016/j.renene.2017.10.017
13. Küçükyavuz, S.: On mixing sets arising in chance-constrained programming. Math. Program. **132**(1–2), 31–56 (2010). https://doi.org/10.1007/s10107-010-0385-3
14. Larsson, T., Patriksson, M., Strömberg, A.B.: Conditional subgradient optimization — theory and applications. Eur. J. Oper. Res. **88**(2), 382–403 (1996). https://doi.org/10.1016/0377-2217(94)00200-2
15. Luedtke, J.: A branch-and-cut decomposition algorithm for solving chance-constrained mathematical programs with finite support. Math. Program. **146**(1–2), 219–244 (2014). https://doi.org/10.1007/s10107-013-0684-6
16. Nedić, A., Ozdaglar, A.: Approximate primal solutions and rate analysis for dual subgradient methods. SIAM J. Optim. **19**(4), 1757–1780 (2009). https://doi.org/10.1137/070708111
17. Ozturk, U.A., Mazumdar, M., Norman, B.A.: A solution to the stochastic unit commitment problem using chance constrained programming. IEEE Trans. Power Syst. **19**(3), 1589–1598 (2004). https://doi.org/10.1109/TPWRS.2004.831651

18. Polyak, B.T.: A general method for solving extremal problems. Dokl. Akad. Nauk SSSR **174**(1), 33–36 (1967)
19. Polyak, B.T.: Subgradient methods: a survey of Soviet research. Nonsmooth Optimization **3**, 5–29 (1978)
20. Polyak, B.T.: Introduction to optimization. Optimization Software (1987)
21. Rockafellar, R.T., Wets, R.J.B.: Scenarios and policy aggregation in optimization under uncertainty. Math. Oper. Res. **16**(1), 119–147 (1991). https://doi.org/10.1287/moor.16.1.119
22. Rokach, L., Maimon, O.: Clustering methods. In: Data mining and knowledge discovery handbook, pp. 321–352. Springer (2005). DOI: https://doi.org/10.1007/0-387-25465-X_15
23. Sashirekha, A., Pasupuleti, J., Moin, N.H., Tan, C.S.: Combined heat and power (CHP) economic dispatch solved using Lagrangian relaxation with surrogate subgradient multiplier updates. Int. J. Electrical Power Energy Syst. **44**(1), 421–430 (2013). https://doi.org/10.1016/j.ijepes.2012.07.038
24. Shor, N.Z.: The rate of convergence of the generalized gradient descent method. Cybern. Syst. Anal. **4**(3), 79–80 (1968). https://doi.org/10.1007/BF01073933
25. Singh, B., Knueven, B.: Lagrangian relaxation based heuristics for a chance-constrained optimization model of a hybrid solar-battery storage system. J. Global Optim. **80**(4), 965–989 (2021). https://doi.org/10.1007/s10898-021-01041-y
26. Singh, B., Morton, D.P., Santoso, S.: An adaptive model with joint chance constraints for a hybrid wind-conventional generator system. CMS **15**(3–4), 563–582 (2018). https://doi.org/10.1007/s10287-018-0309-x
27. Singh, B., Watson, J.P.: Approximating two-stage chance-constrained programs with classical probability bounds. Opt. Lett. **13**(6), 1403–1416 (2019). https://doi.org/10.1007/s11590-019-01387-z
28. Takriti, S., Birge, J.R.: Lagrangian solution techniques and bounds for loosely coupled mixed-integer stochastic programs. Oper. Res. **48**(1), 91–98 (2000). https://doi.org/10.1287/opre.48.1.91.12450
29. Tikhonov, A.N., Arsenin, V.Y.: Solutions of Ill-Posed Problems. Wiley, Washington (1977). https://doi.org/10.2307/2006360
30. Watson, J.P., Wets, R.J.B., Woodruff, D.L.: Scalable heuristics for a class of chance-constrained stochastic programs. INFORMS J. Comput. **22**(4), 543–554 (2010). https://doi.org/10.1287/ijoc.1090.0372
31. Yuan, D., Xu, S., Zhao, H.: Distributed primal–dual subgradient method for multiagent optimization via consensus algorithms. IEEE Trans. Syst. Man Cybern. Part B (Cybernetics) **41**(6), 1715–1724 (2011).https://doi.org/10.1109/tsmcb.2011.2160394

The Customization of the Geodesic Algorithm for Optimal Fastener Arrangement

Julia Shinder⬤, Margarita Petukhova$^{(\boxtimes)}$⬤, Sergey Lupuleac⬤,
and Tatiana Pogarskaia⬤

Peter the Great St. Petersburg Polytechnic University, St. Petersburg 195251, Russia
margarita@lamm.spbstu.ru

Abstract. The search for rational ways to accelerate aircraft assembly seems to be one of the most important tasks of modern aerospace industry. Elimination of the manual operations that follow the temporary fastening can be efficient though the requirements to the final quality are to be preserved. The problem of fastening optimization in its turn is based on the solving of contact problem series that account for mechanical properties of the assembly, dimensions of variation, etc. A special non-iterative geodesic algorithm based on computation of contact forces arising during assembly and geodesic distances between fastener positions is being developed [11] and applied to solving several industrial problems. However, the algorithm includes a set of input parameters that significantly affect the resulting arrangement, i.e. a starting point, weight coefficients for forces and distances, accuracy of the contact problem solution, etc. The further development and application of the algorithm to specific industrial problems requires elaborating of recommendations for the algorithm tuning or choice of parameter values. The present paper discusses the tuning of geodesic algorithm and demonstrates its efficiency on different problems.

Keywords: Aircraft assembly · Fastening optimization · Geodesic distance · Contact problem

1 Introduction

The assembly of basic aircraft structures is a complex and multi-stage process. The main structural elements of an aircraft airframe are first assembled from large scale flexible panels on separate assembly lines and then delivered to the final assembly line. The assembly of almost all components of the airframe is carried out by means of riveting and bolting. Drilling the holes for rivets in predefined positions is carried out directly on the assembly jig. After fixing the parts on the assembly stand, the so-called initial fastening is carried out, that consists of installing 10–20 percent of the fasteners in manual mode in order to reduce the gap between the parts. After that, at the next stage of temporary assembly all the holes are drilled in a semi-automatic mode and the corresponding fasteners are installed.

Installation of fasteners at the initial fastening stage is more expensive than at the subsequent stages. Thus, it is relevant to reduce the number of fasteners installed at this stage while ensuring the smallness of the gap between the parts. In modern aircraft manufacturing at every stage of the assembly the same fastener installation arrangements are used for all products of the series. Therefore, when modeling the assembly process it is also necessary to take into account the deviations of serial products from the nominal value. Thus, a specific optimization problem arises, that is considered in this paper.

There is a number of papers dedicated to the seemingly close problem of optimization the fastener positioning and the design of final riveted joints. One can mention the works of Menassa et al. [8] and Zhang et al. [22], where the finite element analysis is combined with optimization of fastener locations. The optimization of fastener joint topology is discussed by Watkins and Jakiela [19], Thomas et al. [18], Ambrozkiewicz and Kriegesmann [1]. The similar problem of topology optimization for fastener joint is solved by Chickermane et al. [2] by introducing a continuous function of fastener density. Rakotondrainibe et al. [14] combine the optimization of fastener joint with structural optimization of the assembled parts. The optimum fastener arrangement with regard to fatigue is determined by Oinonen et al. [9]. However, there is a fundamental difference between the problems of optimizing the topology of a bolted or riveted joint and the problems of optimizing the fastening assembly process. In the first case, the positions of the fasteners are not predefined. Thus, the unknowns are continuous functions of spatial variables, which allows to consider these problems as continuous optimization ones. In the second case the fasteners can only be installed in predefined positions. In this regard, the problems of optimizing the fastener assembly process are combinatorial optimization problems.

The problem of optimization of fastener arrangement during temporary aircraft assembly was first formulated by Yang, Qu and Ke [20] as a problem of minimization of maximal residual gap after the installation of a certain number of fasteners. The problem was solved with genetic algorithm. Only one initial gap was considered, so the variations of part shape were not taken into account. The optimization of temporary fastener arrangement during the assembly of A320 wing was studied by Lupuleac, Zaitseva et al. [3]. The set of surrogate initial gaps obtained on the base of measurements was used to replicate the deviations of parts on the assembly jig. The close approach was used in [6] for enhancement the assembly process for A350 fuselage part. The close problem of fastening sequence optimization was studied by Pogarskaia et al. [12] for airframe assembly as well as by Tabar et al. [15] for automotive assembly.

The problem of temporary assembly optimization was first formulated in terms of defect probability minimization in [5]. It was solved using several methods including simulated annealing, local variations etc. Optimization of A350 wing-to-fuselage assembly process was considered. It was shown that the optimization of assembly process for industrial large scale structures with standard approaches is applicable only with the use of high performance computing. This prevents assembly process optimization from becoming a common practice in the aircraft manufacturing. The geodesic algorithm was proposed in [11] which allowed to significantly reduce the amount of calculations while maintaining the quality of the results in optimizing the assembly process. This algorithm was applied for optimization of A350 fuselage and wing-to-fuselage fastening process [10,11].

Further modifications of the geodesic algorithm are considered in present paper. Moreover, its performance is thoroughly studied on test examples, and the tuning of the algorithm is discussed.

2   Optimal Fastener Arrangement Problem Statement

2.1   An Approach Used for Assembly Simulation

Let us recall the main aspects of the methodology for simulation of the aircraft assembly process proposed in [4]. The detailed introduction to this technique is provided in [7]. In order to compute the deformed stress-strain state of assembly the contact problem is reformulated into a quadratic programming problem:

$$\min \left(\tfrac{1}{2} u^T K u - f^T u \right),$$
$$Au \leq g, \tag{1}$$

where n is the number of nodal displacements restricted by non-penetration conditions, $u \in \mathbb{R}^n$ is the vector of all restricted displacements in the computational nodes of the junction area (that is a zone of possible contact between parts), $K \in \mathbb{R}^{n \times n}$ is the stiffness matrix reduced to the junction area (positive definite block diagonal matrix with dense blocks), $A \in \mathbb{R}^{m \times n}$ is a linear operator that defines non-penetration condition (full rank sparse matrix), $f \in \mathbb{R}^n$ is the vector of loads from fastening elements, $g \in \mathbb{R}^m$ $(m \leq n)$ is the vector of initial gap between parts.

Here we can classify the input data parameters as follows:

- Invariable parameters: the matrix A provides non-penetration of parts during the contact, and the matrix K reflects the mechanical properties of the parts and the assembly jig. Matrices A and K are unique for the investigated assembly model.
- Variable parameters: the vector of initial gaps g represents the deviations of part geometries caused by manufacturing defects and/or fixations of assembled parts on the jig. It can be obtained by measurements or by means of variation analysis (e.g., see [21]). The vector of external loads f is constructed from the contributions from each fastener acting on the assembly. Thus, the vector of forces is determined solely by the number and location of fasteners. Note that in some circumstances it is required to use a more complex model and also take into account the order of fastener installation [12], but such problems are not considered in the present paper.

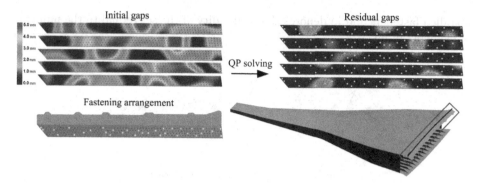

Fig. 1. Assembly problem flow chart

The assembly problem flowchart is shown in Fig. 1. As a result, one can obtain displacements of assembled parts $u_{res} = \arg\min Problem(1)$ and calculate the residual gap:

$$g_{res} = g - Au_{res}. \tag{2}$$

The approaches for fast Problem (1) solving are discussed in [17] and [16]. The quality of the fastening is determined by statistical analysis of the calculated residual gaps.

2.2 Fastener Arrangement Optimization Problem Formulation

There are several typical fastening optimization problems, often arising during serial aircraft assembly, one of them is considered in present paper. It occurs in the temporary assembly stage during the initial fastening.

In the temporary assembly stage the holes in the parts are drilled consequently with installation of temporary fasteners. As a result of this stage, the assembly is completely fastened with temporary fasteners, which will then be replaced with permanent ones. At the beginning of the temporary assembly stage, the so-called initial fastening is carried out in manual mode. It involves the installation of 10–20 percent of the fasteners in order to ensure reliable contact between the joined parts. After completion of the initial fastening, further assembly proceeds in the semi-automatic mode, which significantly reduces the cost of assembly operations. So, it is necessary to minimize the number of installed fasteners during the initial fastening while also minimizing the probability that the residual gap between assembled parts exceeds the given technological tolerances (separately specified for each stage of the assembly process). This is a typical two-objective optimization problem, the detailed formulation of which is presented in the current subsection.

The set of holes for fastener installation is denoted as $H = \{h_i\}_{i=1,n_h}$, where n_h is the total number of holes. The fastener arrangement $H^o = \{h_i^o\}_{i=1,n_f}$ is considered as a subset of set H (see Fig. 2). Here $n_f = |H^o|$ is the number of installed fasteners and $|.|$ denotes the cardinality of the set (equal to the number of elements in it).

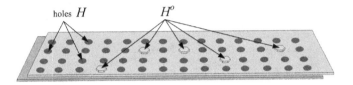

holes H H^o

Fig. 2. Fastener arrangement example

For the fastener arrangement H^o it should be ensured that the residual gap g_{res} does not exceed the predefined value $\hat{g}$ (allowed technological tolerance). Let us define the defect vector $d(H^o) \in \mathbb{R}^m$

$$d_i(H^o, \hat{g}) = \begin{cases} 1, & (g_{res})_i \geq \hat{g} \\ 0, & \text{otherwise} \end{cases} . \tag{3}$$

In other words, the computational node is considered as defective if the residual gap in it exceeds $\hat{g}$.

Cumulative characteristic $P(H^o)$ (following [7], further on it is called the "probability of defect") describes the overall quality of the joint provided by the given fastener arrangement

$$P(H^o) = \frac{\sum\limits_{i=1}^{m} d_i(H^o)}{m}. \tag{4}$$

Thus, the optimization of fastener arrangement is reduced to the simultaneous minimization of number of installed fasteners n_f and determination of the best positions for n_f fasteners among the n_h holes minimizing the probability of defect

$$\min\left(n_f; P(H^o)|_{|H^o|=n_f}\right). \tag{5}$$

The solution of minimization problem (5) is the set $\left\{n_f; \hat{H}^o(n_f)\right\}$ (Pareto frontier). Here the fastener arrangement $\hat{H}^o(n_f)$ is defined as follows:

$$\hat{H}^o(n_f) = \arg\min_{|H^o|=n_f} (P(H^o)). \tag{6}$$

As a rule, starting from a certain number of installed fasteners n_f^{crit}, the probability of defect becomes equal to zero. Optimizing the fastener arrangement for $n_f > n_f^{crit}$ is no use. Therefore, the natural constraint on n_f in the problem (5, 6) is $1 \leq n_f \leq n_f^{crit}$. For industrial aircraft assemblies n_f^{crit} is 10–20% of the total number of holes n_h.

One can see that the stated problem is a combinatorial one, and the number of admissible arrangements of n_f fasteners is equal to $\binom{n_h}{n_f}$ (or $C_{n_h}^{n_f}$). The number of holes in the assembly models used in industry is several hundreds and installing the fasteners in 10% of holes (commonly done during initial fastening) results in enormous number of fastener arrangements (e.g., see Fig. 8 (b) below).

In [5] it was shown that the problem (5, 6) can be solved by standard optimization techniques (simulated annealing, genetic algorithm, method of local variations). However, for full-sized industrial assembly models the fastener arrangement optimization by standard methods is extremely resource-intensive and requires application of high performance computing. The recently proposed geodesic algorithm [11] is designed to drastically reduce the resource intensity of solving the problem (5, 6) by using its specifics. The main principles of geodesic algorithm as well as its modification are discussed in the next section.

3   Geodesic Algorithm and Its Specifics

3.1   Basic Conception

The geodesic algorithm uses the information obtained from the assembly model in order to prevent excessive computations and to reduce the amount of admissible arrangements. Firstly, so called geodesic map is built that connects each hole $h_i \in H$ with the rest of the holes

$$GMap = \{\rho_g\,(h_i, h_j)\}_{h_i, h_j \in H}. \tag{7}$$

Here $\rho_g\,(h_i, h_j)$ is the geodesic distance computed as the length of the shortest path between holes h_i and h_j along the contact surface.

$$\rho_g\,(h_i, h_j) = \min_{\gamma(h_i, h_j)}\,||\gamma(h_i, h_j)||, \tag{8}$$

where γ is any path from h_i to h_j, $||\cdot||$ is the length of curve along the surface.

In assembly problems the surface is the junction area that is defined by computational mesh. The corresponding geodesic distance calculation is described in detail in [13].

Secondly, the necessary external pressure to be applied to the parts in order to keep the residual gap within the specified range is calculated. Let's introduce the parameter δ and state the problem similar to (1)

$$\begin{aligned} \min(\tfrac{1}{2}u^T K u), \\ Au - g \leq 0, \\ (g - Au)_k \leq \delta,\ k \in \{1, ..., m\}. \end{aligned} \tag{9}$$

There are no external loads applied in (9), but additional constraints force the residual gap to be within the limits $[0, \delta]$ in whole junction area. This leads to the appearance of the Lagrange multipliers $\lambda \in \mathbb{R}^m$ that can be interpreted as necessary external pressure in computational nodes making the residual gap smaller or equal than δ. Note that δ has the physical meaning of the tolerance for the allowable residual gap. It is worth mentioning that for a given assembly model the necessary λ vector is determined by the initial gap g and the parameter δ. The number of holes n_h is usually much smaller than the number of computational nodes n (shown by dots in Fig. 3 (a)).

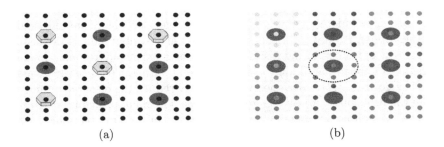

(a)   (b)

Fig. 3. Computational nodes and hole-to-node association

The hole $h_i \in H$ is surrounded by nearby nodes, so that it is possible to split all the nodes into independent, non-intersecting groups S_i with the centers in h_i (see Fig. 3 (b)). According to the S_i, the load λ_x from each node goes to a specific hole. Then hole weight coefficients are calculated according to the given λ

$$w\left(h_i\right) = \sum_{x \in S_i} \lambda_x, \qquad (10)$$

where x is a computational node.

Having the geodesic map $GMap$ and the hole weight coefficients, the following non-iterative procedure for constructing the best fastener arrangement $\hat{H}$ is derived:

- The first fastener is installed in the hole h_1 with maximal weight calculated by (10). Hole h_1 is added to $\hat{H}$.
- Then for each empty hole the minimal geodesic distance is found $r(h) = \min_{\hat{h}_l \in \hat{H}} \rho_g(h, \hat{h}_l)$.
- The next hole where fastener is to be installed provides maximum to the function $\max_{h \in H/\hat{H}} w(h)^\alpha \cdot r(h)$.

The fact that the procedure is non-iterative is understood as the absence of the need to calculate the goal function (that is, to solve the time-consuming contact problem) during the placement of fasteners. This version of the geodesic algorithm was proposed in [11], where a more detailed description can be found. Later in present article, this numerical procedure is referred as basic version of geodesic algorithm.

Parameter α appears when the choice of the next fastener position is being made. It regulates the contribution of hole weight: if α is close to zero, the choice is made mostly based on the distance from already chosen holes. Increasing α makes hole weight dominant over the geometry.

Thus, there are two governing parameters δ and α that affect the result of geodesic algorithm. For given n_f the result is one fastener arrangement $\hat{H}^o(n_f)$ for certain values of δ and α. At the same time the resulting arrangement $\hat{H}^o(n_f)$ heavily depends on these parameters, and the geodesic algorithm requires tuning, which is discussed in the following sections.

3.2   Algorithm Modification

The described above basic version of geodesic algorithm implies computing the necessary external pressure in all the computational nodes and distributing it between the holes H. However, it is possible to obtain the forces in the hole positions necessary to close the gap between parts directly by solving the optimization problem. For this purpose, the statement of (9) is modified. Each hole $h_i \in H$ is assigned to a specific computational node y_i. The condition for the smallness of the residual gap is set only at these nodes, while the non-penetration condition is set at all computational nodes. The reformulated optimization problem has the form:

$$
\begin{aligned}
&\min(\tfrac{1}{2} u^T K u), \\
&Au - g \leq 0, \\
&(g - Au)_{y_i} \leq \delta.
\end{aligned}
\tag{11}
$$

Such transformation leads to the reduction of computational efforts and makes hole weight coefficients' calculation more clear, as $\lambda = \{\lambda_i\}_{i=1,n_h}$ are obtained directly. The algorithm implementation is listed below with both variants of weight coefficient computations.

Let's compare the speed of two versions of the algorithm. Figure 4 (a) shows the results of computational time for solving problems (9) and (11) for different number of computational nodes and $\delta = 0.15$. It can be observed that for the problems with more than 5000 unknowns the difference becomes significant: computations are performed three times faster if the constraints are imposed only in the hole positions.

The next step is to compare the influence of δ value on the computational time. We consider the problem with 7710 nodes and vary the parameter δ for (9) and (11). The problem (11) is solved about three times faster than problem (9) for different δ (see Fig. 4 (b)). Note that Problems (9) and (11) are solved by the Active Set Method (ASM) [17]. The running time of ASM increases with an increase in the number of active constraints. This determines the dependencies shown in the Fig. 4.

This improvement strongly affects the algorithm if it is necessary to find the fastener arrangement that provides low probability of defect $P(H)$ for different input parameters. As it is shown in the next section, the modified algorithm provides better results and requires less time for computations. At the same time, the basic geodesic algorithm described in Sect. 3.1 is more universal, since the

---

Algorithm 1: Determining the optimal fastener arrangement

Data: Geodesic map $GMap$, n_f, parameters δ, α

Result: Set of fastener arrangements $\hat{H}_\delta^\alpha$

if *Basic version* **then**

| Compute hole weight coefficients by solving (9) and using (10)

else

| Compute hole weight coefficients by solving (11)

end

Install the first fastener in the hole with maximal weight $\hat{h}_1 = \arg \max w(h_i)$

$\hat{H}_\delta^\alpha = \{\hat{h}_1\}$

for $i = 2$ *to* n_f **do**

| Calculate $r(h) = \min\limits_{\hat{h}_l \in \hat{H}_\delta^\alpha} \rho_g(h, \hat{h}_l)$

| $\hat{h}_l = \arg \max\limits_{h \in H/\hat{H}_\delta^\alpha} w(h)^\alpha \cdot r(h)$

| $\hat{h}_l \rightarrow \hat{H}_\delta^\alpha$

end

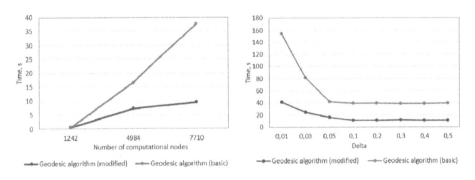

Fig. 4. Comparison of computational costs for different number of nodes and δ

assembly of parts can be carried out not only by applying point loads (installation of fasteners), but also by applying distributed loads (special clamps, etc.).

4   Results

The work of the geodesic algorithm is demonstrated on the examples of two test models. The first one (Model 1) is a simplified version of a typical assembly of two aircraft panels. Each part is made of aluminum alloy, the upper panel is reinforced by two stringers. The finite element model of this assembly is presented in Fig. 5 (a). All displacements are restricted in the nodes marked by red. The initial gap between parts varies from 0.63 to 4.12 mm as it is shown in Fig. 5 (b). The number of computational nodes in the junction area is 1722, the number of holes is 15.

The problem of simultaneous minimization of the number of installed fasteners n_f and determination of the best positions for n_f fasteners minimizing the

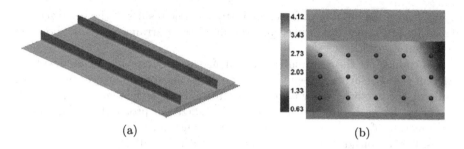

(a)   (b)

Fig. 5. Model 1: finite element model, holes and initial gap (Color figure online)

probability of defect is solved. The allowed technological tolerance $\hat{g}$ defining the defect is set to 0.15 mm. Since the number of holes is small, exhaustive search can be used to find the optimal fastener arrangement. The number of arrangements in exhaustive search is up to 6435 (see Fig. 6, data in color blue).

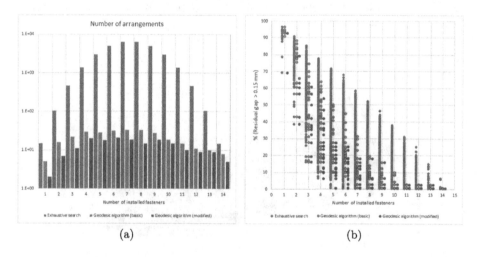

(a)   (b)

Fig. 6. Results for the Model 1: the number of arrangements and the probability of defect (Color figure online)

Figure 6 (b) shows the dependence of the probability of defect on the number of installed fasteners. Blue color corresponds to exhaustive search, orange to the basic geodesic algorithm, and red to the modified geodesic algorithm. Note that already with 5 installed fasteners, the zero value of the defect probability is reached ($n_f^{crit} = 5$). Therefore, a further increase in the number of installed fasteners has no practical meaning and was carried out only to study the features of the algorithms. From Fig. 6 (b) it follows that for any number of installed fasteners there is an optimal arrangement among all the others obtained by the modified geodesic algorithm. The number of arrangements varies from 2 to 21 depending on the number of fasteners installed. The results of the basic geodesic algorithm are somewhat worse, and the number of arrangements is about one and a half times larger.

Depending on the parameters α and δ, the geodesic algorithm results in different fastener arrangements. Although these arguments are continuous, the result of the geodesic algorithm is a finite number of fastener arrangements. Therefore, for each number of fasteners installed the probability of defect (4) as a function of α and δ is a piece-wise constant and takes a finite number of values. Let us consider changing parameters with a small step, δ changes in the range from 0.01 mm to 0.5 mm (which corresponds to the physical parameters of the problem) with a step of 0.01 mm, α varies from 0 to 1 in steps of 0.01. The maximum α is chosen so that its further increase does not lead to new arrangements.

Thus, for each number of fasteners installed, a table containing 50 rows (by δ) and 101 columns (by α) is obtained. The elements of this table are the values of the probability of defect for the corresponding δ and α. The number of distinct values is up to 33 for basic geodesic algorithm and up to 21 for modified one (see Fig. 6 (a)). An example of dependence of the defect probability on the parameters α and δ is presented in Fig. 7.

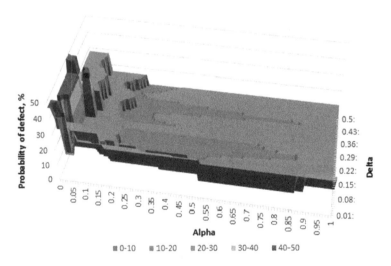

Fig. 7. An example of dependence of the defect probability on α and δ for Model 1 (basic geodesic algorithm, 5 installed fasteners)

For the second numerical test only the modified geodesic algorithm is used (it is denoted as the geodesic algorithm). A full-scale assembly of the wing and fuselage is considered as the Model 2. A finite element model and the set of ten random initial gaps are shown in Fig. 8 (a). The nodes with restricted displacements are marked in red.

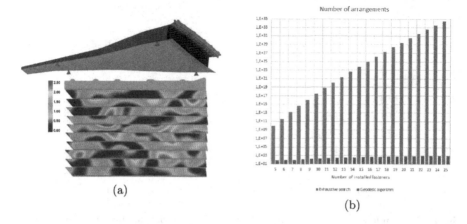

(a)   (b)

Fig. 8. Model 2: finite element model and initial gaps; the number of arrangements (Color figure online)

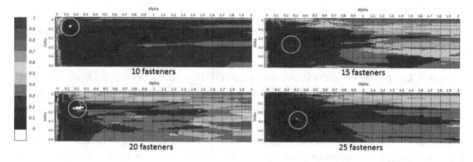

Fig. 9. Normalized probability of defect for different number of the installed fasteners for Model 2 (Color figure online)

For this problem a node is considered defective if the gap in it exceeds $\hat{g} = 0.3\,\text{mm}$. As for the previous model, the problem of two-objective optimization is solved, when the number of installed fasteners and the probability of a defect are simultaneously minimized. Thus, it is required to obtain the Pareto frontier.

The number of computation nodes in the junction area for Model 2 is 3612 and the number of holes is 256. With so many holes, it is impossible to consider the exhaustive search, since even with 10 percent of fasteners installed, the number of possible arrangements reaches 10^{34} (see Fig. 8 (b)).

For this model parameter δ varies in the same range as in the previous case (from 0.01 mm to 0.5 mm with step 0.01 mm) and α varies from 0 to 2 in steps 0.01. The dependence of the normalized defect probability on the parameters α and δ for different number of installed fasteners is presented in Fig. 9. The colors change from blue to red as the probability of defect increases. The zones with minimal probability of defect are marked with white color. Note that for

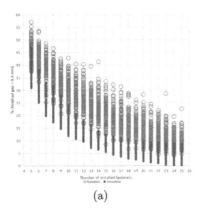
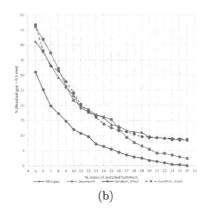

(a)   (b)

Fig. 10. Dependence of defect probability on the number of installed fasteners for Model 2: geodesic and random arrangements; results of comparison of geodesic and other algorithms (Color figure online)

this model, as well as for the previous one, with a different number of installed fasteners the optimal parameters α and δ differ.

In Fig. 10 (a) the probability of defect corresponding to the arrangements obtained by geodesic algorithm are marked in red. The number of installed fasteners varies from 5 to 25. With 25 fasteners installed, the probability of defect is zero ($n_f^{crit} = 25$), so a further increase in the number of installed fasteners is not needed. Since it is impossible to carry out exhaustive search in this case, 100 random arrangements were generated for illustration purposes. The corresponding values of probability of defect are shown in blue in Fig. 10 (a).

The geodesic algorithm was also compared with the other practically used fast algorithms. The first of them, conventionally called the "maximum gap" (green line in Fig. 10 (b), is based on the idea that the next fastener is installed in a hole with the maximum residual gap. In the case of a set of gaps, the average over all gaps in each hole is calculated first and then the maximum is selected. The second algorithm, called "geometric" (violet line in Fig. 10 (b)) does not require solving the contact problem at all. The first fastener is installed in the hole farthest from the rest, and the next ones are installed one-by-one as far as possible from those already installed. All values that the geodesic algorithm gives are located between the solid and dashed red lines. It can be seen from Fig. 10 that the implementations of the geodesic algorithm for all parameter values give quite good results. At the same time, the parameter-optimized geodesic algorithm gives results that are significantly superior to all other optimization algorithms comparable in speed (that is, except for exhaustive search).

In the models under consideration, it is possible to calculate the contact problem with all fastener arrangements obtained by the geodesic algorithm (for different α and δ) for the set of 10 initial gaps. However, for larger models, it is proposed to further consider the methods for optimizing the piece-wise constant objective function with respect to the α and δ parameters.

5 Conclusion

The geodesic algorithm was developed as a very fast and efficient way to optimize the fastener arrangement during the temporary aircraft assembly. As it was shown in [10,11] it provides the results not worse than using classical optimization methods (simulated annealing, genetic algorithm, local variation method), while spending orders of magnitude less computational resources. The geodesic algorithm has already been used to optimize the assembly processes of the A320 and A350 aircraft series. At the same time, present paper shows by examples that the results of the geodesic algorithm notably depend on the choice of parameters. Carrying out tuning by parameters separately for each task can significantly improve the quality of the results. When solving large-scale industrial problems, the geodesic algorithm with varied parameters can produce up to 1000 different arrangements of fasteners, the best of which can be selected using exhaustive search. Further, it is planned to develop a procedure for rapid optimization of the geodesic algorithm in terms of parameters, that will allow it to be quickly tuned without loss of quality.

Acknowledgements. The research was supported by Russian Science Foundation (project No. 22-19-00062, https://rscf.ru/en/project/22-19-00062/). The authors are grateful to Maria Churilova and Vasily Lupuleac for their valuable assistance in preparing the publication.

References

1. Ambrozkiewicz, O., Kriegesmann, B.: Simultaneous topology and fastener layout optimization of assemblies considering joint failure. Int. J. Numer. Methods Eng. **122**(1), 294–319 (2021)
2. Chickermane, H., Gea, H.C., Yang, R.J., Chuang, C.H.: Optimal fastener pattern design considering bearing loads. Struct. Optim. **17**(2), 140–146 (1999)
3. Lupuleac, S., et al.: Combination of experimental and computational approaches to a320 wing assembly. SAE Technical Papers 2017-September (2017)
4. Lupuleac, S., Kovtun, M., Rodionova, O.: Assembly simulation of riveting process. SAE Int. J. Aerosp. **2**(1), 193–198 (2009)
5. Lupuleac, S., Pogarskaia, T., Churilova, M., Kokkolaras, M., Bonhomme, E.: Optimization of fastener pattern in airframe assembly. Assem. Autom. **40**(3), 723–733 (2020)
6. Lupuleac, S., et al.: Optimization of automated airframe assembly process on example of a350 s19 splice joint. In: SAE Technical Paper. SAE International (2019)
7. Lupuleac, S., et al.: Simulation of the wing-to-fuselage assembly process. J. Manuf. Sci. E. T. ASME **141**, 1 (2019)
8. Menassa, R.J., DeVries, W.R.: Optimization methods applied to selecting support positions in fixture design. J. Eng. Ind. **113**(4), 412–418 (1991)
9. Oinonen, A., Tanskanen, P., Björk, T., Marquis, G.: Pattern optimization of eccentrically loaded multi-fastener joints. Struct. Multidiscip. Optim. **40**(1), 597 (2009)

10. Pogarskaia, T., Churilova, M., Bonhomme, E.: Application of a Novel Approach Based on Geodesic Distance and Pressure Distribution to Optimization of Automated Airframe Assembly Process. In: Voevodin, V., Sobolev, S. (eds.) RuSCDays 2020. CCIS, vol. 1331, pp. 162–173. Springer, Cham (2020). https://doi.org/10.1007/978-3-030-64616-5_14

11. Pogarskaia, T., Lupuleac, S., Bonhomme, E.: Novel approach to optimization of fastener pattern for airframe assembly process. In: Procedia CIRP, vol. 93 (2020)

12. Pogarskaia, T., Lupuleac, S., Shinder, J., Westphal, P.: Optimization of the Installation Sequence for the Temporary Fasteners in the Aircraft Industry. J. Comput. Inf. Sci. Eng. **22**(4), 040901 (2022)

13. Popov, N.P., Pogarskaia, T.A.: Geodesic distance numerical computation on compliant mechanical parts in the aircraft industry. J. Phys: Conf. Ser. **1326**, 012026 (2019)

14. Rakotondrainibe, L., Desai, J., Orval, P., Allaire, G.: Coupled topology optimization of structure and connections for bolted mechanical systems. Eur. J. Mech. A. Solids **93**, 104499 (2022)

15. Sadeghi Tabar, R., Wärmefjord, K., Söderberg, R.: Rapid sequence optimization of spot welds for improved geometrical quality using a novel stepwise algorithm. Eng. Optim. **53**(5), 867–884 (2020)

16. Stefanova, M., Baklanov, S.: The relative formulation of the quadratic programming problem in the aircraft assembly modeling. In: Olenev, N., Evtushenko, Y., Jaćimović, M., Khachay, M., Malkova, V., Pospelov, I. (eds.) Optimization and Applications, pp. 34–48. Springer Nature Switzerland, Cham (2022). https://doi.org/10.1007/978-3-031-22543-7_3

17. Stefanova, M., et al.: Convex optimization techniques in compliant assembly simulation. Optim. Eng. **21**(4), 1665–1690 (2020)

18. Thomas, S., Li, Q., Steven, G.: Topology optimization for periodic multi-component structures with stiffness and frequency criteria. Struct. Multidiscip. Optim. **61**(6), 2271–2289 (2020)

19. Watkins, M., Jakiela, M.: Fastener pattern optimization of an eccentrically loaded multi-fastener connection. In: Proceedings of the ASME Design Engineering Technical Conference, vol. 1, pp. 1209–1218 (2010)

20. Yang, D., Qu, W., Ke, Y.: Evaluation of residual clearance after prejoining and pre-joining scheme optimization in aircraft panel assembly. Assem. Autom. **36**(4), 376–387 (2016)

21. Zaitseva, N., Lupuleac, S., Khashba, V., Shinder, Y., Bonhomme, E.: Approaches to initial gap modeling in final aircraft assembly simulation. In: ASME International Mechanical Engineering Congress and Exposition, Proceedings (IMECE), vol. 2 (2020)

22. Zhang, B., Brown, D., St. Pierre, J., Tao, X., Williams, I., Whitehead, G., Wolfe, C., Pillutla, R.: Multi-objectives optimization of fastener location in a bolted joint. SAE Technical Papers 2 (2013)

Author Index

N. Olenev et al. (Eds.): OPTIMA 2023, LNCS 14395, pp. 389–390, 2023.
https://doi.org/10.1007/978-3-031-47859-8